Fundamentals of Electromagnetism

Springer
Berlin
Heidelberg
New York
Barcelona
Hong Kong
London
Milan
Paris
Singapore
Tokyo

A. López Dávalos D. Zanette

Fundamentals of Electromagnetism

Vacuum Electrodynamics, Media, and Relativity

With 72 Figures, 115 Problems,
and CD-ROM for Windows 95/98 and NT

Springer

Professor Dr. Arturo López Dávalos
Centro Atomico Bariloche
Instituto Balseiro
8400 Bariloche, Argentina

Professor Dr. Damián Zanette
Centro Atomico Bariloche
Instituto Balseiro
8400 Bariloche, Argentina

ISBN 3-540-65448-8 Springer-Verlag Berlin Heidelberg New York

Library of Congress Cataloging-in-Publication Data. López Dávalos, Arturo, 1937–. Fundamentals of electromagnetism: Vacuum electrodynamics, media, and relativity/Arturo López Dávalos, Damián Zanette. p. cm. Includes bibliographical references and index. ISBN 3-540-65448-8 (hardcover: alk. paper) 1. Electromagnetism. I. Zanette, Damián, 1963–. II. Title. QC760.L57 1999 537–dc21 98-52461

Typesetting: Data conversion by Satztechnik Katharina Steingraeber, Heidelberg
Cover design: *design & production* GmbH, Heidelberg

SPIN: 10698716 57/3144/ba - 5 4 3 2 1 0 – Printed on acid-free paper

To our teachers
To our students

Preface

The bibliography on electromagnetism is very rich and it may be necessary to justify our effort in writing a new text. The present text arose as a need when lecturing about the subject at Instituto Balseiro, in Bariloche, Argentina. We wanted to have a reference text for the most important topics covered in the course, which would also cover some aspects of interest for the applications of the concepts that the students have learned.

A course on electromagnetism usually follows one in classical mechanics. In the latter a brief introduction to the theory of relativity is normally included. During courses on electromagnetism this theory is traditionally presented again in brief form. This approach has the effect that, unless students later work on related topics, they do not acquire an appropriate knowledge of relativity. In this way, they lose perspective of the importance of one of the most beautiful and fascinating creations of physics in the twentieth century. For this reason, we consider the properties of charges, currents and electromagnetic fields in vacuum, highlighting the relativistic transformation properties of each quantity. In spite of the advantages and elegance of the covariant formulation of electromagnetism, however, we avoid overemphasizing it. It is left for Chap. 9, where we present it after having developed vacuum electrodynamics. In this way, we minimize the risk of negatively impressing those students who are attracted to more concrete problems.

We present the study of electromagnetic fields in material media after electromagnetic phenomena in vacuum have been discussed. We stress the fact that electromagnetic phenomena really take place in the empty space surrounding atoms. The foundations of the electromagnetic theory of material media is presented starting from averages of the microscopic fields. This is absolutely necessary for later applications of the theory to the phenomenology of such media. This presentation – which is not the traditional one – provides a better foundation for each one of the new areas, starting from basic knowledge developed in previous chapters. In this way we prepare the road to a deeper understanding of the formal structure and of the phenomenology which underlie the theory.

As a general rule, we have emphasized the discussion of physical concepts – in connection with the foundations and applications of electromagnetism – leaving aside the details of mathematical derivations. The latter usually divert attention from the more essential aspects. The mathematical derivations are often left as exercises for the reader. Instructors should guide the students in this respect.

Our educational experience teaches us that the solving of problems is the surest way in which students can prove their level of understanding of each topic. For this reason the problems included at the end of each chapter have been chosen to complete the discussion of some topics. It is important to supplement the reading of the text by solving them.

Nowadays, the discussion of many typical problems in electrostatics and magnetostatics can be enriched with numerical solutions. These allow us to appreciate certain subtleties that escape a simply intuitive discussion or an analytical solution of difficult numerical evaluation. We have had the collaboration of Dr. Sergio Pissanetsky, a specialist in the finite element method, who developed the program PhysicSolver specially to accompany the edition of this book. This software is very user-friendly and is based on the author's extensive professional experience. We believe this is a valuable contribution to the pedagogy of physics. Both professors and students will benefit from the use of the solutions to electrostatic and magnetotastic problems generated by PhysicSolver. In the problems proposed at the end of each chapter we have indicated those which can be solved with the software. Teachers and students will be able to suggest many more problems for which there is no simple analytic solution but which can easily be solved with PhysicSolver. Our own experience shows that numerical solutions often surprise the intuition, even of experienced physicists.

The order of presentation of the topics is as follows: Chap. 1 contains a historical introduction. Chapters 2 and 3 are devoted to a review of relativistic kinematics and dynamics. Chapter 4 discusses the properties of electric charges and of the electrostatic field. Chapter 5 contains a brief presentation of the solution of Poisson's and Laplace's equations, including a brief discussion of these equations in other branches of physics and a presentation of the finite element method. Chapter 6 discusses Ampère's law and the Biot–Savart law. The magnetic multipolar expansion, magnetic energy and a variational principle are also discussed here. Maxwell's equations are presented in Chap. 7. Here the symmetries and conservation laws associated with the equations are discussed. Dynamic fields and radiation are treated in Chap. 8. The covariant formulation is discussed in Chap. 9. Chapters 10, 11 and 12 are devoted to the discussion of electromagnetic properties of material media, and in Chap. 13 we present the properties of superconductors from the point of view of electromagnetism.

The order in which topics are presented here has advantages. However, in view of the variety of ways in which a course on electromagnetism can be given, the text can also be useful to those who prefer to follow a different order. For example, after going through Chaps. 1, 2 and 3, if more emphasis on the covariant formulation is wanted, a quick review of Maxwell's equations can be done following Chap. 7. From there one can go on to Chap. 9. In this way one can reach a kind of "summit" from which it is possible to "descend" to treat the remaining topics, continuing the order indicated below:

$$1 \to 2 \to 3 \to 4.1 \to 4.2 \to 6.1 \to 7 \to 9 \to 4 \to 5 \to 6 \to 8$$

$$\Downarrow$$

$$10 \to 11 \to 12 \to 13$$

We believe the chapter on superconductivity is a novelty in this kind of textbook. Our presentation makes it possible for instructors to find a relatively easy way to explain most of the relevant concepts in superconductivity within the framework of a course on classical electrodynamics.

Preliminary versions of the text were sent to many colleagues and we have received suggestions and comments from several of them. In particular, we would like to thank A. Rubens de Castro (University of Campinas) and Alberto Rojo (University of Michigan). One of the authors (ALD) spent a sabbatical semester in 1996 at the Department of Theoretical Physics of Condensed Matter, of the Universidad Autónoma de Madrid, where he completed some parts of the text. It is a pleasure for him to thank Prof. Fernando Flores for his comments and support.

We thank our many colleagues and students at Instituto Balseiro who have contributed to the improvement of the text. We received countless suggestions from those who read sections or certain chapters. We have especially appreciated the collaboration of Néstor Arista, Andrés García and Gerardo Aldazabal who carefully read the complete text, and contributed useful suggestions. Alfredo Tolley, Marcelo Kuperman and Javier Briático cooperated in the selection of the problems. Enzo Dari collaborated with the artwork and Ann Montemayor–Borsinger helped us with the English. All of them deserve our warmest acknowledgement.

San Carlos de Bariloche
February 1999

A. López Dávalos
D.H. Zanette

Contents

Appendix

1. Historical Perspective of Electromagnetism

When we study electromagnetism from a historical perspective, we are confronted with the convergence of three currents of scientific thought originating in seemingly disparate areas. In electromagnetic theory these areas end up being parts of a single conceptual structure. The historical lines of scientific research converging in electromagnetism include the study of

- the structure of the Universe or cosmology, i.e. humankind's desire to know our location in the Universe;
- the phenomena of attraction and repulsion of electric charges and of magnets started by the ancient Greeks, to whom in fact we owe the two parts of the word *electro-magnetism*;
- optical phenomena, i.e. the nature and behavior of light, its decomposition into colors and its velocity of propagation.

The first systematic cosmological vision was elaborated by Ptolemy (Klaudios Ptolemaios, or Ptolemæus, 100–170) of Alexandria and developed in his book *Almagesto* around the year 140. It assumes the Earth to be at the center of the Universe, with the celestial sphere rotating around it. This hypothesis, based on the testimony of the senses, was more welcome than a previous one, proposed by Aristarcus of Samos (280 BC), who assumed the Earth to rotate around the Sun. This hypothesis was in fact considered offensive to the gods.

Ptolemy had a remarkable knowledge of astronomy. In his treaties he demonstrated that the Earth was round. For this he argued that if the polar star rises higher northwards than southwards, the Earth should necessarily be curved in that direction. His argument in favor of the curvature in the east–west direction proved to be an even sharper observation. If clocks are set in time with the Sun, an eclipse of the Moon is found to take place at different times in two places separated in the east–west direction. Since a lunar eclipse is a single objective event to be seen everywhere at the same time, Ptolemy concluded that the clocks were not synchronized, because the bending of the Earth in the east–west direction makes the Sun pass the local meridian of different places at different times.

In spite of this knowledge, Ptolemy was not ready to admit either the Earth's rotation or its translation. According to his vision, rotation of our planet would leave behind everything that is on it, so birds and air could not follow this motion. He also objected to the possibility of the Earth moving through the celestial sphere, because in such a case we would be getting closer to its limits, and some parts would be seen larger than others. These ideas seem to prove that Ptolemy dedicated quite some time to thinking about such possibilities. They also demonstrate that he had little knowledge of mechanical laws. His philosophical and physical ideas were in fact based on those of Aristotle.

Ptolemy's cosmological vision dominated astronomical knowledge during more than a thousand years. Demonstrating a great independence of thought a Polish astronomer, Nicolas Copernicus (Nikolaj Kopernik, 1473–1543), undermined those ideas through a new vision. In 1530 he finished his work *De Revolutionibus Orbium Cœlestium*, in which he claimed that the Earth and the planets rotate around the Sun.

Copernicus' ideas were not based on evidence which would not have been available to Ptolemy. His arguments invoke exclusively reasons of simplicity. He reached the conclusion that if the stars rotate around the Sun the most distant ones should do so at very high speeds, a fact that contradicts common sense. Copernicus also accurately calculated the radii of the planetary orbits and concluded that the Sun is somehow removed from the center of those orbits.

Thanks to his ability to build astronomical instruments Tycho Brahe (1546–1601) opened the way for precise determinations of planetary orbits. His observations were completed by his assistant Johannes Kepler (1571–1630), whose fame surpassed that of his teacher. Kepler worked with Brahe in Prague where the latter had looked for refuge, exiled from his country of birth, Denmark.

Brahe developed the sextant, which allowed him to increase the precision of cosmographic determinations to half a minute of arc. With these instruments, Kepler carried out observations leading him to find the "harmonies" of the planetary system. In 1619 he published his work *Harmonies of the Universe* in which he presents his observations and conclusions. In this book he says:

At last I have found it, and my hopes and expectations are proven to be true that natural harmonies are present in the heavenly movements, both in their totality and in detail – though not in a manner which I previously imagined but in another, more perfect, manner... If you forgive me, I shall be glad; if you are angry, I shall endure it. Here I cast my dice and I write a book to be read by my contemporaries or by the future generations. It may wait long centuries for its reader. But even God himself had to wait six thousand years for those that contemplate His work.

In spite of the magnificence of his work, Kepler was not able to admit the possibility, considered by Giordano Bruno (1548–1600), that the distant stars could be independent solar systems. Soon after the work of Kepler, Galileo Galilei (1564–1642) created a substantial change in the means of astronomical observation by introducing the telescope. Galileo's observations immediately interested Kepler, at the same time that they were subject to jeer or were disregarded by other scientists of the time. About these attitudes it is illustrative to quote one of Galileo's letters to Kepler, where he says:

I am very grateful that you have taken interest in my investigations from the very first glance at them and thus have become the first and almost the only person who gives full credence to my contentions; nothing else could be really expected from a man with your keenness and frankness. But what would you say to the noted philosophers of our University who, despite repeated invitations, still refuse to take a look either at the Moon or the telescope and so close their eyes to the light of truth? This type of people regards philosophy as a book like Æneid or Odyssey and believe that truth will be discovered, as they themselves assert, through the comparison of texts rather than through the study of the world or Nature. You would laugh if you could hear some of our most respectable university philosophers trying to argue the new planets out of existence by mere logical arguments as if these were magical charms.

Galileo's biggest victory was due to his trust in observations.

Isaac Newton (1643–1727) was the great unifier who played the final role in this chapter of cosmology. He combined Copernicus', Kepler's and Galileo's individual discoveries in a coherent system. Newton perceived that Galileo's observations on the free fall of bodies were linked to his astronomical observations. He concluded that the Earth's power of attraction extended also to other planets and that it was a property common to all masses. Newton persevered for more than twenty years to correct an error in his estimates of distances, that prevented him from correctly calculating the radius of the lunar orbit. When he understood the origin of the discrepancy, he was able to set up a harmonic framework. The theory he developed was confirmed through precise calculations of the planetary orbits, and included Copernicus' heliocentric conception, Kepler's laws on the orbits and Galileo's laws on the free fall of bodies.

These concepts established the basis for the theory of relativity in classical mechanics. They formed the system of ideas that physicists had at their disposal at the end of the nineteenth century to interpret optical phenomena, the existence of the ether as a medium in which electromagnetic phenomena took place, and the Earth's motion relative to this medium. Einstein's restricted theory of relativity, presented in Chap. 2, as well as his theory of general relativity, form the basis of modern cosmology.

Although the Greeks had already studied the properties of rubbed amber ($\eta\lambda\varepsilon\kappa\tau\rho o\nu$) and of loadstones of magnetite (FeO–Fe_2O_3), a mineral found in the region of Magnesia in Turkey, it is not until the eighteenth century that research on these topics acquired momentum. Thus, although no satisfactory explanation from first principles was given in that century, at least a coherent phenomenological description emerged. The early history of electricity and of magnetism is plagued with empirical observations and a lot of imprecision, blended with a few really important achievements. The relative ease of experimenting with magnets, compared to the difficulty of obtaining reproducible results in electrostatics, made progress in the study of magnetism relatively fast when compared to that of electricity.

William Gilbert, born in 1544, published in 1600 his monumental treatise *De Magnete*. In this treatise he summarizes all the experiments carried out by him in the seventeen years prior to the publication of this book. Before that, some specialists in magnetism like the Italian scientist Giambattista della Porta (1535–1615), founder of the first scientific academy, tried to prove the lack of support of some popular beliefs. One of these was the superstition of seamen claiming that garlic and onions affected magnets in such a way that it was dangerous to toss one's breath to the compass after having eaten some of those vegetables.

Besides sensing that the Earth behaves like a large magnet, similar to the spherical magnets with which he had experimented, Gilbert gave quite a precise description of the magnetic field:

Rays of magnetic virtue spread out in every direction in an orbe; the center of this orbe is not at the pole (as Porta reckons) but at the center of the stone and of the terella.

Although a clear description of the experiments was given, Gilbert lacked a theoretical framework which would give a coherent vision of what he saw or experienced. His theory of magnetism was based on the belief that magnets had a soul inherited from the Earth, which he believed should also have one.

Experimentation with static electricity was boosted after the construction by Francis Hauksbee (1666–1713) in 1706 of a machine for electrostatic generation and accumulation of charge with great efficiency. In 1745 the Dutch Pieter Musschenbroeck (1692–1761) of Leyden discovered that dielectrics increase the capacity to accumulate charge. He fixed a wire through the cork of a metallic container filled with water, which allowed him to charge the *condenser* he had just invented. The first manifestation of the increased capacity of the device took place when an assistant accidentally touched the wire and suffered a strong discharge. Musschenbroeck perfected this system in 1706 linking it to a rubbing electric generator. It was popularized as the *Leyden Bottle*.

A current of thought setting the basis for modern developments of the theory of electromagnetic phenomena appeared in the second half of the eighteenth century. In a way similar to what happened with heat theories, it had been initially postulated that electric and magnetic effects were transmitted through fluids. In 1729 Stephen Gray (1670–1736) showed that electric charge could be transmitted from one body to another through metals or other materials. Benjamin Franklin (1706–1790) postulated that the entity transmitting charge was a single fluid whose excess or defect produced accumulation of charge of one sign or the other. In 1759, a treatise applying the theory of the single fluid both to electricity and magnetism was published in Saint Petersburg. Its author, Franz Maria Aepinus, exposed very clearly Franklin's ideas and fostered their diffusion.

In 1733 Charles François de Cisternay du Fay showed the existence of two types of electricity which he called resinous and vitreous, because they appeared when rubbing resin with skin, and glass with silk, respectively. As this idea competed strongly with that of Franklin, many experiments were performed trying to prove one or the other. Meanwhile, the Swede Johan C. Wilcke and, independently, the Dutch Anton Brugmans presented a two fluid theory in the area of magnetic phenomena in 1778.

In 1785, Charles Augustin Coulomb (1736–1806) published his first experimental results on the attraction of electric and magnetic charges. His measurements, carried out with the torsion balance he had invented, allowed him to reach precise conclusions and to make quantitative determinations on the laws governing the forces between charges and magnets. Coulomb was a positivist and, as such, restricted his descriptions to what he observed, without taking the risk of elaborating any theory or of advancing any interpretation. Precise observation and detailed experimentation in Galileo's style played in his experiments a fundamental role. In this sense the torsion balance, as a tool for the advancement of knowledge, can be compared with Tycho Brahe's sextant or with Galileo's telescope. Coulomb's work marks the beginning of a very intense activity in this subject. The mathematical description of his observations was developed by Siméon Denis Poisson (1781–1840), who however did not participate in the later progress of the theory.

Further research was encouraged by the invention of the electric pile in 1800. Alessandro Volta (1745–1827) put at the disposal of other scientists a compact and easily handled physicochemical source of electricity. Volta himself made important contributions to the physiological effects of electricity and, in particular, determined the consequences of electric stimuli in different parts of the body. Volta proved that conductors applied to his eyes gave place to an excitation of the optical nerve that made him *"see the stars."*

In 1820 Hans Christian Oersted (1777–1851) was looking for a connection between electric and magnetic phenomena when he discovered that electric currents were able to deflect magnetized needles. This result, connecting two phenomena initially considered independent, produced a revolution in scientific circles. It immediately gave rise to a series of experiments allowing this connection to be quantified. The most outstanding works were those of Jean Baptiste Biot, Félix Savart, Dominique François Arago and André-Marie Ampère.

Arago (1786–1853) was the first to inform the Academy of Sciences of Paris, on September 11th, 1820, of results proving that an electric current behaves in many respects like a magnet. Together with Ampère (1775–1836), he carried out the first magnetization experiments made with a solenoid. A month later, in England, Sir Humphrey Davy (1778–1829) presented results similar to those of Ampère and Arago.

One week after Arago's presentation Ampère put forward his hypothesis that ferromagnetism could be due to internal electric currents flowing perpendicularly to the axis of the magnet. After his death, letters by Augustin Jean Fresnel (1788–1827) dating back to 1821 were found among his papers suggesting that the currents causing ferromagnetism should be molecular and not macroscopic in origin. The fact that these currents do not produce heating, as macroscopic currents do, was among the reasons given by Fresnel for this hypothesis. These ideas were corroborated almost a hundred years later through the study of atomic structure.

Michael Faraday was born in 1791 and died in 1867. In 1813, while he worked as an artisan printer, he attended lectures by Sir Humphrey Davy. The latter took him on as his assistant after Faraday presented him with the notes he had taken of his classes. Starting in 1821, he carried out countless experiments in electricity: he experimented with electromagnetic induction, found the opening and closing currents of circuits and introduced the concept of lines of force. As a consequence of his work *"on the rotation of a current around a magnet and the rotation of a magnet around a current,"* as well as the liquefaction of chlorine, he was chosen as a member of the Royal Society.

Faraday conceived the idea of the cage bearing his name as a shielding against electromagnetic induction. He worked in electrochemistry and enunciated the law of equivalence relating the elementary electric charge with the number of atoms per cubic centimeter of a substance. In this way he facilitated the first determination of the elementary charge and suggested the atomic character of electricity. He discovered the optical effect that bears his name which shows the action of a magnetic field on the propagation of light in a material medium. Of all his achievements, surely the one which most impacted subsequent developments of electromagnetism was the concept of a field. In Maxwell's own words:

Faraday, with his mind's eye saw lines of force traversing all space, where the mathematicians saw centers of force attracting at a distance: Faraday saw a medium where they saw nothing but distance... I also found that several of the most fertile methods of research discovered by the mathematicians could be expressed much better in terms of ideas derived from Faraday than in their original form.

From the most remote antiquity humankind has been interested in the phenomena associated with light. Empedocles (490–430 BC) and Euclid (c. 300 BC) left the first known writings on optical phenomena. In these they discuss the nature of light, and observations are mentioned on rectilinear propagation, reflection and refraction. René Descartes (1596–1650) considered light to be a kind of pressure transmitted through a medium filling all transparent bodies, the ether. According to this hypothesis, the origin of colors was due to rotational motions with different velocities of the particles of this medium.

The law of refraction was discovered by Willebord Snell (1591–1626) in 1621. In 1657 Pierre de Fermat (1601–1665) enunciated the principle of minimum time according to which light goes from one place to another in the smallest possible time. According to this idea, refraction is due to the resistance that different media present to the passage of light.

Interference phenomena were discovered independently by Robert Boyle (1627–1691) and by Robert Hooke (1635–1703). Hooke was the first to postulate that light consists of vibrations spreading instantaneously or with high speed. In fact, decades before Hooke's birth, Galileo had tried to measure the velocity of light, coming to the conclusion that it was too high for his experimental method. The confirmation that light has a finite propagation velocity was due to Ole Römer (1644–1710) who based his conclusions on the observations of the eclipses of Jupiter's moons.

When Newton demonstrated that white light could be broken down by means of a prism and that each color had an index of refraction of its own, he revealed a fundamental property of colors. In spite of this, the reservations he expressed towards the undulatory theory produced a general rejection of this theory, until the interference experiments of Thomas Young (1773–1829) in 1802 gave new support to it. Previously, the undulatory theory had been backed by Christian Huygens (1629–1695). He enunciated the principle that bears his name, which he applied to explain all known optical phenomena.

Supporters of the corpuscular theory, who assumed light to consist of particles propagating in straight lines, proposed an explanation of diffraction phenomena for the annual prize of 1818 of the Paris Academy of Sciences. Instead, the Academy rewarded Augustin Fresnel's work which, based on the undulatory hypothesis, was the first of a series that would defeat the corpuscular theory. Among those defeated in the competition for the prize of 1818 were Pierre Simon de Laplace (1749–1827) and Jean Baptiste Biot (1774–1862). The essence of Fresnel's work was a combination of Huygens' principle with Young's interference principle. With these elements he showed that he could explain both the rectilinear propagation of light and diffraction effects. Arago made a fundamental contribution to the wave theory when he experimentally confirmed that at the center of the shade of a small disk there appears a circular luminous point.

Fresnel began studies on the nature of light coming from stellar sources, comparing it with light from terrestrial sources. Arago was able to show that, for all practical purposes, light from both types of source behaves in a similar way. On the basis of these results Fresnel developed a theory of the partial drag of the ether by the Earth. This hypothesis was confirmed by Armand Hippolyte Louis Fizeau (1819–1896) in an interference experiment, using light propagating in a moving liquid.

Fresnel and Arago also demonstrated that two light rays polarized at ninety degrees to one another do not interfere. This result, incompatible with longitudinal light waves, made Young postulate that the vibrations were transverse. Study of the elastic properties of the ether was based on this hypothesis. Scientists who also set up the theory of elasticity such as Claude Louis Marie Henri Navier (1785–1836), Augustin Louis Cauchy (1789–1857), Young himself and others took part in these developments.

Until the last decades of the 19th century, the development of optics was totally independent of the evolution of electromagnetic theory. The unification of both areas in a single structure was due to James Clerk Maxwell (1831–1879) who was able to summarize the results of the phenomenology of electricity and magnetism in the group of equations that bears his name. Maxwell's work was not limited to compiling the laws describing such phenomena, but he also introduced some very important concepts. In particular, he formalized the notion of electric and magnetic fields. He also put forward the principle of conservation of electric charge – that was to find its more rigorous form in the theory of relativity – and modified the laws of electromagnetism in order to satisfy this principle.

Maxwell stood out already in his childhood as an extremely inquisitive boy. At the age of ten he attended Edinburgh Academy as a boarder. Before he turned fifteen, a work of his on a method to build ellipses was read at the Royal Society of Edinburgh. At sixteen he entered Edinburgh University, where he studied for three years. At seventeen he used birefringence to locate the tensions undergone by a transparent material, a technique later transformed into an important tool in engineering. His conclusions delighted Nicol who in recognition presented him with two of his polarizing prisms. Shortly afterwards Maxwell entered Cambridge to attend the famous Trinity College. He not only cultivated the study of sciences, but was also interested in philosophy and developed an ability to write both in prose and in verse. After concluding his studies, he returned to research publishing a work on the lines of force proposed by Faraday, with whom he developed a personal friendship.

In his first works on electromagnetism Maxwell constantly appeals to mechanical models. In 1864, however, he presented to the Royal Society his work *Dynamic Theory of the Electromagnetic Field* in which he does without the mechanical model, and makes a description of the field by means of twenty differential equations. His theory was expounded in several publications and summarized in his *Treatise of Electricity and Magnetism*, published in 1865. This book is not easy to read, and modern versions of the theory are due to revisions by Heaviside and by Hertz. Maxwell, however, stuck to the idea that a medium was necessary for electromagnetic phenomena to occur. In the closing lines of the *Treatise* he says:

Hence all these theories lead to the conception of a medium in which the propagation takes place, and if we admit this medium as an hypothesis, I think it ought to occupy a prominent place in our investigations, and that we ought to endeavour to construct a mental representation of all the details of its action, and this has been my constant aim in this treatise.

One very important result of electromagnetic theory was the prediction of the possibility of propagation of electromagnetic waves with a velocity which could be derived from purely electrical and magnetic measurements. Rudolph Kohlrausch (1809–1858) and Wilhelm Weber (1804–1894) performed these measurements in 1856 and established that the velocity of propagation of electromagnetic waves was the same as that of light. As a consequence, Maxwell postulated that light consists of electromagnetic waves. In 1888, Heinrich Hertz (1857–1894) generated electromagnetic waves starting from oscillating electric circuits.

In spite of its success in the description of optical and electromagnetic phenomena, Maxwell's theory was initially resisted because, as we said above, in its original formulation it used mechanical elements for which a satisfactory model was lacking. However the idea was accepted gradually that electromagnetic fields cannot be reduced to simpler elements and that the mechanical elements postulated by Maxwell are not essential. By showing that electricity and magnetism are nothing but two different manifestations of a single phenomenon, Maxwell's formulation constitutes a first example of a unified description of physical interactions.

Besides his contributions to electromagnetism, Maxwell made contributions of great importance to the kinetic theory of gases, setting down the basis for the foundation of statistical mechanics. He was also the first director of the Cavendish Laboratory. In his inaugural conference on experimental physics, he rejected the idea that progress in physics was coming to an end, and that, as some of his colleagues mantained, the only thing left was to devote oneself to a refined determination of well-known fundamental constants. In that same conference he stated that the new laboratory should coordinate its work with other important centers in the country and abroad to assure a swifter advancement of knowledge.

The search for a relativity principle for the description of electromagnetic phenomena, as well as the experimental confirmation of the nonexistence of an absolute reference system, fixed to the ether, relative to which Maxwell equations were valid, led to the formulation of the theory of special relativity by Albert Einstein (1879–1955) in 1905. Abandoning the idea of a privileged reference frame had cosmological implications comparable to the formulation of the Copernican heliocentric theory. In 1916 Einstein demonstrated that the theory of relativity could be generalized to include systems in arbitrary motion. This led him to unify the description of accelerated systems with that of gravitational forces through the so-called *equivalence principle.* General relativity allowed the formulation of a cosmological theory based on first principles.

At the time of Maxwell's formulation, electromagnetic forces and gravity were the only two forces known in Nature. Einstein dreamt all his life of unifying the description of electromagnetic and gravitational phenomena. As with electromagnetic forces, making Maxwell's theory compatible with the principles of quantum mechanics gave rise to quantum electrodynamics, a theory capable of predicting with high precision the behavior of electrons and photons. We know today that besides these forces there are others governing the behavior of the *elementary particles*, the last constituents of matter.

Elementary particles are divided into three groups: leptons, quarks and intermediate bosons. Leptons comprise the electron, the muon, the tauon and their three neutrinos. Quarks, in six varieties, are in turn the constituents of hadrons encompassing the baryons (proton, neutron, lambda, sigma, etc.) and the mesons (pions, kaons, etc.). The intermediate bosons, i.e. the photon, the W and Z bosons and the gluons, are the particles that mediate the electromagnetic, weak and strong forces, respectively. Weak interactions characterize the forces between leptons and quarks. Strong interactions manifest themselves in the forces between quarks and, in an indirect way, between hadrons. The electromagnetic interaction describes the force among charged particles. Finally, gravitation, mediated by a not-yet-found graviton, is always present.

Between 1964 and 1967 Abdus Salam (1926–1996) and, independently, Steven Weinberg formulated a unified theory of quantum electrodynamics and of the weak interactions, called electroweak theory. More recently, the theory of strong interactions called quantum chromodynamics has been unified with the electroweak theory into a single theory called grand unified theory. In this description the three forces – electromagnetic, weak and strong – turn out to be three aspects of a single interaction, just as electricity and magnetism are two aspects of the force originating in the electric charge. In this case the unification happens at distances of the order of 10^{-28} cm. If two particles, whatever their type at larger distances, approach at distances smaller than the unification range, the apparent distinction between them disappears and their interaction is due to a single force.

The desire to unify all forces, partially described above, is supplemented with the more ambitious idea, similar to Einstein's, of including gravity in a single theory. This theory, still in its speculative stages, would show, if experimentally verified, the connection between the study of the forces between elementary particles at microscopic distances on one hand and the cosmological description of the structure and origin of the Universe on the other.

Further Reading

M. Born and E. Wolf: *Principles of Optics* (Pergamon, New York, 1959).
J.M. Jauch: *The Trial of Galileo Galilei* (CERN, Genève, 1964).
D.C. Mattis: *The Theory of Magnetism* (Harper, New York, 1965).
H. Reichenbach: *From Copernicus to Einstein* (Philosophical Library, New York, 1942).
A. Sommerfeld: *Electrodynamics – Lectures on Theoretical Physics Vol. III* (Academic Press, New York, 1952).

2. Relativistic Kinematics

A fundamental characteristic of the theory of electromagnetic phenomena is its incompatibility with the laws of classical mechanics. Indeed, application of Galilean transformations and of Newton's equations to the evolution of systems governed by electric and magnetic interactions leads to results in contradiction with experimental observations. Since the dynamics of material particles must frequently be included in the study of electromagnetic phenomena, the inconsistency between electromagnetism and classical mechanics became one of the major problems of physics at the end of the nineteenth century. The solution to this problem was found through the formulation of the special theory of relativity by Albert Einstein in 1905. This theory modifies the concepts of space and time of classical physics, and replaces Newtonian mechanics in the description of dynamical processes. The theory of special relativity, which is consistent with Maxwell's formulation of electric and magnetic phenomena, constitutes the fundamental basis for describing the evolution of mechanical systems subject to electromagnetic interactions. It is therefore convenient to begin the study of electromagnetism with an analysis of the principles of special relativity and the most important consequences derived therefrom. This chapter and the following one are aimed at this purpose.

2.1 Relativity in Classical Mechanics

Although the principle of relativity is usually associated with the theory developed by Einstein, it is important to emphasize that the laws of classical mechanics satisfy a similar principle. This is stated by the fact that Newton's laws have the same form in any inertial reference frame. This property of physical laws is called *covariance.*

This means that the three laws:

- Principle of inertia,
- $\boldsymbol{F} = \dot{\boldsymbol{p}} = m\boldsymbol{a}$,
- Principle of action and reaction,

have the same form in all inertial systems.

The covariance of the principle of inertia becomes clear by keeping in mind the definition of an *inertial reference system.*

Newton's second law can be written in each of its components as:

$$F_j = m\frac{\mathrm{d}^2 x_j}{\mathrm{d}t^2}. \tag{2.1}$$

Study of the covariance of these equations leads us to analyze how the three quantities involved in this law transform under a change of reference system. Mass is introduced in classical mechanics as an invariant quantity, characteristic of a moving body. In a new reference system the mass m' is thus identical to m:

$$m' = m. \tag{2.2}$$

For the acceleration we observe, without loss of generality, that if the new system moves with respect to the original one with velocity v in the x direction, the transformations of space and time between both systems are the so-called Galilean transformations

$$\begin{aligned} t' &= t, \\ x' &= x - vt, \\ y' &= y, \\ z' &= z, \end{aligned} \tag{2.3}$$

From here it follows that

$$\frac{\mathrm{d}^2 x_j}{\mathrm{d}t^2} = \frac{\mathrm{d}^2 x'_j}{\mathrm{d}t'^2}, \tag{2.4}$$

that is to say, $\boldsymbol{a}' = \boldsymbol{a}$. Finally, since the relationship (2.1) can be considered as a definition of force, the Newton's second law is covariant.

From this definition of *force* we can also derive the covariance of the principle of action and reaction: since acceleration and mass are invariant, forces among interacting bodies do not vary when being considered from different inertial reference frames.

2.2 Prerelativistic Observations

Optical and electromagnetic phenomena are explained by the laws enunciated by Maxwell in the second half of the nineteenth century. Since these laws are not invariant under transformations of type (2.3), before Einstein proposed his theory of special relativity it was assumed that the enunciation of a principle of relativity encompassing electromagnetic phenomena was impossible.

In particular, the existence of a privileged reference frame, with respect to which Maxwell equations were valid, was postulated. The idea of a privileged system was also associated with another different concept. The existence of electromagnetic waves – first foreseen theoretically and then verified experimentally – required, in the classical mind, the existence of a material medium. This would serve as support for the electric and magnetic fields – "*a subject was needed for the verb to undulate.*" This medium was called *ether*, and the reference system privileged by electromagnetic laws was recognized as fixed to it. The ether hypothesis gave place to several experiments and observations aimed at detecting its state of motion relative to the Earth.

Here we describe three of these observations, with different results in each case.

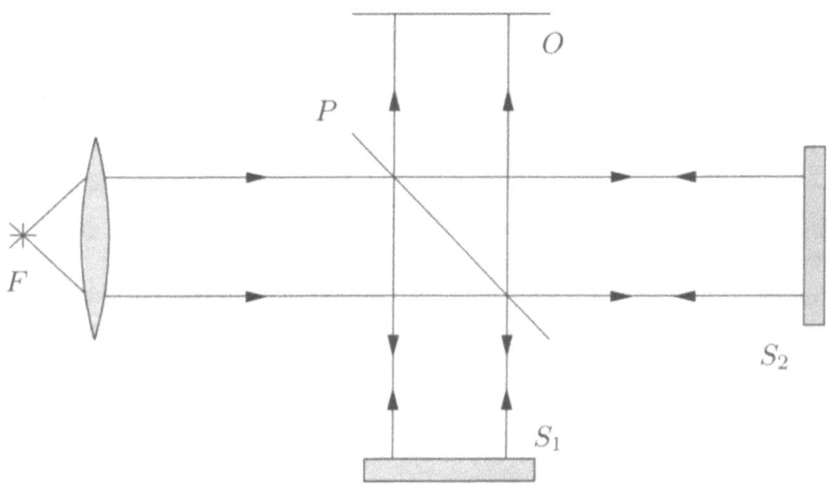

Fig. 2.1. Michelson–Morley interferometer

2.2.1 The Michelson–Morley Experiment (1881–1904)

Not being a privileged body in the Universe, the Earth cannot be expected to be fixed to the ether. Since our planet moves around the Sun at a speed of 30 km/s, it was assumed that its velocity relative to the ether would be at least of that magnitude. To detect this motion Albert Abraham Michelson (1852–1931) and Edward William Morley (1834–1923) used an interferometer like the one shown in Fig. 2.1. The experiment was carried out, for the first time, in 1881, and subsequently repeated with several improvements up to 1904 [2.1]. Light from the source F is divided in two by the half-silvered plate P. These beams are reflected at the mirrors S_1 and S_2, and return through the same plate P to the observation point O, where an interference pattern of fringes is obtained. Let us suppose the apparatus to move relative to the ether with a velocity v parallel to S_1P. According to Galilean transformations, the velocity of light adds to that of the interferometer, and therefore the difference in optical paths in the segments PS_1P and PS_2P, given by the difference in the times spent by light in traversing them, is

$$\Delta = \frac{2}{c\sqrt{1-\beta^2}}\left(\frac{l_1}{\sqrt{1-\beta^2}} - l_2\right). \tag{2.5}$$

Here l_1 and l_2 are the distances between the plate P and the corresponding mirrors, c is the speed of light, and $\beta = v/c$.

Rotating the instrument by 90°, l_1 and l_2 interchange their roles, and we have

$$\Delta' = \frac{2}{c\sqrt{1-\beta^2}}\left(l_1 - \frac{l_2}{\sqrt{1-\beta^2}}\right). \tag{2.6}$$

The expected change in the interference pattern is proportional to the difference between Δ and Δ', given by

$$\Delta' - \Delta = (l_1 + l_2)\beta^2. \tag{2.7}$$

All attempts carried out over two decades of work showed, however, no measurable difference between the quantities Δ and Δ'.

This result allows us to conclude that **the ether is fixed to the Earth** or that, since it is impossible to detect its motion, it is at least dragged by our planet.

2.2.2 The Aberration of Fixed Stars (1728)

Once sufficiently fine astronomical observations became possible, it was observed that the apparent position of a star changes over the year [2.2]. For a star located perpendicularly to the plane of the Earth's orbit this angular displacement is a maximum, and reaches 20.5″ of arc with respect to the average position. This phenomenon, called stellar aberration, was discovered by James Bradley (1693–1762) in 1728. Bradley explained it by observing that the superposition of the motion of the Earth, at a velocity $v = 30$ km/s, with that of light coming from the star, results in light arriving at an angle α with respect to the real direction of the star (Fig. 2.2), given by

$$\tan\alpha = \frac{v}{c}. \tag{2.8}$$

The effect is similar to the inclination with which we see rain falling, on a windless day, from a moving car's window. The result (2.8) would be different if the star were in a different location with respect to the ecliptic. In any case, it can be explicitly calculated, and it explains the observed value for aberration quite precisely. In fact, this phenomenon was used by Bradley as an alternative method to measure the speed of light. As a conclusion, we can say that **the ether is at rest with respect to the fixed stars, and the Earth moves relative to it.**

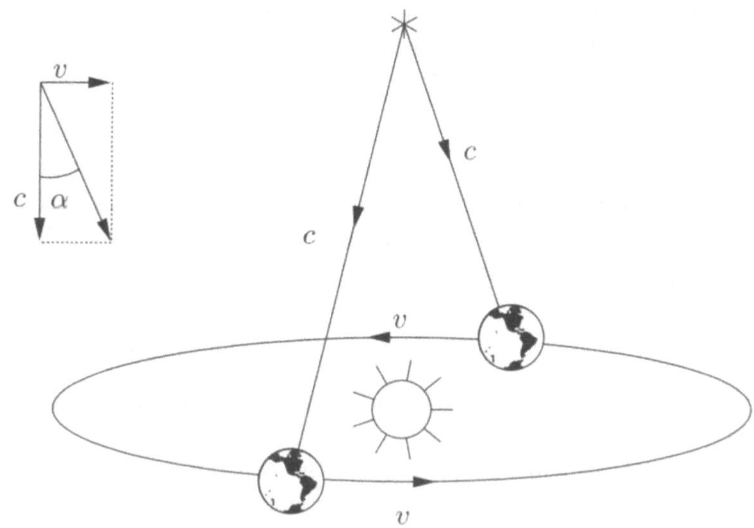

Fig. 2.2. Aberration of fixed stars due to the motion of Earth

2.2.3 Fizeau's Experiment (1865)

By means of an interferometric method of coherent beams which traverse a liquid current (Fig. 2.3), Armand Fizeau measured the speed of light in a moving material medium.

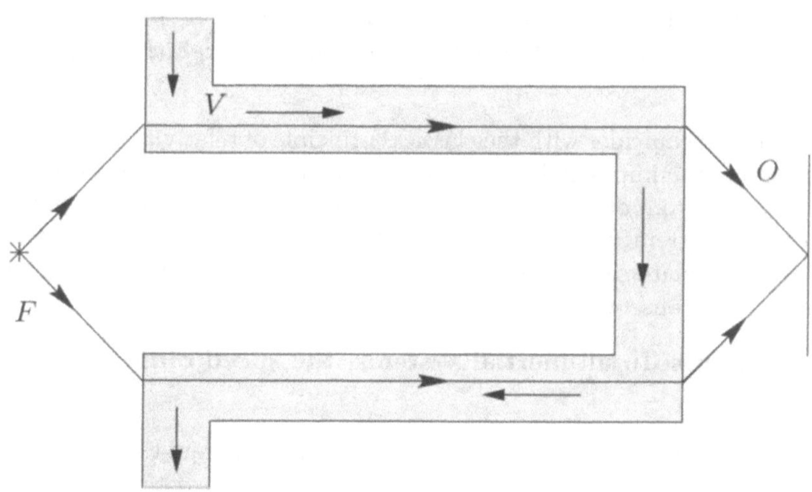

Fig. 2.3. Fizeau's interferometer

The result obtained indicated that the velocity of the material medium adds to the velocity of light, not in all its magnitude but affected by a drag coefficient:

$$v = \frac{c}{n} \pm kV. \tag{2.9}$$

Here c/n is the speed of light in the medium at rest, n the index of refraction of the medium, and V the velocity of the liquid in the interferometer. The drag coefficient is given approximately by

$$k \approx 1 - \frac{1}{n^2}. \tag{2.10}$$

Then, according to this experiment, **the ether is partially dragged along by the material medium**.

Towards the end of the nineteenth century, the contradictory results of these three observations claimed for an urgent explanation.

2.3 The Special Theory of Relativity

The explanation of these seemingly contradictory results on the state of motion of the ether was proposed by Albert Einstein in his work *Zur Elektrodynamik bewegter Körper*[1] [2.3]. Einstein enunciated two principles that constitute the basis of the special theory of relativity, which deeply modified classical ideas about space and time.

- First principle: **The laws expressing the evolution of physical systems are the same in all reference systems in relative uniform motion.**

This enunciation coincides with the classical principle of relativity. It is equivalent to stating the impossibility of detecting the motion of uniform translation of a system from observations made in that same system. Only relative velocities of a system with respect to another are relevant. Because of its nature, this postulate cannot be verified experimentally in a direct form but only through its consequences.

- Second principle: **In all inertial systems, the speed of light in vacuum is** c.

That is to say, all observers in uniform relative motion measure the same value for the speed of light. This second postulate eliminates the existence of a privileged system associated with the propagation of light and, consequently, with the validity of Maxwell's equations. Together with the first principle, it leads

[1] *On the Electrodynamics of Moving Bodies*

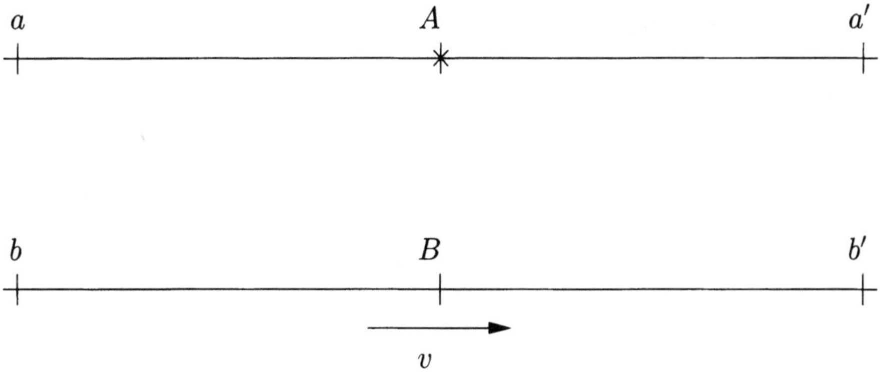

Fig. 2.4. Reference frames in relative motion

to the conclusion that Galileo's classical transformation laws (2.3) should be replaced by the Lorentz transformations, introduced by Hendrik Antoon Lorentz (1853–1928). Under these transformations Maxwell's equations are covariant.

In order to understand better the need for modifying the notions of space and time implied by the principles of the theory of relativity, let us imagine the following *thought experiment*, first proposed by Einstein in his original article. Let A be a reference system considered at rest, and let B be another system moving relative to A with velocity v (Fig. 2.4). In fact, according with the first postulate we cannot assign any meaning to the absolute velocities of each system and we can only speak of their relative velocity. Let us assume we choose the origin of time as the moment at which the space origins of both systems, points A and B, coincide. Let us suppose in addition that, at that moment, a light source placed at point A flashes. For an observer at A the spherical wave spreading with velocity c arrives simultaneously at the equidistant points a and a'. As seen from the other system, according to the second principle, the spherical wave is centered at B and should reach simultaneously the equidistant points b and b'. In particular, an observer at B sees the arrival of the luminous front at points a and a' at different times.

We conclude that the relativistic postulates force us to revise the concept of simultaneity. According to this experiment, the simultaneity of two events occurring at different points in space is relative to the reference system from which one observes them. This conclusion invalidates the classical concept of absolute time. In what follows, therefore, for a reference system S with space coordinates (x, y, z), we assign a characteristic timescale t in such a way that in a second system S′, with coordinates (x', y', z') the events are ordered according to a timescale t', different from t.

2.4 Reference Systems and Lorentz Transformations

Given a reference system, every *event* takes place at a given time at a certain point in space. The space location is determined by a system of Cartesian coordinates (x, y, z), so that the position of the event is established knowing the distances to the planes associated with each coordinate. These distances are measured with a standard ruler.

The definition of the time location of an event requires us to be more cautious. We assume that there is a clock at the origin of coordinates, and a set of clocks synchronized with the first one at each point in space. The time coordinate t of an event taking place at a given point is the time measured by the clock located at the point where the event happens. The postulate of invariance of the speed of light allows us to synchronize clocks by sending light signals from the origin to each point in space. The transformations between systems of coordinates refer to the space and time locations of an event in two systems S and S′, similar in every respect, in relative motion of uniform translation.

We assume that the Cartesian axes of both systems, (x, y, z) and (x', y', z'), are parallel, and that the relative motion is along the x axis with velocity v. The derivation of the transformations takes the following postulates implicitly into account:

- Whatever the relationship obtained between t and t', the sense of time flow is invariant. In other words, the thermodynamic laws of evolution toward equilibrium are valid in both systems. Moreover causality is invariant, so that events causally linked must take place in the same time order in both systems.
- Whenever a measurement of a physical quantity – for example, the wave length of an spectral line – is carried out, the result should be the same for any observer at rest with respect to the experiment. This postulate can be called *invariance principle of the proper quantities.*

The relationship between the space–time coordinates (t, x, y, z) and (t', x', y', z') can be obtained by analyzing a series of ideal experiments.

2.4.1 Experiment I

Let us imagine a couple of rods OP and $O'P'$ of identical proper length, at rest with respect to S and S′, respectively (Fig. 2.5). As a result of this experiment, it is observed that at the instant at which the origins of both systems coincide points P and P' also coincide.

In fact, let us admit for a moment that this were not so. We could assume then that the length of the rod $O'P'$, measured by S, is $O'P' = \alpha(v)OP$, since v is the only parameter involving the relationship between the reference systems. However, due to the equivalence between both systems, we must

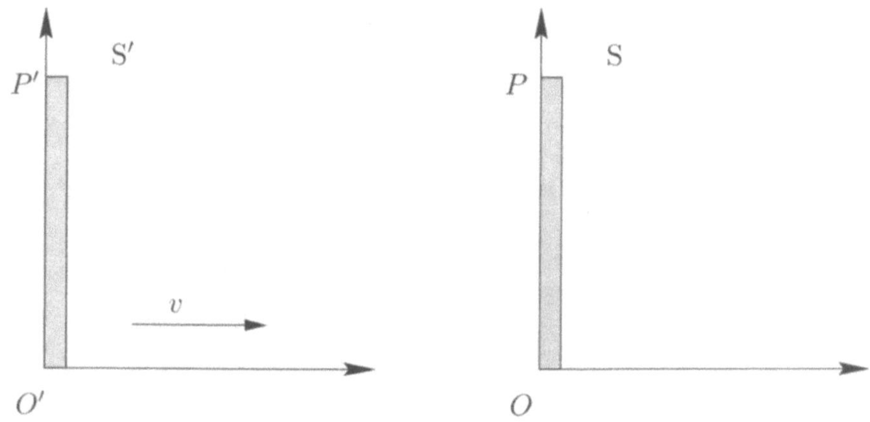

Fig. 2.5. Experiment I

have, according to what is observed in S', $OP = \alpha(-v)O'P'$. This implies $\alpha(v)\alpha(-v) = 1$. Now, performing the experiment with S' moving with respect to S at velocity $-v$, we would obtain $\alpha(v) = \alpha(-v)$. From here we conclude that $\alpha(v) = 1$. Because of the relativity principle, the only acceptable result is $\alpha = 1$. As a consequence, $O'P' = OP$ for both systems.

2.4.2 Experiment II

Due to the relativity of the concept of simultaneity, a clock in S cannot be compared with one in S', since they are not mutually stationary. On the other hand, we can compare the clock in S' with each one of the clocks in S located at the points where that clock passes by. Let us consider a light source at the origin of S', where the clock is also placed, such that it emits a signal perpendicularly to the relative velocity $\boldsymbol{v}$. This signal is reflected on a mirror and returns towards the clock.

For an observer in S, the time interval between the start and return of the signal is

$$\Delta t' = \frac{2y_0'}{c}, \tag{2.11}$$

(Fig. 2.6). For an observer at rest with respect to S the time interval Δt between the same pair of events can be measured with two synchronized clocks, separated a distance $v/\Delta t$ along the x axis.

Since c is invariant, this interval is given by

$$\Delta t = \frac{2y_0}{c}\frac{1}{\sqrt{1-\beta^2}}, \tag{2.12}$$

where $\beta = v/c$. According to Experiment I, $y_0 = y_0'$. Then from (2.11) and (2.12) we obtain

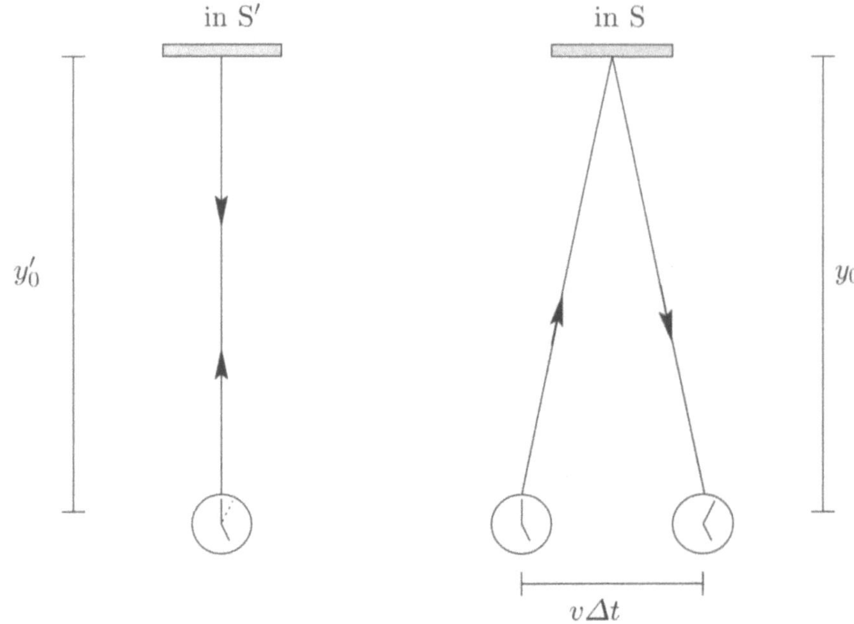

Fig. 2.6. Experiment II

$$\Delta t = \frac{\Delta t'}{\sqrt{1-\beta^2}}. \tag{2.13}$$

Then, the *proper time interval* $\Delta t'$, between two events that happen at the same point in space, corresponds in S to a longer interval Δt.

This phenomenon, unknown in classical physics and called the *dilation of proper time*, has received experimental confirmation: the disintegration of mesons captured in an absorbent shows that the half life of a meson at rest is $\tau' = 1.6 \times 10^{-6}$ s. On the other hand, measurements of the absorption in the atmosphere of mesons produced by cosmic radiation and whose velocity is close to c, show that these survive journeys of the order of 20 km. This implies that the half life of moving mesons is $\tau \approx 20\,\text{km}/c \approx 7 \times 10^{-5}$ s. From here it turns out that $\tau/\tau' \approx 50$ and $v \approx c\sqrt{1 - 1/2500} \approx 0.9998\,c$. This result is compatible with the measured energy of cosmic mesons.

2.4.3 Experiment III

Let us consider a rod of proper length x_0', parallel to the x axis at rest in S′, and let us try to establish its length in the system S. As a method of measuring lengths, a light beam is sent from one end of the rod towards the other, where it is reflected and returns toward its origin. Seen from S′ , the time elapsed between the start and the arrival of the beam is

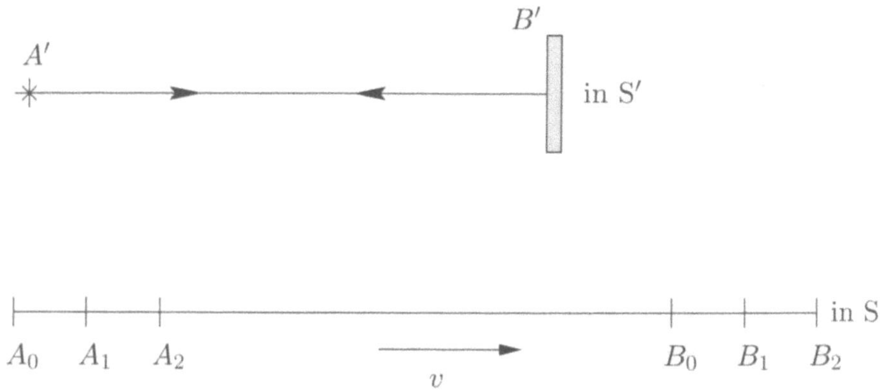

Fig. 2.7. Experiment III

$$\Delta t' = \frac{2x'_0}{c}. \tag{2.14}$$

In the system S, light was emitted in A_0, reflected at point B_1 and received on its way back at A_2 (Fig. 2.7). We know that while light travels from A_0 to B_1 with speed c, B_0 has moved up to B_1 with velocity v and therefore $A_0B_1/B_0B_1 = c/v$. Similarly, while light travels along the segment B_1A_2, A_1 moves up to A_2, and we have $B_1A_2/A_1A_2 = c/v$. Taking into account that the total time measured by S is $\Delta t = (A_0B_1 + B_1A_2)/c$, we obtain

$$\Delta t = \frac{2x_0}{c(1-\beta^2)}. \tag{2.15}$$

Finally, since $\Delta t'$ is a proper time, it is related to Δt according to (2.13), and so the lengths x_0 and x'_0 are linked by the relationship:

$$x_0 = x'_0\sqrt{1-\beta^2}. \tag{2.16}$$

This other fundamental consequence of the principles of relativity is called *Lorentz length contraction.* It was postulated by H.A. Lorentz in his paper *Electromagnetic Phenomena in a System Moving with any Velocity less than that of Light* [2.4], as a possible way of explaining the Michelson–Morley experiments, understanding it as an elastic phenomenon in the arms of the interferometer. Its correct interpretation, in the framework of special relativity, is due to Einstein. The Lorentz contraction implies that the length of an object measured by a moving observer in the direction of the motion is smaller than when the observer and the object are at mutual rest. As an example, let us consider the electrons of the linear accelerator LINAC of Centro Atómico Bariloche, which consists essentially of a 3 m long tube formed by a succession of resonant cavities where a microwave accelerates electrons. These electrons soon reach an energy of 25 MeV and, therefore, acquire a velocity $v \approx 0.9998\,c$. At this velocity, $\sqrt{1-\beta^2} \approx 1/50$, so that the 3 m of the LINAC measure, for an observer fixed to the electrons, about 6 cm.

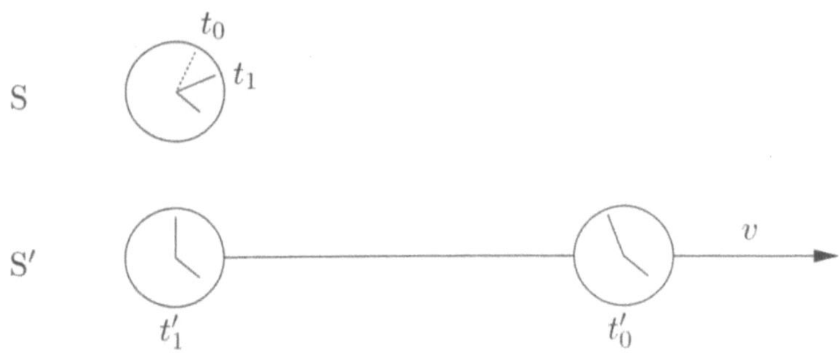

Fig. 2.8. Experiment IV

2.4.4 Experiment IV

Let us consider now the change in the synchronization of clocks when passing from one reference system to another. Assume that two clocks fixed in S′ indicate times t'_0 and t'_1 when they face a clock fixed in S which travels between them. This clock indicates, for those same events, the times t_0 and t_1, respectively. The interval $t_1 - t_0$ is a proper time, because it refers to two events with the same space coordinates, so that the following relationship holds:

$$t'_1 - t'_0 = \frac{t_1 - t_0}{\sqrt{1-\beta^2}}. \qquad (2.17)$$

For the observer in S, due to time dilation, the clocks in S′ are fast, but will also appear desynchronized (Fig. 2.8). Indeed, if it were not so, the observer in S could conclude that

$$t_1 - t_0 = \frac{t'_1 - t'_0}{\sqrt{1-\beta^2}}, \qquad (2.18)$$

by application of the principle of relativity to the system S′.

This relationship, however, contradicts (2.17). Equation (2.18) should thus be modified, assuming the existence of a desynchronization δ between the clocks of S′ relative to S:

$$t_1 - t_0 = \frac{t'_1 - t'_0 + \delta}{\sqrt{1-\beta^2}}. \qquad (2.19)$$

This equation, together with (2.17), allows us to determine δ as

$$\delta = -\frac{x'_0\beta^2}{v}, \qquad (2.20)$$

where x'_0 is the distance between the clocks in S′. The sign of δ indicates, according to (2.19), that a series of synchronized clocks fixed to S′ successively facing a clock in S appear, to the observer in S, successively fast.

The conclusions of these four experiments allow us to outline the *Lorentz transformations*, linking space–time coordinates of an event in two systems

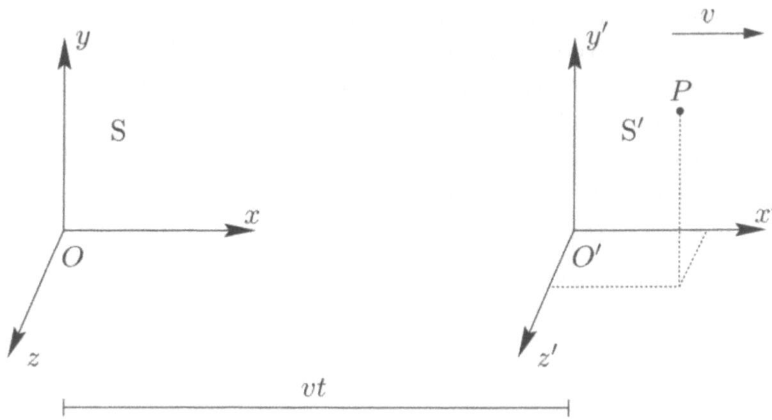

Fig. 2.9. Coordinate transformation between reference systems

of coordinates in relative motion of uniform translation (Fig. 2.9). For simplicity, and without loss of generality, we consider the origins of both systems coincident at $t = t' = 0$.

Given an event with coordinates (t', x', y', z') in S', we observe in the first place that, in agreement with Experiment I, the coordinates transverse to the velocity, y' and z', are not modified. As a consequence, $y = y'$ and $z = z'$. The coordinate in the direction of motion is given by $x = vt + x'_{\mathrm{S}}$, where x'_{S} indicates the distance between the origin and x' as seen from system S. Length contraction implies that this distance is $x'_{\mathrm{S}} = x'\sqrt{1-\beta^2}$. Therefore,

$$x = vt + x'\sqrt{1-\beta^2}. \tag{2.21}$$

Moreover, keeping in mind that at time $t = t' = 0$ clocks at the origins of both systems are synchronized, and that for S the clock associated with the event considered is fast by an amount $\delta = x'v/c^2$ (cf. (2.20)), it follows that

$$t = \frac{t' + x'v/c^2}{\sqrt{1-\beta^2}}. \tag{2.22}$$

Combining these last equations, we obtain the Lorentz transformations:

$$\begin{cases} t' = (t - xv/c^2)/\sqrt{1-\beta^2}, \\ x' = (x - vt)/\sqrt{1-\beta^2}, \\ y' = y, \\ z' = z. \end{cases} \tag{2.23}$$

The inverse transformations are

$$\begin{cases} t = (t' + x'v/c^2)/\sqrt{1-\beta^2}, \\ x = (x' + vt')/\sqrt{1-\beta^2}, \\ y = y', \\ z = z', \end{cases} \tag{2.24}$$

which are obtained from (2.23) by solving for (t, x, y, z) as functions of the primed variables. The inverse transformations can also be obtained, as was to be expected according to the principle of relativity, by simply changing v for $-v$. We note also that the Lorentz transformations can be written in a highly symmetrical form, replacing the time coordinate by ct, making the units for all coordinates homogeneous:

$$\begin{cases} ct' = (ct - \beta x)/\sqrt{1-\beta^2}, \\ x' = (x - \beta\, ct)/\sqrt{1-\beta^2}, \\ y' = y, \\ z' = z. \end{cases} \tag{2.25}$$

2.5 Properties of the Lorentz Transformations

From the mathematical point of view, the most important property of the Lorentz transformations is the fact that they constitute a one-parameter group, given by the relative velocity $\boldsymbol{v}$ between the reference systems. Indeed, as we have shown in the previous section, the transformations have an inverse. One can also prove that the composition or successive application of two Lorentz transformations is also a Lorentz transformation. The parameter of the composition of two transformations corresponding to collinear velocities v_1 and v_2 is given by

$$v = (v_1 + v_2)\left(1 + \frac{v_1 v_2}{c^2}\right)^{-1}. \tag{2.26}$$

This result represents the relativistic *velocity addition law*, that can be obtained from the transformations (2.23). Let us consider a particle moving with velocity $\boldsymbol{u} = (u_x, u_y, u_z)$ with respect to S. Differentiation of the space components in (2.23) with respect to the corresponding time component, allows us to relate $\boldsymbol{u}$ to the velocity of the particle in system S$'$. The result is

$$\begin{aligned} u'_x &= (u_x - v)/(1 - vu_x/c^2), \\ u'_y &= u_y\sqrt{1-\beta^2}/(1 - vu_x/c^2), \\ u'_z &= u_z\sqrt{1-\beta^2}/(1 - vu_x/c^2). \end{aligned} \tag{2.27}$$

It must be noticed that, when applied to the motion of a light beam, this velocity addition law leaves invariant the modulus of this velocity. In the limit $v/c \to 0$, Lorentz transformations are identical to those of Galileo, (2.3), and the addition law (2.27) reduces to the classical expression. This limit shows that the theory of special relativity reproduces classical results when the velocities involved in the systems studied are much smaller than the speed of light.

Given two events A and B with coordinates $(t_{\mathrm{A}}, x_{\mathrm{A}}, y_{\mathrm{A}}, z_{\mathrm{A}})$ and $(t_{\mathrm{B}}, x_{\mathrm{B}}, y_{\mathrm{B}}, z_{\mathrm{B}})$, the quantity

$$\Delta s^2 = c^2(t_{\mathrm{A}} - t_{\mathrm{B}})^2 - (x_{\mathrm{A}} - x_{\mathrm{B}})^2 - (y_{\mathrm{A}} - y_{\mathrm{B}})^2 - (z_{\mathrm{A}} - z_{\mathrm{B}})^2 \tag{2.28}$$

is invariant under changes of reference system, namely

$$\begin{aligned} &c^2(t_{\mathrm{A}} - t_{\mathrm{B}})^2 - (x_{\mathrm{A}} - x_{\mathrm{B}})^2 - (y_{\mathrm{A}} - y_{\mathrm{B}})^2 - (z_{\mathrm{A}} - z_{\mathrm{B}})^2 = \\ &c^2(t'_{\mathrm{A}} - t'_{\mathrm{B}})^2 - (x'_{\mathrm{A}} - x'_{\mathrm{B}})^2 - (y'_{\mathrm{A}} - y'_{\mathrm{B}})^2 - (z'_{\mathrm{A}} - z'_{\mathrm{B}})^2. \end{aligned} \tag{2.29}$$

The quantity Δs is called the *invariant interval* between the events A and B. For infinitely close events, the differential invariant interval is

$$\mathrm{d}s^2 = c^2\mathrm{d}t^2 - \mathrm{d}x^2 - \mathrm{d}y^2 - \mathrm{d}z^2. \tag{2.30}$$

This invariance property has its parallel in classical physics. Indeed, Galilean transformations leave invariant separately the space distance between two events $|\boldsymbol{r}_{\mathrm{A}} - \boldsymbol{r}_{\mathrm{B}}|$, and the time distance $t_{\mathrm{A}} - t_{\mathrm{B}}$. Let us assume now the invariant interval between two infinitely close events is such that $c^2\mathrm{d}t^2 < \mathrm{d}x^2 + \mathrm{d}y^2 + \mathrm{d}z^2$, that is to say, $\mathrm{d}s^2 < 0$. The space distance between the events is larger than the path traveled by light in the time separating them. In this case, it is possible to find a reference system in which both events are simultaneous. In fact, from the Lorentz transformations we get

$$c\ \mathrm{d}t' = \frac{c\ \mathrm{d}t - \beta\ \mathrm{d}x}{\sqrt{1-\beta^2}}. \tag{2.31}$$

We can first make a space rotation to eliminate $\mathrm{d}y$ and $\mathrm{d}z$. We can then find a value of $\beta < 1$ such that $\mathrm{d}t'$ is zero, that is, $\beta = c\ \mathrm{d}t/\mathrm{d}x$. An interval such that $\mathrm{d}s^2 < 0$ is said to be *space-like*: there exists a *proper system* in which the distance between the events is purely spatial.

If, on the other hand, the interval satisfies $c^2\mathrm{d}t^2 > \mathrm{d}x^2 + \mathrm{d}y^2 + \mathrm{d}z^2$, we have $\mathrm{d}s^2 > 0$. In this case it is possible to find a reference system such that the space distance between the events vanishes. By first making $\mathrm{d}y$ and $\mathrm{d}z$ zero by means of a rotation, it is enough to take $v = \mathrm{d}x/\mathrm{d}t$ to have $\mathrm{d}s^2 = c^2\mathrm{d}t'^2$. Such an interval is said to be *time-like*.

A third possibility consists in that $c^2\mathrm{d}t^2 = \mathrm{d}x^2 + \mathrm{d}y^2 + \mathrm{d}z^2$, such that $\mathrm{d}s^2 = 0$. If $\mathrm{d}t = 0$, the events are simultaneous and coincident in space, and they have these properties as seen from any other reference system. If on the other hand $\mathrm{d}t \neq 0$, we deal with two events that can be joined by means of a light signal.

As a conclusion of the previous analysis, we can say that, in the framework of special relativity, *the propagation of physical actions with velocity higher than that of light is impossible.* By physical action we understand a connection capable of linking the two events causally. Indeed, let us assume that two events A and B are causally linked by a physical action which propagates with a velocity greater than c. In this case, the interval between the events satisfies $c^2\mathrm{d}t^2 < \mathrm{d}x^2 + \mathrm{d}y^2 + \mathrm{d}z^2$, and the velocity of the physical agent is

$u^2 = (\mathrm{d}x^2+\mathrm{d}y^2+\mathrm{d}z^2)/\mathrm{d}t^2 > c^2$. According to (2.31), it is possible in this case to find a system where the time order, and therefore the causal relationship, is reversed. It is enough to take $\beta > c\,\mathrm{d}t/\mathrm{d}x$ for the sign of $\mathrm{d}t'$ to be opposite to that of $\mathrm{d}t$.

2.6 Minkowski Diagrams

We present now a geometric representation of the Lorentz transformations. We start by taking a reference system S (one-dimensional in space) with coordinates (t, x) which can be represented in a plane by means of orthogonal axes x and ct (Fig. 2.10).

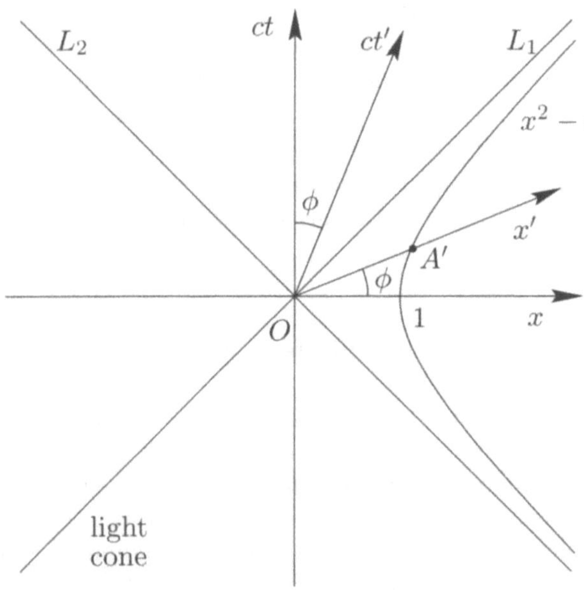

Fig. 2.10. Minkowski diagram

As we already indicated, the factor c in the time axis is introduced to homogenize units. A representation of the coordinate x of a moving point as a function of t in this graph is called the *world line* of the moving point. So, the world lines of light rays going through the origin at time zero are given by the straight lines OL_1 and OL_2. The world line of a point going through the origin at $t = 0$ and moving with constant velocity v is determined by the equation

$$x = vt = \frac{v}{c}\,ct. \tag{2.32}$$

This is a straight line making an angle ϕ with the ct axis, such that $\tan\phi = v/c$. If this moving point is interpreted now as the origin of coordinates of a second reference system S′, its world line is nothing but the ct' axis of the system S′. Indeed, the point is at rest in S′, so that events happening there are spatially coincident, although in general they take place at different times. The axis of the space coordinate x' determines also an angle ϕ with x, such that the straight line OL_1 bisects the angle between the axes ct' and x', as required by the invariance of the speed of light. Notice that plotting the axes of S as orthogonal is an arbitrary choice. The principle of relativity demands that any system can be represented with its axes orthogonal; the axes of other reference systems determine different angles, according to their relative velocities.

The geometrical representation of the Lorentz transformations is completed by fixing the measure scales to be used on the axes of each reference system. To do this it is necessary to notice that the hyperbola $x^2 - c^2t^2 = 1$ is the locus of events that in the original system are separated from the event at the origin by a unitary interval. Due to the invariance of the interval, these events are also at unitary distance from O in S′.

The hyperbola cuts the x' axis at the point A' of coordinates $x' = OA'$ and $ct' = 0$ so that the interval between A' and the origin should be unitary. Therefore, the unit of distance on the x' axis is the segment OA', given by

$$U = OA' = \sqrt{\frac{1+\beta^2}{1-\beta^2}}. \tag{2.33}$$

Considering now the hyperbola $c^2t^2 - x^2 = 1$, we see that, because of the symmetry of the space and time coordinates, OA' is also the unit of measure on the ct' axis. The coordinates in S′ of any event P can now be determined according to the usual convention. To find its coordinate x' for example, a line parallel to the axis ct' is drawn through P and its intersection with the x' axis is determined. The distance between this intersection and the origin, divided by the unit of measure U, is the coordinate.

This graphical representation of the Lorentz transformations is called the *Minkowski diagram.* Such a geometrical interpretation provides a very convenient method for studying properties of the transformations. For example, the world lines of light rays going through the origin – determining the so-called *light cone* – divide the whole of space–time into regions whose intervals relative to the event at the origin have different characteristics.

The events inside the light cone, above and below the origin, are separated from O by time-like intervals. In the upper cone we find the events corresponding to the absolute future of O, and in the lower cone we have those belonging to its absolute past. The time axes of any other system S′ are in this way divided in two half-axes: the positive one, corresponding to the future, and the negative one to the past. We also see from here that any event P separated from O by a time-like interval can be *brought to rest.* This means that it can be seen as a coincident event with O, in a conveniently chosen system, such that in the new system both events occur at the same place. The time axis ct' of such system is precisely OP.

Events with a space-like separation with respect to O are found outside the light cone. These events can be transformed in such a way that they appear as simultaneous with the event at the origin. This happens in a system whose space-axis joins the origin and the event in question. Events located in these areas cannot be causally connected with the origin, because the time order of O and one of such events may be altered by changing the reference system.

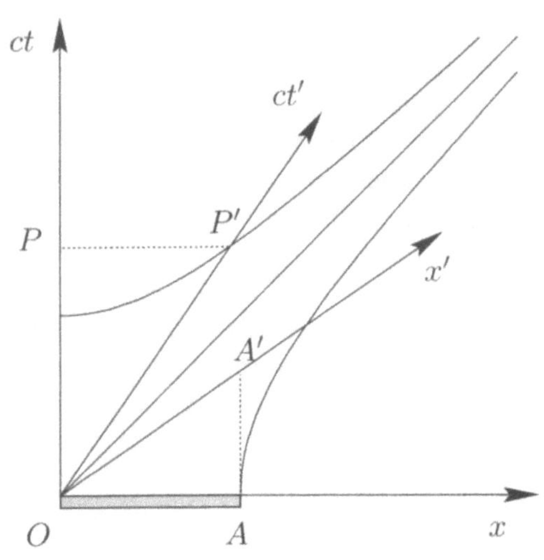

Fig. 2.11. Lorentz contraction and time dilation

Using Minkowski diagrams we can analyze in a qualitative geometric form the phenomena of length contraction and time dilation (Fig. 2.11). In the first place, we consider a ruler OA of unit length, at rest in a system S whose axes are drawn orthogonal. The world lines of the ends of the ruler are parallel to the ct axis. The space distance in S′ between two simultaneous events on this pair of lines is then OA'. As can be seen from the figure, OA' is smaller than the length unit in S′, determined by the intersection of the x' axis with the hyperbola $x^2 - c^2t^2 = 1$. Therefore the moving ruler appears shorter to an observer in S′.

In the same way let us assume a clock at rest at the origin of S′. Its world line is the axis ct'. Let P' be an event separated from the origin by a unit of time in S′. The time component of that event in S is the segment OP, larger than unity. The moving clock therefore appears to be slow.

Minkowski diagrams allow us to rederive the Lorentz transformations. Indeed, let us consider an event P with space and time coordinates (x, ct)

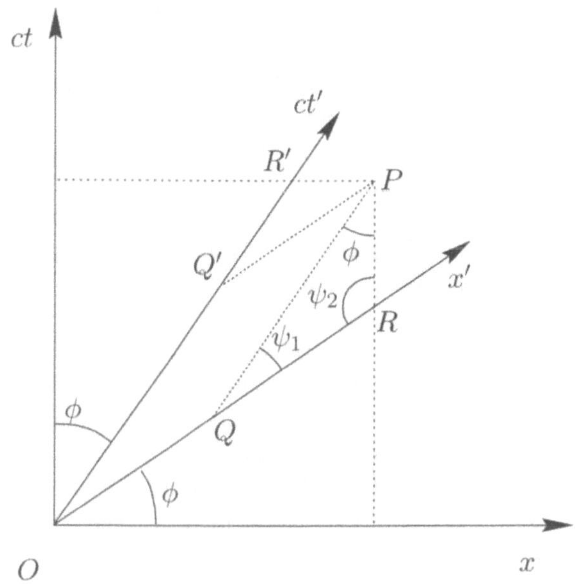

Fig. 2.12. Rederivation of the Lorentz transformations

(Fig. 2.12). In the primed system its coordinates are $x' = OQ/U$ and $ct' = OQ'/U$, where U is the unit of measure on the primed axis, given by (2.33). Let us consider now the triangle PQR. We notice that $\psi_1 = \pi/2 - 2\phi$, $\psi_2 = \pi/2 + \phi$ and $\tan\phi = v/c$. Applying the sine theorem we obtain

$$\frac{PQ}{\sin\psi_2} = \frac{PR}{\sin\ \psi_1} \Rightarrow \frac{OQ'}{\cos\phi} = \frac{ct - vx/c}{\cos 2\phi}. \tag{2.34}$$

As we noticed above, the unit of measure U relates OQ' to ct'. Replacing the trigonometric functions of ϕ by their expression in terms of the tangent, we find

$$ct' = \frac{ct - vx/c}{\sqrt{1-\beta^2}}. \tag{2.35}$$

This is the Lorentz transformation for the time variable. For the position x', the result is totally symmetric due to the likeness of the triangles $PQ'R'$ and PQR. We leave the explicit calculation of x' as an exercise for the reader.

The classical limit of the Minkowski diagrams was introduced by Hermann Minkowski (1864-1909) in his original works. These representations of the Galilean transformations, discussed in the exercise solved in the problem section, could be called *Galileo diagrams*.

2.7 Interpretation of Prerelativistic Experiments

We have shown that classical physics is unable to give a coherent explanation of all of the results of the prerelativistic experiments described in Sect. 1.2. We show now how we can correctly interpret these observations within the framework of the Lorentz transformations.

The Michelson–Morley experiment can be explained as being a consequence of the principle of relativity and of the invariance of c. This last postulate, in particular, implies that the speed of light is the same along both arms of the interferometer, thus making it impossible to notice any displacement of the interference pattern as the apparatus is rotated.

Stellar aberration can be simply interpreted starting from the velocity addition law. Let us assume that a light ray approaches the Earth perpendicularly to the direction of its translational motion. For an observer fixed to the source, light propagates with velocity $\boldsymbol{u} = (0, -c, 0)$. On Earth, the velocity is given by $\boldsymbol{u}' = (-v, -c\sqrt{1-\beta^2}, 0)$, where v is the velocity of the Earth, and $\beta = v/c$ (note that $u = |\boldsymbol{u}| = u' = |\boldsymbol{u}'| = c$). The angle of incidence θ' with respect to the normal to the ecliptic is then

$$\tan\theta' = \beta/\sqrt{1-\beta^2}. \tag{2.36}$$

Since $v \ll c$, this result is, in practice, numerically equivalent to the classical one.

The relativistic velocity addition law also explains correctly Fizeau's experiment. Since light moves relative to the liquid medium at rest with speed c/n, the speed of light in a system fixed with respect to the laboratory is

$$u = \left(\frac{c}{n} \pm v\right)\left(1 \pm \frac{v}{nc}\right)^{-1} \approx \frac{c}{n} \pm v\left(1 - \frac{1}{n^2}\right). \tag{2.37}$$

This result agrees with the experimental observation.

2.8 The Speed of Light: 1675–1983

The speed of light has been measured with growing accuracy since in 1675 the Danish astronomer Ole Römer concluded that the irregularities in the eclipses of Jupiter's satellites had their origin in the finite speed of light. Starting from the different duration of the periods of visibility of these satellites, as the distance from the Earth to Jupiter varied over the year, Römer estimated the speed of light to be about 2.25×10^{10} cm/s.

In 1728 James Bradley found the prerelativistic explanation discussed above for the phenomenon of stellar aberration, and concluded that light should propagate at approximately 2.85×10^{10} cm/s. In 1849 Fizeau argued that the Doppler effect, explained in 1842 by the Austrian physicist Christian Johann Doppler (1803–1853) in connection with sound, could also occur with light. He postulated that the color of light of a given source would change if the source approached or moved away from the observer.

This idea was not verified until quite some time later, and in fact the Doppler shift toward the red in the galactic spectra is one of the strongest evidences in favor of the expansion of the Universe. In 1849 Fizeau carried out the first determination of the speed of light measuring distances and times on the surface of Earth. For this he used a jagged wheel and measured the rotation speed of the wheel needed for light, passing between two of the teeth of the wheel and reflected a certain distance away, to be either intercepted by one of the teeth of the wheel or to pass through the next interval between the teeth. Jean Bernard Léon Foucault (1819–1868) was an assistant of Fizeau, who perfected the method, replacing the wheel by two revolving mirrors. The result obtained in 1862 for the speed of light was of 2.97665×10^{10} cm/s. The distance between the revolving mirrors was of only 20 m. With this method the first determinations of the speed of light in a dielectric, water, were made. Alfred Cornu (1841–1902) improved the method of the jagged wheel and established in 1876 a value of 2.99990×10^{10} cm/s.

In 1882 Michelson made determinations of the speed of light, and established the value of 2.9978888×10^{10} cm/s. Michelson's interferometric determinations, that began in 1878, implied a determination of differences in speeds. His measurements of the speed of light should be considered as a different result from the experiments which bear his name. In 1927 Michelson reproduced Foucault's method, considerably increasing the precision, and obtained a value of 2.99798×10^{10} cm/s.

Starting from the decade of 1940, and thanks to technological advances in the area of microwaves, a series of precise determinations of the speed of light were made. These experiments involved electromagnetic radiation with wavelengths of the order of centimeters in resonant cavities working at frequencies of the order of the 10^{10} Hz. In October 1972 a group directed by the American physicist Kenneth Evenson, using a set of laser chains, obtained the value $2.932923750031 \times 10^{10}$ cm/s.

The Convention of the General Conference on Weights and Measures established in 1967 the present definition of the second: one second is 9, 192, 631, 770 times the period of the radiation emitted in the transition between the two hyperfine levels of the ground state of the cesium atom ^{133}Ce. In 1983 the same General Conference adopted the speed of light as a reference constant for the definition of the units of length and, in this way, the meter was defined as the distance traveled by light in a 299 792 458th of a second. The British units of measure define the inch and the foot with reference to this definition of the meter. With this, the speed of light came to occupy a definitive role as universal constant, being used as the reference for the definition of the units with which until then it had been measured.

Problems

2.1 Three light sources at rest in a three-dimensional reference system S flash simultaneously. Another reference system S$'$ moves with respect to S. Identify the points P' in S$'$ from which the light sources are seen to flash simultaneously. Calculate the distances measured in S$'$ from P' up to the light sources, and the time it takes for light to reach P'. Compare these quantities with the corresponding ones measured in S. What would happen if there were N light sources? What if we lived in D space dimensions?

2.2 A rod at rest in a reference system S makes an angle α with the x axis. What angle will this rod make with the x' axis as seen from a reference system S$'$ moving with respect to S with velocity v in the positive x direction?

Hint: Transform the position of a generic point along the rod from S to S$'$, taking into account that to define the angle in S$'$ the whole rod has to be observed simultaneously.

2.3 Two trains which run on the same rail are initially separated a distance D, and start simultaneously, as seen from the platform, one towards the other. Both have velocities of modulus v with respect to the platform. Calculate, for a system fixed to Earth and for another system fixed to one of the moving trains, the time needed for the collision to occur. Do these times satisfy relativistic time dilation? Why? A fly travels with velocity $V > v$ between the front ends of both trains, touching them successively without stopping to rest. Calculate, in both systems, the distance traveled by the fly before it perishes squashed in the collision. Do these distances satisfy the Lorentz contraction? Why? Plot in a Minkowski diagram the trajectories of the trains and the insect.

Hint: Write equations for the trajectories of the trains and the insect in one of the systems, and transform them to the other.

2.4 An isotropic source at rest emits particles of velocity v in pulsed form and period T. Each pulse determines a spherical particle front whose radius increases with time. Find the form of the particle front and the distance between successive fronts, measured by observers with respect to which the source moves with velocity V.

Hint: Apply the hint to Problem 2.2.

2.5 The *rapidity* of a particle moving with velocity u can be defined as $\phi\,(u) = \arg\tanh\,(u/c)$. Prove that for particles moving with collinear velocities, the rapidity is an additive quantity.

2.6 An object having velocity v and acceleration a is observed from a one-dimensional reference system S. Determine the acceleration of that object as seen from a system S$'$ moving relative to S with velocity V.

2.7 A cube with its faces parallel to the coordinate axes moves with velocity V along the x axis of an inertial system S. Determine the appearance of this cube in an instantaneous picture taken with a camera located on the z axis a certain distance away from the origin.

Hint: Take into account the finiteness of the speed of light.

2.8 Obtain the classical limit of the Minkowski diagrams and represent the Galilean transformations in a form equivalent to that of Minkowski.

Solution: To obtain the classical limit we must let the speed of light tend towards infinity: $c \to \infty$. We stress that the presence of the factor c in the time axis was introduced by symmetry and dimensional homogeneity arguments in the relativistic space–time coordinates. Thus, for clarity when taking this limit, it is convenient, first, to transform the time variable from ct to $v_0 t$, where v_0 is an arbitrary reference velocity smaller than that of light.

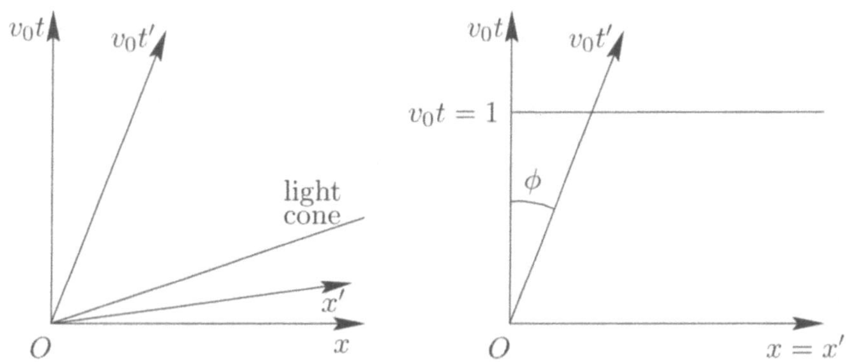

Fig. 2.13. Galileo diagram as limit of a Minkowski diagram

With this change, the light cone (Fig. 2.13) and the x' axis have slopes v_0/c and $v_0 v/c^2$, respectively, while the time axis of both systems has a slope equal to v/v_0.

We observe that for $c \to \infty$, both the light cone and the x' axis tend to coincide with the x axis, in such way that the slope of the x' axis is always smaller than that of the light cone. Let us study what happens to the invariant hyperbolas in the classical limit. We notice that, in this limit, these curves are not longer symmetric, since one of them closes on the x axis while the other opens up. In the new variables, the equation determining the hyperbola that crosses the x axis at $x = 1$, $(x^2 - c^2t^2 = 1)$ is written as $x^2 - (c/v_0)^2 v_0^2 t^2 = 1$, from which we obtain

$$v_0 t = \pm \frac{v_0}{c}\sqrt{x^2 - 1}.$$

Taking the limit $c \to \infty$, this relationship implies

$$v_0 t = 0 \text{ for } |x| \geq 1.$$

In a similar way we can analyze the limit of the equation determining the hyperbola that cuts the v_0t axis at $v_0t = 1$, given by

$$v_0^2t^2 - v_0^2x^2/c^2 = 1.$$

In the limit $c \to \infty$, this reduces to the equation of the horizontal straight line

$$v_0t = \pm 1.$$

This straight line cuts the time axes of the different reference systems at different distances from the origin. Since time is an invariant under Galilean transformations, the intersection of that line with the t' axis determines the unit of measure U on that axis. From Fig. 2.13 it follows that

$$U = \sqrt{1 + v^2/v_0^2}.$$

On the space axis, on the other hand, the unit of measure is independent of the reference system. This result shows that for Galileo diagrams, the unit of measure depends on the relative velocity between the systems. Minkowski diagrams add to this an additional variation of scale characteristic of Lorentz transformations. The symmetry loss when passing from the Lorentz transformations to Galilean transformations is evident comparing Figs. 2.10 and 2.13.

References

2.1 A.A. Michelson: *The Relative Motion of the Earth and the Luminiferous Ether*, Am. J. Sci. **22**, 120 (1881); A.A. Michelson and E.W. Morley: *On the Relative Motion of the Earth and the Luminiferous Ether*, Am. J. Sci. **34**, 333 (1887).

2.2 A. Stewart: *The Discovery of Stellar Aberration*, Sci. Am. **210** (3), 100 (1964).

2.3 A. Einstein: *Zur Elektrodynamik bewegter Körper*, Ann. Phys. Lpz. **17**, 891 (1905).

2.4 H.A. Lorentz: *Electromagnetic Phenomena in a System Moving with any Velocity less than that of Light*, Proc. Acad. Sci. Amsterdam **6** (1904).

Further Reading

P. Couderc: *La Relativité* (Presses Universitaires de France, Paris, 1941).

H.A. Lorentz et al.: *The Principle of Relativity* (Dover, New York, 1952).

J. Wheeler and L. Taylor: *Space–Time Physics* (Freeman, New York, 1966).

3. Relativistic Dynamics

The modifications introduced by the theory of relativity in the concepts of classical kinematics make it necessary to consider a reformulation of the laws of mechanics, in order to bring them into agreement with relativistic postulates. It can be expected that this reformulation modifies in a substantial way the laws of mechanics, since these respond to a Galilean principle of relativity.

Classically, a particle can be indefinitely accelerated until reaching a velocity as high as required. This is not possible in the framework of relativity due to the existence of the velocity of light as a limit of speed. As a consequence, we should expect relativistic mechanics to provide a means of increasing the inertia of a particle as it is accelerated, so that it cannot exceed the speed of light. Inertia is measured by the mass, an invariant in classical mechanics, and it is thus to be expected that relativity forces a revision of its definition. In addition, from a formal point of view, classical laws are invariant under Galilean transformations, as was shown in Chap. 2, and are not invariant under Lorentz transformations. Finally, the law of action and reaction involves the concept of simultaneity in the action of forces not necessarily coincident in space; as we have seen, simultaneity ceases to be an invariant concept in the theory of relativity, so this principle requires revision.

In this chapter we discuss the relativistic formulation of mechanics for systems free of the action of gravity. Gravitational forces should be included in a more general formulation called, in fact, the *general theory of relativity*.

3.1 Conservation Laws. Relativistic Momentum

Given a particle of mass m moving with velocity $\boldsymbol{v}$, we define its momentum as

$$\boldsymbol{p} = m\boldsymbol{v}, \tag{3.1}$$

by analogy with the corresponding classical quantity. We assume this quantity to have additive properties and postulate the *law of conservation of momentum*:

- The momentum of an isolated (force-free) system is a constant of motion.

We also introduce the *law of conservation of mass*, which states that

- The total inertial mass of an isolated system remains constant in time.

Let us study the consequences of these conservation laws, when considered from different reference systems. Let us assume that in a system S′ two identical particles of mass m move with opposite velocities u' and $-u'$, parallel to the x' axis, so that they collide plastically and remain united and at rest. From a system S moving relative to S′ with velocity $-v$ in direction of x', the particles have initial velocities u_1 and u_2 – given by the addition law – and masses m_1 and m_2 which *a priori* cannot be considered as identical.

According to the conservation principles which we assume valid in S, we have

$$\begin{aligned} m_1 + m_2 &= M, \\ m_1 u_1 + m_2 u_2 &= MV, \end{aligned} \tag{3.2}$$

where M is the sum of the masses immediately after the collision as measured in S. The velocity V can be related to u_1 and u_2 through the law of addition of velocities

$$u' = \frac{u_1 - V}{1 - u_1 V/c^2}, \qquad -u' = \frac{u_2 - V}{1 - u_2 V/c^2}. \tag{3.3}$$

Starting from (3.2) it is possible to show that $m_1/m_2 = -(u_2 - V)/(u_1 - V)$. Had we used the classical law for addition of velocities this relationship would have implied $m_1/m_2 = 1$. On the other hand, applying (3.3), we obtain

$$\frac{m_1}{m_2} = \frac{\sqrt{1 - u_2^2/c^2}}{\sqrt{1 - u_1^2/c^2}}. \tag{3.4}$$

Thus, the masses of two identical particles vary as a function of their velocity u inversely proportional to $\sqrt{1 - u^2/c^2}$. If we choose V so that $u_2 = 0$, we can write

$$m = \frac{m_0}{\sqrt{1 - u^2/c^2}}, \tag{3.5}$$

where m_0 is a constant, called *rest mass* or *proper mass* of the particle, since $m = m_0$ when $u = 0$.

Notice that, in the process we have considered, the mass after the collision is larger than the sum of the rest masses of each one of the particles, $M > 2m_0$. We shall show that this difference is associated with the elastic energy involved in the collision. On the other hand, we notice that in the limit $c \to \infty$ the results obtained coincide with the classical ones. In particular, in this limit the mass does not depend on velocity, just as in classical mechanics.

The conservation of momentum and mass together with the Lorentz transformations require therefore the dependence of the mass of a particle on its velocity. This property was predicted by Lorentz in 1904, analyzing the action of the electromagnetic force on the motion of a particle, in his paper *Electromagnetic Phenomena in a System Moving with any Velocity less than that of Light* [3.1]. It has been checked quantitatively by measuring the charge to mass ratio for electrons in relativistic motion emitted from a radioactive source of electrons (Bucherer experiment). Later on, we show that the electric charge is invariant under Lorentz transformations. The Bucherer experiment thus allows for a direct measurement of the electron mass as a function of velocity. The results are in total agreement with (3.5).

3.2 Relativistic Force, Work and Energy

The concept of *force*, which is involved in the principle of conservation of momentum, has to be generalized to the relativistic case. According to the results of the previous section, the two possible classical definitions of force, i.e. the product of mass times acceleration and the time derivative of momentum, are not equivalent quantities in the theory of relativity. In view of the form given to the law of conservation of momentum, we prefer the second definition:

$$\boldsymbol{F} = \frac{\mathrm{d}\boldsymbol{p}}{\mathrm{d}t}. \tag{3.6}$$

The equations of motion for a particle of rest mass m_0 on which a specified force $\boldsymbol{F}$ acts is

$$\frac{\mathrm{d}}{\mathrm{d}t}\left(m_0 u_i/\sqrt{1-u^2/c^2}\right) = F_i, \tag{3.7}$$

where the index i runs over the three Cartesian coordinates and $u^2 = u_x^2 + u_y^2 + u_z^2$. The difference between the Newtonian definition of force through the acceleration and the relativistic definition given above becomes explicit if we observe that the time derivative of momentum can be written as

$$\boldsymbol{F} = \frac{\mathrm{d}\boldsymbol{p}}{\mathrm{d}t} = m\frac{\mathrm{d}\boldsymbol{u}}{\mathrm{d}t} + \boldsymbol{u}\frac{\mathrm{d}m}{\mathrm{d}t}. \tag{3.8}$$

From here it follows that force and acceleration are not in general parallel.

The transformation laws for force, derived by applying Lorentz transformations to the velocities and time involved in the definition (3.6), can be used for a detailed study of dynamics. If the system S′ moves with velocity v in the x direction relative to the original system S, the corresponding components of the force are related through

$$\begin{aligned} F_x &= F'_x + \left(u'_y F'_y + u'_z F'_z\right) v/(c^2 + u'_x v), \\ F_y &= c^2\sqrt{1-v^2/c^2}\,F'_y/(c^2 + u'_x v), \\ F_z &= c^2\sqrt{1-v^2/c^2}\,F'_z/(c^2 + u'_x v). \end{aligned} \tag{3.9}$$

The *work* carried out by a force $\boldsymbol{F}$ along a differential displacement $\mathrm{d}\boldsymbol{r}$ is given by

$$\mathrm{d}W = \boldsymbol{F} \cdot \mathrm{d}\boldsymbol{r}. \tag{3.10}$$

According to the law of energy conservation, this work must equal the change in kinetic energy of the particle:

$$\mathrm{d}W = \mathrm{d}T = \frac{\mathrm{d}}{\mathrm{d}t}(m\boldsymbol{u}) \cdot \mathrm{d}\boldsymbol{r} = m_0 u \frac{\mathrm{d}u}{\mathrm{d}t}(1 - u^2/c^2)^{-3/2}. \tag{3.11}$$

Therefore, $\mathrm{d}T$ depends on the rate of velocity change. From here we obtain

$$\mathrm{d}T = c^2 \mathrm{d}m = \mathrm{d}\left(m_0 c^2 / \sqrt{1 - u^2/c^2}\right), \tag{3.12}$$

that integrated over the velocity from 0 to u gives

$$T = (m - m_0)c^2. \tag{3.13}$$

Inspired by this relationship, and taking an important conceptual step forwards, Einstein formulated the general equivalence between mass and energy. This equivalence, which Einstein discussed in his paper *Ist die Trägheit eines Körpers von seinem Energiegehalt abhängig?* [1] [3.2], has been verified in a variety of situations. Equation (3.13) describes the kinetic energy as the difference between a quantity dependent on the velocity of the particle u and the same quantity evaluated at $u = 0$, namely, as a difference of masses. Then, the total energy of a free particle is defined as

$$E = mc^2 = m_0 c^2 + T, \tag{3.14}$$

that is to say, as the sum of the kinetic energy T and the energy associated with the rest mass, $m_0 c^2$. Let us observe that from (3.14) we conclude that the law of conservation of mass is equivalent to the law of conservation of energy.

Expanding expression (3.14) in series in powers of u/c, we obtain the total energy in the limit of low velocities

$$E \cong m_0 c^2 + m_0 u^2/2. \tag{3.15}$$

This consists of two contributions. The first one, $m_0 c^2$, is the *rest energy* associated with the particle mass. The second one coincides with the classical form of the kinetic energy.

Rewriting the first equation of (3.2) in the center of mass system we have

$$\frac{m_0}{\sqrt{1 - u_1^2/c^2}} + \frac{m_0}{\sqrt{1 - u_2^2/c^2}} = M_0. \tag{3.16}$$

Here, it is clear that the total mass M_0 of the two particles after the collision is larger than the sum $2m_0$ of the original particles at rest. This fact can be interpreted by assuming the mass equivalence of the elastic energy of deformation.

[1] *Is the Inertia of a Body Dependent on its Energy Content?*

3.3 Tensor Formulation of the Lorentz Transformations

The principle of relativity expressing the covariance of all physical laws finds its most elegant and precise mathematical expression through *tensor algebra and tensor calculus* (Appendix C). These allow us to enunciate relationships in a way independent of the coordinate system. This formulation notably simplifies the notation and makes apparent the covariant character of mechanical laws. Consequently, it is useful to study it in further detail.

In the first place, in order to get a homogenous notation for space–time coordinates, we define $x^0 = ct$, $x^1 = x$, $x^2 = y$ and $x^3 = z$. The Lorentz transformations (2.25) can then be written as

$$x'^{\nu} = a^{\nu}{}_{\mu} x^{\mu} \qquad (\nu = 0, 1, 2, 3), \tag{3.17}$$

where we omit the sum over repeated indices like μ (Einstein convention, see Appendix C). The coefficients $a^{\nu}{}_{\mu}$ form a matrix whose explicit form is

$$a^{\nu}{}_{\mu} = \begin{pmatrix} \gamma & -\gamma\beta & 0 & 0 \\ -\gamma\beta & \gamma & 0 & 0 \\ 0 & 0 & 1 & 0 \\ 0 & 0 & 0 & 1 \end{pmatrix}, \tag{3.18}$$

where $\gamma = (1 - \beta^2)^{-1/2}$ and the indices ν and μ run over rows and columns, respectively. The inverse transformation is of the form $x^{\nu} = b^{\nu}{}_{\mu} x'^{\mu}$, where the coefficients $b^{\nu}{}_{\mu}$ can be obtained by changing β to $-\beta$ in the direct transformation.

The transformations (3.17) form a group and are applied to a four-dimensional pseudoEuclidian manifold, with coordinates (x^0, x^1, x^2, x^3) and line element given by

$$\mathrm{d}s^2 = g_{\mu\nu} \mathrm{d}x^{\mu} \mathrm{d}x^{\nu}, \tag{3.19}$$

where

$$g_{\nu\mu} = \begin{pmatrix} 1 & 0 & 0 & 0 \\ 0 & -1 & 0 & 0 \\ 0 & 0 & -1 & 0 \\ 0 & 0 & 0 & -1 \end{pmatrix} \tag{3.20}$$

is called the *metric tensor*. The line element coincides, in fact, with the differential invariant interval.

In this formulation physical phenomena take place at points of a four-dimensional space, the *space–time continuum*. A reference system is a particular choice of a system of coordinates in that space. A transformation of the Lorentz group corresponds to a change between systems of coordinates. The principle of relativity implies that the quantities involved in the formulation of a physical law have a defined tensor character under Lorentz transformations.

The most elementary tensors are tensors of zero rank or *scalars*. These are invariant quantities under the transformations, as for example the interval ds^2, the rest mass, and the velocity of light.

Tensors of first rank or *four-vectors* are combinations of four quantities, with well defined transformation rules. The *contravariant* four-vectors, denoted by A^ν ($\nu = 0, 1, 2, 3$), transform like the coordinates, $A'^\nu = a^\nu{}_\mu A^\mu$. The *covariant* four-vectors B_ν, transform with the inverse matrix, $B'_\nu = b^\mu{}_\nu B_\mu$. Each four-vector physical quantity, like the four-dimensional position or velocity, has two tensor representations of rank one, one covariant and one contravariant related through the metric tensor (3.20):

$$A_\mu = g_{\mu\nu} A^\nu. \tag{3.21}$$

Four-tensors of rank n are labeled by n indices, and transform as products of n four-vector components. In this case, given a tensor, it is necessary to specify the indices relative to which it is contravariant or covariant. For example, a twice-contravariant tensor of rank two, $T^{\mu\nu}$, transforms according to

$$T'^{\mu\nu} = a^\mu{}_\rho a^\nu{}_\sigma T^{\rho\sigma}. \tag{3.22}$$

An example of a second rank tensor is the metric tensor $g_{\mu\nu}$. As we show later on, the electromagnetic field is also a tensor of rank two.

3.4 Covariant Formulation of Mechanics

The tensor formulation of relativistic mechanics makes it necessary to redefine the quantities of interest, such as velocity, momentum or force, in terms of tensor entities, so that they have well-defined transformation properties under Lorentz transformations.

Because of (3.17) we know that the coordinates x^ν form a contravariant four-vector which we call the *position four-vector*. This has as a first component ($\nu = 0$), the time coordinate ct. The three remaining components are the space coordinates, so that $x^\nu = (ct, x, y, z)$, which is notated concisely as $x^\nu = (ct, \boldsymbol{x})$. By extension, the first coordinate of any four-vector is usually called the *time component*, and the remaining ones *space components*.

Starting from the position four-vector and differentiating each of its components with respect to the proper time τ in the history of the particle, which is a scalar because of the relation $ds^2 = c^2 d\tau^2$, we can build the four-vector

$$V^\nu = \frac{dx^\nu}{d\tau}. \tag{3.23}$$

These four quantities constitute the *velocity four-vector* or *Minkowski velocity*. The interval of proper time $d\tau$ is related to dt according to $d\tau = dt\sqrt{1 - v^2/c^2}$. It follows that

$$V^\nu = \frac{1}{\sqrt{1 - v^2/c^2}} \frac{\mathrm{d}x^\nu}{\mathrm{d}t}. \tag{3.24}$$

The components of the four-vector velocity are therefore

$$V^\nu = \frac{1}{\sqrt{1 - v^2/c^2}} (c, \boldsymbol{v}), \tag{3.25}$$

where $\boldsymbol{v}$ indicates the three components of the velocity in its usual definition. We note that the latter does not have definite tensor properties, since each one of its components is the ratio between the space part of the position four-vector, and the fourth component of the same four-vector. The modulus of the velocity four-vector, which is a scalar, is $|V|^2 = g_{\mu\nu} V^\mu V^\nu = V_\nu V^\nu = c^2$.

The product of a four-vector times a scalar is also a four-vector. We can thus introduce the momentum four-vector as

$$p^\nu = m_0 V^\nu = m(c, \boldsymbol{v}) = (E/c, \boldsymbol{p}), \tag{3.26}$$

where $\boldsymbol{p}$ is the momentum in the usual sense. The modulus of the momentum four-vector is

$$p_\nu p^\nu = -p^2 + E^2/c^2 = m_0^2 V_\nu V^\nu = m_0^2 c^2, \tag{3.27}$$

from which it follows that $E^2 = c^2 p^2 + m_0^2 c^4$.

The derivative of p^ν with respect to the proper time τ defines a four-vector associated with the force, called the *Minkowski force*,

$$F^\nu = \frac{\mathrm{d}p^\nu}{\mathrm{d}\tau} = \frac{1}{\sqrt{1 - v^2/c^2}} \left(\frac{1}{c} \frac{\mathrm{d}E}{\mathrm{d}t}, \boldsymbol{F} \right), \tag{3.28}$$

where $\boldsymbol{F}$ is the ordinary force. From this it follows that the covariant form of the equation of motion for a particle under the action of an external force is

$$F^\nu = \frac{\mathrm{d}p^\nu}{\mathrm{d}\tau}. \tag{3.29}$$

This equation is written in the same way in all inertial systems, since all quantities involved have a well defined tensor character. In particular, the four-vector structure of F^ν allows us to reobtain the transformation laws for force, (3.9).

The scalar product of the force four-vector and the velocity four-vector is a scalar given by

$$F_\nu V^\nu = m_0 \frac{\mathrm{d}V_\nu}{\mathrm{d}\tau} V^\nu = \frac{1}{2} m_0 \frac{\mathrm{d}}{\mathrm{d}\tau} (V_\nu V^\nu) = 0, \tag{3.30}$$

since the modulus of V^ν is also a scalar. Replacing the explicit form of the components of each four-vector, this equation allows us to obtain

$$\frac{\mathrm{d}E}{\mathrm{d}t} = \boldsymbol{F} \cdot \boldsymbol{v}. \tag{3.31}$$

This is the usual relationship between the change of energy and the work done by the force on a particle.

3.5 Relativistic Analytical Dynamics

By analogy with classical analytical dynamics it is possible to build a covariant version of the *principle of least action* consistent with the theory of relativity, and to derive from it the relativistic equations of motion. This generalization completes the extension of the formalism of classical mechanics to its relativistic covariant version.

Given a one-particle system, we postulate the existence of a function $\mathcal{L}$, called the *Lagrangian*, which is a function of x^ν and V^ν, that is to say, of the coordinates and of the velocity. Starting from the Lagrangian we define the *action* of the particle in a trajectory going between the proper times τ_1 and τ_2 as

$$\mathcal{S} = \int_{\tau_1}^{\tau_2} \mathcal{L}\, \mathrm{d}\tau. \tag{3.32}$$

This quantity is defined using the proper time τ in order to build a covariant formalism. For the same reason, we require that both the Lagrangian $\mathcal{L}$ as well as the action $\mathcal{S}$ are scalars, invariant under Lorentz transformations. It is important to point out that the Lagrangian is not unique, since the only condition it should satisfy is that the Euler–Lagrange equations derived from it coincide with the Newtonian equations.

We enunciate now the *Hamilton principle* by requiring that the motion of the particle is such that the action reaches an extremum

$$\delta\mathcal{S} = 0. \tag{3.33}$$

From this equation and from (3.32) we can obtain, in complete analogy to the classical case, the Euler–Lagrange equations

$$\frac{\mathrm{d}}{\mathrm{d}\tau}\frac{\partial\mathcal{L}}{\partial V^\nu} - \frac{\partial\mathcal{L}}{\partial x^\nu} = 0. \tag{3.34}$$

We notice that in this case the number of independent coordinates or degrees of freedom of the particle is four. The reason is that now time is no longer a free parameter in (3.32). The equations of motion give the relationship between the coordinate x^0 and the proper time τ. In these equations, the quantity $\partial\mathcal{L}/\partial V^\nu$ representing the canonical momentum in Lagrangian theory is a covariant four-vector. By analogy with the classical Lagrangian formulation, we associate it with the canonical momentum four-vector

$$p_\nu = \frac{\partial\mathcal{L}}{\partial V^\nu}, \tag{3.35}$$

and we require therefore that

$$p_\nu = \frac{\partial\mathcal{L}}{\partial V^\nu} = m_0 V_\nu = m_0 g_{\mu\nu} V^\mu. \tag{3.36}$$

Integrating this equation with respect to velocity, we can find the general form of the Lagrangian:

$$\mathcal{L} = \frac{1}{2} m_0 V^\nu V_\nu + f(x^\nu), \tag{3.37}$$

where f is an arbitrary scalar function of the coordinates. It describes the interaction of the particle with an external potential.

For a free particle, because of the isotropy of space–time, we assume that the Lagrangian $\mathcal{L}$ does not depend on the coordinates, and therefore set $f = 0$. The Lagrangian for a free particle is then

$$\mathcal{L}^{(0)} = \frac{1}{2} m_0 V^\nu V_\nu. \tag{3.38}$$

Compare this result with the classical Lagrangian $\mathcal{L} = mv^2/2$. Being a scalar, the Lagrangian of a free particle has an invariant value, equal to $m_0 c^2/2$. However, for the formulation of the equations of motion it is necessary to leave explicit the dependence on the components of the velocity four-vector.

To consider the case of a particle in an external potential we take a Lagrangian of the form $\mathcal{L} = \mathcal{L}^{(0)} + \mathcal{L}^{(1)}$, where $\mathcal{L}^{(1)}$ describes the potential. This part of the Lagrangian should be a scalar. Therefore its form depends on the tensor character of the quantities which define the external potential. The only interactions that can be described in a simple form in the framework of the covariant formulation of special relativity are the electromagnetic ones. Indeed forces of elastic type, for example, require the formulation of relativistic fluid mechanics; gravitational forces on the other hand, as we have already mentioned, require a generalization of the theory of relativity.

It is possible to show that electromagnetic fields can be described by means of a *four-vector potential* A^ν whose space components are the *vector potential* $\boldsymbol{A}$, from which the magnetic field is derived, and the time component is the *electrostatic potential* ϕ. We thus have

$$A^\nu = (\phi, \boldsymbol{A}). \tag{3.39}$$

The electric and magnetic fields are given by

$$\boldsymbol{E} = -\nabla\phi - \frac{1}{c}\partial_t \boldsymbol{A}, \tag{3.40}$$

$$\boldsymbol{B} = \nabla \times \boldsymbol{A}. \tag{3.41}$$

We show next that the Lagrangian

$$\mathcal{L} = \frac{1}{2} m_0 V^\nu V_\nu + \frac{q}{c} A^\nu V_\nu \tag{3.42}$$

gives rise to the canonical equations governing the motion of a particle of charge q in the presence of the fields $\boldsymbol{E}$ and $\boldsymbol{B}$. Note that this form of the Lagrangian is not identical with (3.37); indeed, the interaction term depends on the velocities. We study the consequences of this fact as follows.

Due to the dependence of $\mathcal{L}$ on the velocity, the canonical momentum

$$p^\nu = \frac{\partial \mathcal{L}}{\partial V_\nu} = m_0 V^\nu + \frac{q}{c} A^\nu \tag{3.43}$$

is not equal to the kinetic momentum defined in (3.26). In fact, the components of the canonical momentum are

$$p^0 = \frac{m_0 c}{\sqrt{1 - v^2/c^2}} + \frac{q}{c} A^0, \tag{3.44}$$

$$\boldsymbol{p} = \frac{m_0 \boldsymbol{v}}{\sqrt{1 - v^2/c^2}} + \frac{q}{c} \boldsymbol{A}. \tag{3.45}$$

The first of these equations implies a relationship between the potential energy $q\phi = qA^0$, the kinetic energy $m_0 c^2/\sqrt{1 - v^2/c^2}$, and the time component of the canonical momentum, in the form

$$E = c p_0 = \frac{m_0\, c^2}{\sqrt{1 - v^2/c^2}} + q\phi, \tag{3.46}$$

where E is the total energy. Equation (3.46) indicates that when the particle moves keeping the total energy constant, its kinetic energy increases or decreases at the expense of the potential energy. This interpretation can be extended to the other components of the momentum four-vector in the case in which a given component of $\boldsymbol{p}$ is conserved. According to (3.45), the total momentum $\boldsymbol{p}$ is made up of the kinetic momentum $m_0 \boldsymbol{v}/\sqrt{1 - v^2/c^2}$ and of the *potential momentum* $q\boldsymbol{A}/c$. In the case in which a component of momentum is conserved, the corresponding component of kinetic momentum increases or diminishes at the expense of potential momentum $q\boldsymbol{A}/c$. This allows us to give a physical interpretation to the vector potential in connection with particle dynamics in the presence of electromagnetic fields.

The equations of motion obtained from (3.34) for the Lagrangian of electromagnetic interactions can be written as

$$\frac{\mathrm{d}}{\mathrm{d}\tau}(m_0 V_\mu) = \frac{\partial}{\partial x^\mu}(\frac{q}{c} A_\nu V^\nu) - \frac{q}{c}\frac{\mathrm{d}A_\mu}{\mathrm{d}\tau}. \tag{3.47}$$

This equation has the form of the relativistic Newton's second law associated to a force four-vector

$$F_\mu = \frac{\partial}{\partial x^\mu}(\frac{q}{c} A_\nu V^\nu) - \frac{q}{c}\frac{\mathrm{d}A_\mu}{\mathrm{d}\tau}. \tag{3.48}$$

Considering the usual three-dimensional quantities, we can check that this four-vector corresponds to a force

$$\boldsymbol{F} = \frac{\mathrm{d}(m\boldsymbol{v})}{\mathrm{d}t} = \frac{q}{c}\nabla(\boldsymbol{A}\cdot\boldsymbol{v}) - q\nabla\phi - \frac{q}{c}\frac{\mathrm{d}\boldsymbol{A}}{\mathrm{d}t}. \tag{3.49}$$

Since the time variation of $\boldsymbol{A}$ is due both to its intrinsic time variation and to that of the particle coordinates, we have

$$\frac{\mathrm{d}\boldsymbol{A}}{\mathrm{d}t} = \partial_t \boldsymbol{A} + (\boldsymbol{v}\cdot\nabla)\boldsymbol{A}, \tag{3.50}$$

where ∂_t indicates partial differentiation with respect to time. Using some vector identities, we arrive at the simplified form

$$\boldsymbol{F} = q\left(-\nabla\phi - \frac{1}{c}\partial_t \boldsymbol{A}\right) + \frac{q}{c}\boldsymbol{v} \times (\nabla \times \boldsymbol{A}). \tag{3.51}$$

According to (3.40) and (3.41), we identify the gradient of ϕ in the first term of (3.51) as the electric field $\boldsymbol{E}$, and the curl of $\boldsymbol{A}$ in the second term as the magnetic field $\boldsymbol{B}$. The electromagnetic force on a charge q, called the *Lorentz force*, is then

$$\boldsymbol{F} = q\left(\boldsymbol{E} + \frac{\boldsymbol{v} \times \boldsymbol{B}}{c}\right). \tag{3.52}$$

The Lorentz force separates into two well-differentiated parts. The first part, independent of velocity, is the *electric force.* The second part, proportional to v/c, is the *magnetic force.* According to our presentation, it is clear that this result – which was well-known prior to the formulation of relativity theory – fits in a completely consistent way in the covariant formalism of relativistic dynamics.

The covariant formulation of the Lorentz force requires the definition of a four-vector whose space components are precisely that force. This, in turn, requires introducing the electromagnetic fields in the relativistic formalism. We study this problem in Chap. 9, after presenting the phenomenology and the classical formulation of electromagnetism.

Equation (3.47) allows us to write

$$\frac{\mathrm{d}}{\mathrm{d}\tau}\left(m_0 V_\nu + \frac{q}{c}A_\nu\right) = \frac{\partial}{\partial x^\nu}\left(\frac{q}{c}A_\alpha V^\alpha\right). \tag{3.53}$$

Expressed in its components, this is

$$\frac{\mathrm{d}}{\mathrm{d}\tau}\left(mc + \frac{q}{c}\phi\right) = \frac{1}{c}\partial_t\left(q\frac{\phi - \boldsymbol{v}\cdot\boldsymbol{A}/c}{\sqrt{1 - v^2/c^2}}\right), \tag{3.54}$$

$$\frac{\mathrm{d}}{\mathrm{d}\tau}\left(m\boldsymbol{v} + \frac{q}{c}\boldsymbol{A}\right) = -\nabla\cdot\left(q\frac{\phi - \boldsymbol{v}\cdot\boldsymbol{A}/c}{\sqrt{1 - v^2/c^2}}\right). \tag{3.55}$$

These relations give the rate of variation of total energy and momentum in terms of the variations or gradients of the *interaction function*

$$qA_\nu V^\nu/c = q(\phi - \boldsymbol{v}\cdot\boldsymbol{A}/c)/\sqrt{1 - v^2/c^2}. \tag{3.56}$$

The analogy between classical and relativistic results can be extended to give a Hamiltonian formulation of mechanics. The Hamiltonian of a particle is defined as $\mathcal{H}(p^\mu, x^\mu) = p^\nu V_\nu - \mathcal{L}$. From here the tensor analog of Hamiltonian equations follows. Note that this tensor version of the Hamiltonian is a scalar and, in particular, it cannot coincide with the energy of the particle. We leave the reader the task of writing the equations of Hamiltonian dynamics explicitly and of finding the relationship between the quantities defined here and the energy.

Problems

3.1 A particle of mass $2m$ at rest disintegrates giving rise to two particles of mass m which move in opposite directions with velocity v, so that the classically defined momentum is conserved. Show that if the particle of mass $2m$ moves initially with velocity $u \neq 0$, the relativistic addition law for velocities implies that the classical momentum is not conserved in the disintegration.

3.2 A nuclear reactor of 600 MW of electric power converts thermal energy of fission into electric power with an efficiency of 30%. If the fuel is natural uranium containing 0.7 % of U^{235}, which is the fissile isotope, determine what amount of fuel is used per day to generate such power. Each fissioned U^{235} atom liberates about 200 MeV. The answer, in the terms outlined here, differs by a factor of the order of 10 from the real result, because to carry out a more precise estimate the energy contribution of plutonium generated from U^{238} should be included in the calculation. The mass of an uranium atom is 3.9×10^{-22} g.

3.3 Two beams of particles of energy E collide head-on in a particle accelerator. Calculate the energy E' of a particle in one of the beams in the rest system of the other beam. Compare with the classical result. Calculate E' if $E = 30\,\text{GeV}$ and the colliding particles are protons (for protons, $Mc^2 = 1\,\text{GeV}$).

Hint: Compare the particles in each beam with the trains of Problem 2.3.

3.4 By means of a radioactive source, positrons are introduced in a solid material. Inside the material a positron at rest is annihilated with an electron emitting two photons (why two?). Calculate the momentum of the electron in terms of the angle formed by the two emitted photons. This method is used to study the momentum distribution of conduction electrons in metals.

3.5 Study the motion of a particle of rest mass m_0, initially at rest, under the action of a constant force $\boldsymbol{F}$. Compare with the classical result. In what limit would this latter result be reobtained?

Hint: Write and solve the equation of motion. To analyze the classical limit, compare the velocity of the particle with c.

3.6 If a particle of rest mass M decays into a group of particles whose total mass is smaller than M by an amount ΔM, show that the maximum kinetic energy of the i-th particle (of mass m_i) is

$$T_i^{\max} = \Delta M c^2 \left(1 - \frac{m_i}{M} - \frac{\Delta M}{2M}\right).$$

Hint: If the i-th has its maximum kinetic energy, what is the energy of the remaining particles?

3.7 A constant force $\boldsymbol{F} = F\hat{\boldsymbol{y}}$ acts on a particle moving in the (x, y) plane, whose initial velocity is $\boldsymbol{v}(0) = v_0\hat{\boldsymbol{x}}$. (a) Find the momentum of the particle as function of time. (b) Find its velocity as a function of time. (c) Make a schematic plot of each of the components of the velocity as function of t. (d) How is it possible that, even if no forces act along the x direction, the velocity in that direction **does** change with time?

Hint: Write and solve the four components of the covariant equation of motion.

3.8 A relativistic charged particle moves in the presence of a uniform magnetic field, $\boldsymbol{B} = B\ \hat{\boldsymbol{z}}$. (a) Prove that if the initial velocity of the particle is normal to the field its motion is circular and uniform in the plane perpendicular to $\boldsymbol{B}$. Calculate the (cyclotron) frequency and the radius of the orbit. Plot schematically both quantities as functions of the modulus of the velocity. (b) Show that if the particle has a velocity component initially parallel to the field, a motion at constant velocity in the direction of $\boldsymbol{B}$ is superimposed on the circular motion (Fig. 3.1). How does this influence the velocity dependence of the frequency and of the radius of the orbit perpendicular to the field?

Hint: Take into account the Lorentz force.

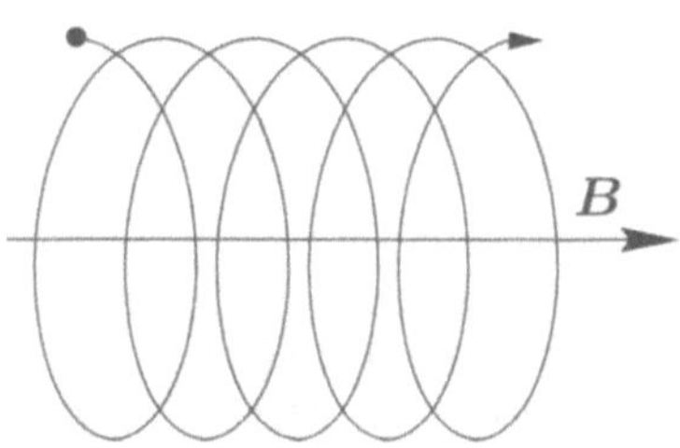

Fig. 3.1. Point charge moving in a uniform magnetic field

3.9 Show that the Hamiltonian of a particle in the presence of an electromagnetic field, obtained from the Lagrangian given in the text, is $\mathcal{H} = p^\nu p_\nu/2m_0 - qA^\nu p_\nu/cm_0$. Using the Hamilton canonical equations (a) obtain the equations of motion for the particle in terms of the Lorentz force and (b) show that the total energy can be written as $E = m_0c^2/\sqrt{1 - v^2/c^2} + q\phi/c$.

3.10 Consider a body of mass M_0 at rest in a reference system S, which emits an amount of energy E_0 during a finite time interval in the form of radiation. Assume that the momentum associated with this radiation is zero as, for example, in the case of the emission of two photons moving in opposite directions or of a spherical wave. Study the phenomenon from a system S′ with respect to which the body moves with velocity V. Explain why the body maintains its uniform motion in system S′, in spite of the fact that the emitted radiation has, in this system, a net momentum.

Solution: The momentum four-vector associated with radiation is

$$p^{\nu} = (E/c, \boldsymbol{p}).$$

In system S it is given by $p_0^{\nu} = (E_0/c, \mathbf{0})$. According to the transformation law for p^{ν}, (3.21), in S′ we have

$$p'^{\nu} = (E'/c, \boldsymbol{p}') = \left(E_0/c\sqrt{1 - V^2/c^2}, E_0\boldsymbol{V}/c^2\sqrt{1 - V^2/c^2}\right).$$

Since in S the body remains at rest, its velocity in S′ must remain constant, even when it transfers a momentum $\boldsymbol{p}' = E_0\boldsymbol{V}/c^2\sqrt{1 - V^2/c^2}$ to the radiation. This fact can be explained admitting that, after the emission, the body changes its rest mass to a value M_0', so that the relation

$$\frac{M_0\boldsymbol{V}}{\sqrt{1 - V^2/c^2}} - \frac{M_0'\boldsymbol{V}}{\sqrt{1 - V^2/c^2}} = \frac{E_0\boldsymbol{V}}{c^2\sqrt{1 - V^2/c^2}}.$$

holds. From this we conclude that

$$E_0 = (M_0 - M_0')c^2.$$

The radiated energy is therefore emitted at the expense of change in the mass of the body. This observation was made by Einstein in 1957 and suggests that the mass–energy equivalence must be a general law of Nature.

References

3.1 H.A. Lorentz: *Electromagnetic Phenomena in a System Moving with any Velocity less than that of Light*, Proc. Acad. Sci. Amsterdam **6** (1904).
3.2 A. Einstein: *Ist die Trägheit eines Körpers von seinem Energiegehalt abhängig?*, Ann. der Physik **17** (1905).

Further Reading

A. Einstein: *The Meaning of Relativity* (Princeton University Press, Princeton, 1953).
L.D. Landau and E.M. Lifshitz: *The Classical Theory of Fields* (Pergamon, Oxford and Addison–Wesley, Reading, Mass. 1971).
J. Wheeler and L.W. Taylor: *Space–Time Physics* (Freeman, New York, 1966).

4. Electrostatics

As mentioned in Chap. 1, electromagnetism is a theory based on a series of quantitative experiments carried out over many years, starting from the eighteenth century. These experiments were compiled into a consistent theory of electric and magnetic phenomena by James Clerk Maxwell in 1865. Until then electric and magnetic interactions were considered two independent phenomena. Maxwell's equations,

$$\begin{aligned} &\nabla \cdot \boldsymbol{E} = 4\pi\rho, \\ &\nabla \times \boldsymbol{B} - \tfrac{1}{c}\partial_t \boldsymbol{E} = \tfrac{4\pi}{c}\boldsymbol{j}, \\ &\nabla \cdot \boldsymbol{B} = 0, \\ &\nabla \times \boldsymbol{E} + \tfrac{1}{c}\partial_t \boldsymbol{B} = 0, \end{aligned} \tag{4.1}$$

provide the first unification of two apparently different kinds of interactions. These equations allow for the electric and magnetic fields $\boldsymbol{E}$ and $\boldsymbol{B}$ to be given as functions of position and time, in terms of the *charge density* ρ and the *current density* $\boldsymbol{j}$ producing them. Equations (4.1) constitute the basis of a theory describing in a unified way electric, magnetic and optical phenomena. Their validity has been checked from atomic to astronomical scales. At the atomic level it must be made compatible with quantum-mechanical laws.

In this chapter we study the properties of the electric field. On the basis of the foundations of the theory of relativity discussed in the previous chapters, we begin discussing the essential properties of the electric charge. This physical entity is the origin of all electromagnetic phenomena. Next, we analyze in some detail the properties of the electrostatic field and the electrostatic potential, discussing some particular examples and introducing the multipole expansion. We introduce the concept of electrostatic energy and present a variational principle linked to it. In addition to its intrinsic importance, this principle provides the basis for numerical calculation algorithms. Finally, we study an application of the concept of electrostatic energy to the calculation of the so-called classical electron radius. This was the first historical attempt at formulating a theory of elementary particles.

4.1 Properties of the Electric Charge

The electric charge is the property of matter giving rise to electromagnetic interactions. It appears in nature in two symmetrical forms, called positive and negative. The sign convention is arbitrary and was established by Benjamin Franklin. He for the first time called *positive* the charge of glass rubbed with silk and *negative* the opposite one.

The fundamental properties of the electric charge, derived from experience, are:

- *Interaction* with other charges, in the form of a force given – for charges at rest – by Coulomb's law. Two charges of the same sign at rest repel each other with a force in the direction of their relative position. Two charges of opposite signs, on the other hand, are similarly attracted. For two charges q_1 and q_2 at rest at positions $\boldsymbol{r}_1$ and $\boldsymbol{r}_2$ Coulomb's law states that the force produced by q_1 on q_2 is

$$\boldsymbol{F}_{21} = q_1 q_2 \frac{\boldsymbol{r}_{21}}{r_{21}^3}, \tag{4.2}$$

where $\boldsymbol{r}_{21} = \boldsymbol{r}_2 - \boldsymbol{r}_1$. This expression, in fact, defines the value of the electric charge. Notice that $\boldsymbol{F}_{21} = -\boldsymbol{F}_{12}$. Throughout this book we use the cgs Gaussian system of units. In this system, the unit of electric charge is usually called the *electrostatic unit of charge* and is denoted by esu. The force between two charges $q_1 = q_2 = 1\,\text{esu}$ at $1\,\text{cm}$ from each other equals $1\,\text{dyne}$. In the international system of units (SI) Coulomb's law is written

$$\boldsymbol{F}_{21} = \frac{1}{4\pi\varepsilon_0} q_1 q_2 \frac{\boldsymbol{r}_{21}}{r_{21}^3}. \tag{4.3}$$

In this case the force is measured in newtons (N), and the numerical value of the factor $1/4\pi\varepsilon_0$ equals $10^{-7}\,c^2$, where c is the speed of light. In SI the unit of charge is defined starting from the force between currents (see Chap. 6) once the unit of current is defined. The resulting unit of charge is the coulomb, $1\,\text{C} = 2.998 \times 10^9\,\text{esu}$. Note that this proportionality factor coincides numerically with a tenth of the speed of light measured in the cgs system. In order to illustrate the relative magnitude of the electrostatic force in comparison with the gravitational force, we note that the ratio of the electric to the gravitational interaction between the electron and the proton in a hydrogen atom is given by $e^2/Gm_\text{e}m_\text{p}$, where G is the gravitational constant, m_p is the mass of the proton, m_e is the mass of the electron and e is its charge. The value of this ratio is approximately 2×10^{39}, indicating that, at the atomic level, the electrostatic force is comparatively much more intense than the gravitational one.
- A *superposition principle* for electromagnetic forces holds. This implies that the total force produced by a system of charges on a test charge equals the sum of the forces produced by each of them.

- *Conservation* in all physical processes. This property implies that, even in processes where particles are created or annihilated, the total initial charge equals the total final charge.
- *Quantization*, according to which any electric charge is an integer multiple of an elementary charge. This property was discovered by Faraday in electrolysis experiments. He linked the amount of charge passing through an electrolytic cell to the amount of matter deposited on the electrodes. Robert Andrews Millikan (1868–1953), in a famous series of experiments performed between 1906 and 1916, verified this quantization [4.1]. Until the decade of 1970 it was believed that the elementary charge was that of the electron. Although it has not yet been observed in a free state, the smallest quantity of charge currently known is the charge of a quark, which is a third of the electron charge. The latter is negative, and its value is $e = -4.803 \times 10^{-10}$ esu.
- *Relativistic invariance*, which establishes that the total charge of a given physical system is the same for all inertial observers. In this sense, it differs from the other fundamental property of matter, the mass, whose transformation laws have been studied in previous chapters.

Although the electric charge appears in quantized form only, for our level of description and for convenience in the mathematical formulation charge distributions are usually described by means of a charge density $\rho(\boldsymbol{r}, t)$ which depends on position and time, and such that the total charge Q in a volume V is given by

$$Q = \int_V \rho(\boldsymbol{r}, t)\, \mathrm{d}^3 r. \tag{4.4}$$

Point charges can also be represented by means of charge densities using the *Dirac delta distribution* (Appendix A). For example, a point charge q, located at the point $\boldsymbol{r}_0(t)$, is described by a density $\rho(\boldsymbol{r}, t) = q\delta(\boldsymbol{r} - \boldsymbol{r}_0(t))$. Charge conservation requires that the time variation of the total charge contained in a volume V is determined by the charges leaving or entering that volume:

$$\partial_t Q = \partial_t \int_V \rho(\boldsymbol{r}, t)\, \mathrm{d}^3 r = -\int_{S(V)} \boldsymbol{j}(\boldsymbol{r}, t) \cdot \mathrm{d}\boldsymbol{s}. \tag{4.5}$$

Here $S(V)$ is the surface enclosing the volume V, $d\boldsymbol{s}$ is the outward surface element, and $\boldsymbol{j}(\boldsymbol{r}, t)$ is the electric current density. The integral of the electric current density over a surface defines the *electric current*, i.e. the amount of charge crossing the surface per unit time. The unit of electric current in the cgs system is $\mathrm{esu\,s^{-1}}$ also called statampere (statA); in SI it is the ampere (A), equivalent to 3×10^9 statamperes. Current density is measured in $\mathrm{statampere\,cm^{-2}}$ in the cgs system and in $\mathrm{A\,m^{-2}} = 3 \times 10^{-2}\,\mathrm{statA\,cm^{-2}}$ in SI.

The conservation of electric charge, (4.5), can be expressed in *differential* or *local form* using the divergence theorem to transform the surface integral

(4.5) into a volume integral, and observing that it should be valid for any volume V. In this way we obtain the *equation of continuity*

$$\partial_t \rho + \nabla \cdot \boldsymbol{j} = 0. \tag{4.6}$$

This relation expresses the conservation of charge in terms of the charge and current densities. We show next that, in its relativistic covariant form, the equation of continuity establishes a natural link between conservation and invariance of charge: two experimental facts which at first glance appear as independent. To understand this linking, let us analyze the following result.

Theorem: Consider a four-dimensional "volume" V limited by the (three-dimensional) "surfaces" Σ_0, Σ_1 and Σ_2 (Fig. 4.1). Let V^μ be a four-vector depending on the coordinates x^ν, vanishing for $x^\nu \to \pm\infty$ and such that its four-dimensional divergence is zero:

$$\partial_\nu V^\nu = 0. \tag{4.7}$$

In the limit in which the surfaces Σ_1 and Σ_2 extend up to infinity in four-dimensional space we thus have

$$\int_{\Sigma_1} V^\nu \, \mathrm{d}\sigma_\nu = \int_{\Sigma_2} V^\nu \, \mathrm{d}\sigma_\nu, \tag{4.8}$$

where $\mathrm{d}\sigma_\nu$ are the components of the outgoing normal to each of the surfaces.

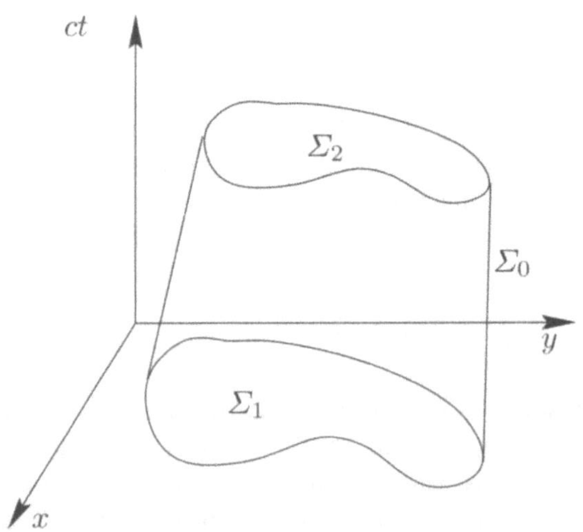

Fig. 4.1. Four-dimensional "volume" bounded by three-dimensional "surfaces"

The proof of this theorem is a direct application of the four-dimensional version of the divergence theorem:

$$\int_V \partial_\nu V^\nu \mathrm{d}^4x = \int_{S(V)} V^\nu \mathrm{d}\sigma_\nu. \tag{4.9}$$

In this case, $S(V) = \Sigma_0 + \Sigma_1 + \Sigma_2$. In addition, according to the hypotheses about V^ν, the integral over Σ_0 vanishes in the limit in that Σ_1 and Σ_2 extend to infinity. Taking into account the appropriate sign for the normal to each surface, the result (4.8) is obtained.

Let us now take Σ_1 to be the three-dimensional space given by the spatial coordinates of a reference system S. Its normal has the direction of the time axis of S. Similarly, let Σ_2 be the manifold given by the space coordinates of a system S′, with normal parallel to its time axis. The thesis of the theorem, (4.8), having been proved, it can now be written as

$$\int V^0 \mathrm{d}^3r = \int V'^0 \mathrm{d}^3r', \tag{4.10}$$

where the integrals are carried out over the space coordinates in each system. This expression represents a relativistic invariance law, since it indicates that the quantity obtained after integrating V^0 over space is the same for all inertial observers, namely, a scalar. This theorem can be generalized to four-tensors of any rank n with zero divergence. In this case it can be shown that the integrals over the surfaces Σ_1 and Σ_2 give rise to four-tensors of rank $n-1$.

To apply this result to the electric charge density we notice that the total charge $Q = \int_V \rho(\boldsymbol{r},t)\, \mathrm{d}^3r$ is a relativistic invariant. This suggests that $\rho(\boldsymbol{r},t)$ could be the time component of a four-vector. The question then naturally arises of whether a four-vector exists whose time component is proportional to ρ. According to the hypotheses of the theorem, this four-vector should have vanishing four-divergence. But we notice that the continuity equation (4.6) has the form of a four-divergence set equal to zero, where the charge density appears differentiated with respect to time. Consequently the quantities differentiated with respect to the space coordinates, that is to say, the components of the current density, are the right candidates to complete the four-vector we look for. Indeed, as a consequence of the conservation of charge, the group of four quantities

$$J^\nu = (c\rho, j_x, j_y, j_z) = (c\rho, \boldsymbol{j}), \tag{4.11}$$

are the components of a four-vector with zero four-divergence.

The theorem implies that the integral of the time component, namely the total charge, is a relativistic invariant. This result appropriately expresses the connection between two different experimental results, i.e. conservation and relativistic invariance of the electric charge.

4.2 Electric Field

From now on we concentrate on the study of systems of charges at rest. The interaction between such charges is specified by Coulomb's law (4.2) and the superposition principle, so that the force exerted by a group of charges q_i located at points $\boldsymbol{r}_i$ on a charge q located at $\boldsymbol{r}$ is

$$\boldsymbol{F}(r) = q \sum_i q_i \frac{\boldsymbol{r} - \boldsymbol{r}_i}{|\boldsymbol{r} - \boldsymbol{r}_i|^3}. \tag{4.12}$$

For the charge distribution q_i, the *electric field* produced by these charges at the point $\boldsymbol{r}$ is defined as the force acting on a unit charge located at $\boldsymbol{r}$:

$$\boldsymbol{E}(r) = \sum_i q_i \frac{\boldsymbol{r} - \boldsymbol{r}_i}{|\boldsymbol{r} - \boldsymbol{r}_i|^3}. \tag{4.13}$$

The electric field is thus fully determined by the charge configuration q_i, usually called *sources* of the field. The definition of electric field implies an important conceptual step regarding expression (4.12) for the Coulomb force. In fact, the latter could also have been attributed to an *action at a distance.* The concept of a field produced by the sources, occupying the space surrounding them, implies that the charges modify the "state of the vacuum" and create physical conditions capable of being detected by a test charge.

The notion of electric field can be generalized to the case in which it is produced by a continuous charge distribution with density $\rho(\boldsymbol{r})$. In fact the charge in a differential volume element is $\rho(\boldsymbol{r})\mathrm{d}^3 r$ and so:

$$\boldsymbol{E}(r) = \int \rho(\boldsymbol{r}') \frac{\boldsymbol{r} - \boldsymbol{r}'}{|\boldsymbol{r} - \boldsymbol{r}'|^3} \, \mathrm{d}^3 r'. \tag{4.14}$$

In the cgs system, the electric field is measured in dyn/esu, a combination of units called a gauss. In SI units, the electric field is measured in volt/meter, equal to 0.33×10^{-4} gauss (G).

A result equivalent to Coulomb's law – since it also characterizes the electric field of a charge distribution in a unique way – is *Gauss's theorem.* This states that the flux of the electric field through a closed surface is proportional to the total charge contained within that surface. To prove this theorem, we calculate the divergence of the field produced by a point charge q. Without loss of generality, we consider the charge to be located at the origin, so that

$$\nabla \cdot \boldsymbol{E} = q \, \nabla \cdot \frac{\boldsymbol{r}}{r^3} = \frac{q}{r^2} \partial_r \left(r^2 \frac{1}{r^2} \right) = 0, \tag{4.15}$$

except for $\boldsymbol{r} = 0$. The divergence has been calculated in spherical coordinates, where the field has only a radial component. We see from (4.15) that the electric field of a point charge has zero divergence over all space, except possibly at the position of the charge, since $\boldsymbol{E}$ is singular there. For the field produced by an arbitrary charge distribution, the superposition principle implies that

$$\nabla \cdot \boldsymbol{E} = 0 \tag{4.16}$$

at all points not occupied by charges.

To show what happens in the points occupied by charges, let us consider again a point charge q, surrounded by an arbitrary closed surface S, and by a small spherical surface S_0, of radius ε, concentric with the charge (Fig. 4.2).

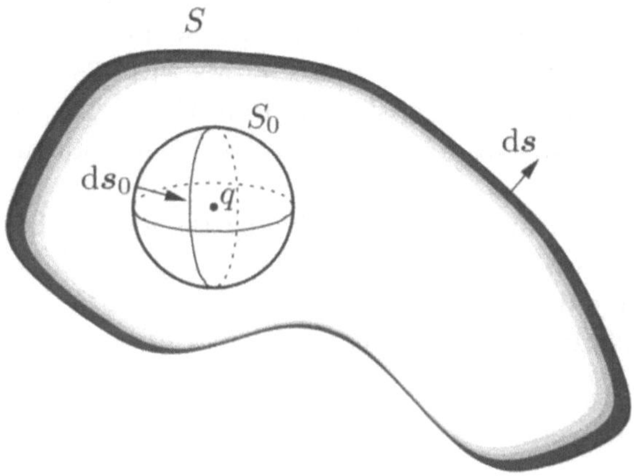

Fig. 4.2. Gauss's theorem

Because of the divergence theorem, the integral of $\nabla \cdot \boldsymbol{E}$ in the volume V_0 bounded by the surfaces S and S_0 is equal to the flux of the field through these surfaces. In our case,

$$\int_S \boldsymbol{E} \cdot \mathrm{d}\boldsymbol{s} + \int_{S_0} \boldsymbol{E} \cdot \mathrm{d}\boldsymbol{s}_0 = \int_{V_0} \nabla \cdot \boldsymbol{E} \; \mathrm{d}^3 r = 0. \tag{4.17}$$

Now, on the surface of the sphere S_0 the field has constant modulus $E = q/\varepsilon^2$ and is normal to the surface. As a consequence, $\int_{S_0} \boldsymbol{E} \cdot \mathrm{d}\boldsymbol{s}_0 = -4\pi\varepsilon^2 q/\varepsilon^2 = -4\pi q$. Since this result is independent of ε, we can rewrite (4.17) as

$$\int_S \boldsymbol{E} \cdot \mathrm{d}\boldsymbol{s} = 4\pi q. \tag{4.18}$$

Applying the superposition principle once more, this identity can be generalized to any charge distribution surrounded by the closed surface S, where q is now the total charge inside S. Gauss's theorem is thus proved.

Notice that the total charge enclosed by S can be expressed as the integral of the charge density in the volume V determined by that surface. Using Gauss's theorem and the divergence theorem we can write

$$4\pi q = 4\pi \int_V \rho(\boldsymbol{r}) \; \mathrm{d}^3 r = \int_S \boldsymbol{E} \cdot \mathrm{d}\boldsymbol{s} = \int_V \nabla \cdot \boldsymbol{E} \; \mathrm{d}^3 r. \tag{4.19}$$

Since these identities should be valid for arbitrary volumes, we obtain

$$\nabla \cdot \boldsymbol{E} = 4\pi\rho. \tag{4.20}$$

This is the differential form of Gauss's theorem. This result generalizes the divergence of the electric field to points that do contain charge.

It is worth emphasizing that the differential and integral form of Gauss's theorem, as well as Coulomb's law, are rigorously equivalent results if the condition $\nabla \times \boldsymbol{E} = 0$ is added (see next section). This is a direct consequence of the dependence of the Coulomb force on r^{-2}. Such a dependence implies that the Coulomb force falls off with distance from the source in the same way as the area of the surface surrounding the source grows with distance. Here we have demonstrated Gauss's theorem starting from Coulomb's law. We suggest as an exercise for the reader to rederive Coulomb's law starting from Gauss's theorem.

Equation (4.18) is particularly appropriate for finding the electric field under conditions of high symmetry, since it avoids the explicit calculation of the integrals over the charge distributions required to determine $\boldsymbol{E}$ from Coulomb's law. As for the differential form, we notice that (4.20) is the first of Maxwell's equations (4.1).

4.3 Electrostatic Potential

As in the case of the gravitational force, the electric field can be derived from a potential,

$$\boldsymbol{E} = -\nabla\phi \tag{4.21}$$

where the function $\phi(\boldsymbol{r})$ is called *electrostatic potential.* In the cgs system, the units of the electric potential are gauss cm, called statvolt. In the SI system the potential is measured in volts (V). The relation between both units is $299.8\,\mathrm{V} = 1\,\mathrm{statvolt}$.

When the field sources are limited in space, we can take $\phi(\boldsymbol{r}) \to 0$ for $r \to \infty$, so the electrostatic potential is given by

$$\phi(\boldsymbol{r}) = \int \rho(\boldsymbol{r}')\,\frac{\mathrm{d}^3 r'}{|\boldsymbol{r} - \boldsymbol{r}'|}. \tag{4.22}$$

in terms of the charge density.

Since the electric field can be derived from a potential, it turns out that the electrostatic force is *conservative.* In fact, the line integral of the field between two arbitrary points is given by

$$\int_1^2 \boldsymbol{E} \cdot \mathrm{d}\boldsymbol{l} = -[\phi(2) - \phi(1)]. \tag{4.23}$$

According to the Stokes theorem this implies that the electric field is irrotational:

$$\nabla \times \boldsymbol{E} = 0. \tag{4.24}$$

Equation (4.24) does not correspond to any of Maxwell's equations, and holds for electrostatic problems only. Together with Gauss's theorem (4.20), (4.24) forms a system of two first-order equations giving the electrostatic field in terms of the charge distribution.

Equations (4.20) and (4.24) can be combined to give a second-order equation for ϕ:

$$\nabla^2\phi = -4\pi\rho. \tag{4.25}$$

This is *Poisson's equation*, whose solution in free space – with the condition of being zero at infinity – is (4.22). In the absence of charge, the Poisson equation becomes

$$\nabla^2\phi = 0, \tag{4.26}$$

called *Laplace's equation.* In the next chapter we study some methods to solve both of these equations.

4.4 Examples of Potentials and Fields. Multipole Expansion

In this section we illustrate the calculation of potentials and electrostatic fields in some simple situations. These examples are of importance because of their applications.

In the first place, let us evaluate the potential along the axis of a uniformly charged disk (Fig. 4.3). Let us assume the disk has radius a, surface charge density σ, and is centered at the origin of coordinates, with its plane perpendicular to the z axis.

The space charge density of the disk can be expressed as

$$\rho(x,y,z) = \begin{cases} \sigma\delta(z) & \text{for } x^2+y^2<a^2 \\ 0 & \text{for } x^2+y^2>a^2, \end{cases} \tag{4.27}$$

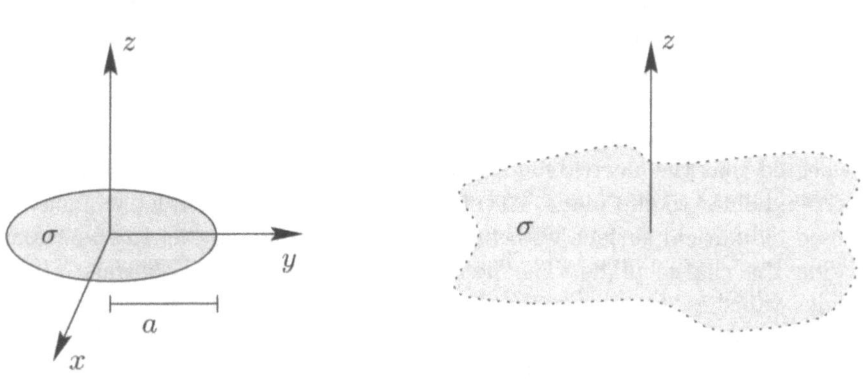

Fig. 4.3. Uniformly charged disk and infinite plane

since it must vanish for all $z \neq 0$ and for $x^2 + y^2 > a^2$. The potential on the z axis, calculated from (4.22), turns out to be

$$\begin{aligned} \phi(0,0,z) &= 2\pi \int_0^a r'\mathrm{d}r' \int_{-\infty}^{+\infty} \mathrm{d}z' \, \frac{\sigma\,\delta(z')}{\sqrt{r'^2 + (z-z')^2}} \\ &= 2\pi \int_0^a r'\mathrm{d}r' \, \frac{\sigma}{\sqrt{r'^2 + z^2}} = 2\pi\sigma\left(\sqrt{z^2+a^2} - |z|\right), \end{aligned} \tag{4.28}$$

where the integral is evaluated in cylindrical coordinates.

It is instructive to analyze the limits of this result for $z \to \infty$ and for $z \to 0$. These limits are obtained by expanding ϕ in powers of a/z and z/a, respectively. In the first case, we obtain

$$\phi(0,0,z) = \pi a^2 \sigma \left(\frac{1}{|z|} - \frac{a^2}{4|z|^3} + \frac{a^4}{8|z|^5} - \frac{5a^6}{64|z|^7} + \cdots \right). \tag{4.29}$$

The leading term in this expansion is $\phi = \pi a^2 \sigma / |z|$. For $z \gg a$ the potential ϕ coincides with the potential generated by a point charge of the same magnitude as the total charge of the disk, $\pi a^2 \sigma$. Accordingly, sufficiently far away from the disk the details of the space distribution of the charges generating the potential are lost. This is a general property of all bounded charge distributions and, as shown in the next section, can be used to obtain an approximate description of the potential at large distances from the sources.

In the limit $z \to 0$ the potential is $\phi \approx 2\pi\sigma(a - |z|)$. This result implies that the electric field near the center of the disk is

$$\boldsymbol{E} = 2\pi\sigma \,\mathrm{sign}(z)\hat{\boldsymbol{z}}. \tag{4.30}$$

This coincides with the field generated by an infinite plane with uniform surface charge density σ placed at $z = 0$. To prove this, however, we cannot go through the calculation of the potential of the infinite plane, as in (4.28), since the integral involved diverges for $a \to \infty$, as implied by the additive constant proportional to a. A possible solution to this problem consists in calculating the field directly by means of Coulomb's law, as in (4.13).

The symmetry properties of the charged infinite plane, however, allow us to carry out this calculation in a simpler way, using Gauss's law. Indeed, it is expected that the electric field depends neither on x nor on y, and that it is perpendicular to the plane. Accepting *a priori* these facts let us consider a closed cylindrical surface with base area s and axis parallel to the z axis, crossing the charge plane. The field flux across the lateral surface of the cylinder vanishes. On the bases the field is normal to the surface and has constant modulus E. The product of E times the surface $2s$ of the bases should be equal to the total charge contained inside, $4\pi s\sigma$. In this way the result (4.30) is rederived.

Let us consider now a charge distribution which is extremely important in several applications, namely, the *dipole*. It is defined as a system of two equal and opposite charges, q and $-q$, separated by a distance d, in the limit $d \to 0$ and $q \to \infty$ such that $qd = p$ remains constant. This system represents a distribution of zero total charge, where the positive charge is slightly displaced with respect to the negative charge. It is a good approximation to the charge distribution in an atom or in a molecule.

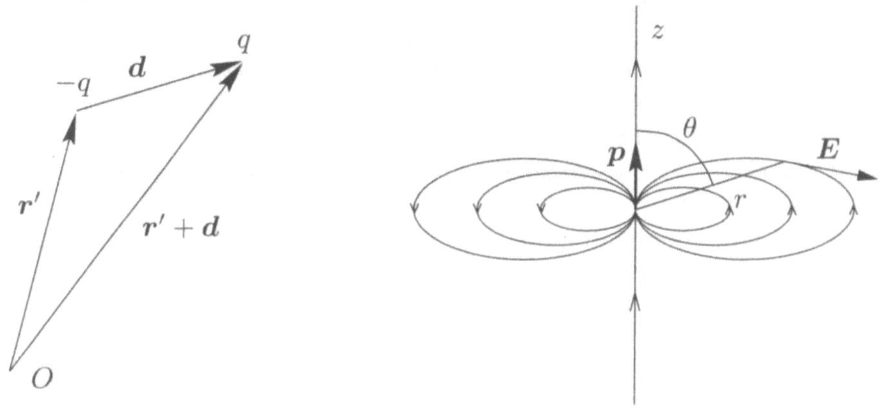

Fig. 4.4. Electric dipole

Assume that the negative charge $-q$ is placed at the point $\boldsymbol{r}'$, and the positive one at $\boldsymbol{r}' + \boldsymbol{d}$, where $\boldsymbol{d}$ is a vector of modulus d (Fig. 4.4). According to the superposition principle, the potential generated by these charges at a point $\boldsymbol{r}$ is given by

$$\phi(\boldsymbol{r}) = \frac{-q}{|\boldsymbol{r} - \boldsymbol{r}'|} + \frac{q}{|\boldsymbol{r} - \boldsymbol{r}' - \boldsymbol{d}|}. \tag{4.31}$$

This expression can be expanded in series in powers of d, so that in the limit $d \to 0$, the only non-vanishing term is

$$\phi(\boldsymbol{r}) = q\,\boldsymbol{d} \cdot \nabla' \frac{1}{|\boldsymbol{r} - \boldsymbol{r}'|} = \frac{\boldsymbol{p} \cdot (\boldsymbol{r} - \boldsymbol{r}')}{|\boldsymbol{r} - \boldsymbol{r}'|^3}. \tag{4.32}$$

Here ∇' is the gradient with respect to $\boldsymbol{r}'$ and $\boldsymbol{p} = q\boldsymbol{d}$ is the *moment* of the dipole.

If the dipole is placed at the origin, and its moment is parallel to the z axis, it follows from (4.32) that, in spherical coordinates, the potential is

$$\phi(\boldsymbol{r}) = \frac{p}{r^2} \cos\theta. \tag{4.33}$$

Starting from (4.32), the electric field of a dipole placed at $\boldsymbol{r}' = 0$, can be written as

$$\boldsymbol{E}(\boldsymbol{r}) = \frac{3\hat{\boldsymbol{r}}(\boldsymbol{p}\cdot\hat{\boldsymbol{r}}) - \boldsymbol{p}}{r^3}, \tag{4.34}$$

where $\hat{\boldsymbol{r}}$ is the unit vector in the radial direction. The dipole field has azimuthal symmetry, and decays at large distances as r^{-3}, i.e. faster than the field of a point charge. The *lines of force*, parallel at each point to the electric field, have been plotted in Fig. 4.4.

The concept of the electric dipole can be used to give an approximate description of the potential – and, therefore, of the electric field – produced by an arbitrary charge distribution. Its generalization leads to the *multipole expansion* of the potential of a charge configuration. The multipole expansion expresses the total potential generated by a limited distribution of charges as a sum of potentials due to *multipoles* of increasing order. The potential of the multipole of order 2^l ($l = 0, 1, 2...$) is characterized by a decay law proportional to r^{-l-1}, corresponding to an arrangement of 2^l point charges. For a dipole, $l = 1$.

In Cartesian coordinates, the multipole terms can be obtained by expanding the function $|\boldsymbol{r} - \boldsymbol{r}'|^{-1}$ appearing in the expression of the potential for an arbitrary charge distribution (4.22), as a series in powers of r':

$$\frac{1}{|\boldsymbol{r} - \boldsymbol{r}'|} = \frac{1}{r} - \sum_i \frac{\partial(1/r)}{\partial x_i} x_i' + \frac{1}{2}\sum_{i,j} \frac{\partial^2(1/r)}{\partial x_i \partial x_j} x_i' x_j' + \cdots, \tag{4.35}$$

where the summations run from 1 to 3, and $x_1 \equiv x$, $x_2 \equiv y$, $x_3 \equiv z$ are the components of $\boldsymbol{r}$, with similar definitions for the primed variables. It should be noted that the differentiations have been carried out with respect to the non-primed coordinates.

Explicitly evaluating the derivatives and introducing (4.35) into (4.22), the electrostatic potential generated by a charge distribution with density $\rho(\boldsymbol{r})$ can be written as

$$\phi(\boldsymbol{r}) = \frac{Q}{r} + \sum_i p_i \frac{x_i}{r^3} + \frac{1}{2}\sum_i \sum_j q_{ij} \frac{3x_i x_j - \delta_{ij} r^2}{r^5} + \cdots, \tag{4.36}$$

where

$$\begin{aligned} Q &= \int \rho(\boldsymbol{r}')\,\mathrm{d}^3 r', \\ p_i &= \int \rho(\boldsymbol{r}')\,x_i'\,\mathrm{d}^3 r', \\ q_{ij} &= \int \rho(\boldsymbol{r}')\,x_i'\,x_j'\,\mathrm{d}^3 r'. \end{aligned} \tag{4.37}$$

This series in negative powers of r can be considered as an asymptotic approximation to the potential at a given distance from the charge distribution. In fact, a truncation of the series provides a good approximation to the value of $\phi(\boldsymbol{r})$ for sufficiently large r.

Let us analyze the physical meaning of the coefficients Q, p_i and q_{ij}. The first one is the total charge of the distribution. This coefficient is in the leading term of the multipole expansion for large values of r. This implies, as is intuitively expected, that the potential behaves as $\phi \approx Q/r$ for sufficiently large r so that the details of the charge distribution become irrelevant and only the total charge is of interest. Higher order coefficients contain information about those details. The constants p_i are the Cartesian components of the vector

$$\boldsymbol{p} = \int \rho(\boldsymbol{r}')\boldsymbol{r}'\mathrm{d}^3r'. \tag{4.38}$$

Comparing the second term of (4.35) with (4.32), we note that this contribution to the potential has exactly the form of the potential of a dipole $\boldsymbol{p}$, given by (4.38). The vector $\boldsymbol{p}$ is called the *dipole moment* of the distribution.

The *quadrupole moments* q_{ij} form a symmetrical tensor of second rank and therefore only six of them are independent. However, starting from the form of the second order term in (4.36), it can be shown that the quadrupole potential does not change if we add a constant k to the diagonal terms of the tensor. In fact,

$$\begin{aligned} & q_{xx}\frac{3x^2-r^2}{r^5} + q_{yy}\frac{3y^2-r^2}{r^5} + q_{zz}\frac{3z^2-r^2}{r^5} \\ &= (q_{xx}+k)\frac{3x^2-r^2}{r^5} + (q_{yy}+k)\frac{3y^2-r^2}{r^5} + (q_{zz}+k)\frac{3z^2-r^2}{r^5}. \end{aligned} \tag{4.39}$$

This property allows us to redefine the components of the tensor so that it has zero trace. The new components are

$$\bar{q}_{ij} = \int \rho(\boldsymbol{r}')\left(x_i'x_j' - \frac{1}{3}r'^2\delta_{ij}\right)\mathrm{d}^3r', \tag{4.40}$$

in such a way that only five of the quadrupole moments are independent.

The form of $\bar{q}_{ij}$ is such that they vanish for a spherical and homogeneous charge distribution. The quadrupole moment of a given distribution is a measure of its departure from a uniformly charged sphere. The quadrupole moments play an important role in the microscopic theory of electromagnetic phenomena in material media, and are of frequent use in nuclear physics.

The remaining terms in (4.36), giving contributions of higher order to the potential, can be interpreted in a similar way in terms of higher order moments of the charge distribution. The multipole moments to all orders are given by the charge distribution; conversely, it can be shown that the set of all moments uniquely determines $\rho(\boldsymbol{r})$.

The multipole moments of a distribution can be expressed in more compact form – with only two indices – if the multipole expansion is expressed in spherical coordinates. When $r' < r$, the function $|\boldsymbol{r} - \boldsymbol{r}'|^{-1}$ can be written as a series in powers of r'/r in the form

$$\frac{1}{|\boldsymbol{r}-\boldsymbol{r}'|} = 4\pi \sum_{l=0}^{\infty} \sum_{m=-l}^{l} \frac{1}{2l+1} \frac{r'^l}{r^{l+1}} Y^*_{lm}(\theta',\varphi')\, Y_{lm}(\theta,\varphi), \tag{4.41}$$

where θ and φ are the spherical angular coordinates of $\boldsymbol{r}$ and the asterisk indicates complex conjugation. The functions $Y_{lm}(\theta,\varphi)$, studied in Appendix B, are called *spherical harmonics* and are given by

$$Y_{lm}(\theta,\varphi) = \sqrt{\frac{2l+1}{4\pi}\frac{(l-m)!}{(l+m)!}}\, P_l^m(\cos\theta)\, \exp(im\varphi), \tag{4.42}$$

where $P_l^m(\cos\theta)$ are the *associated Legendre functions*:

$$P_l^m(x) = \frac{(-1)^m}{2^l l!}(1-x^2)^{m/2} \frac{d^{l+m}}{dx^{l+m}}(x^2-1)^l. \tag{4.43}$$

Given a bounded distribution of charge with density $\rho(\boldsymbol{r})$ the corresponding electrostatic potential is

$$\phi(\boldsymbol{r}) = \sum_{l,m} \frac{4\pi}{2l+1} Q_{lm}\, Y_{lm}(\theta,\varphi)\, r^{-l-1}, \tag{4.44}$$

where

$$Q_{lm} = \int \rho(r')\, r'^l\, Y^*_{lm}(\theta',\varphi')\, \mathrm{d}^3 r', \tag{4.45}$$

are the *spherical multipole moments* of the distribution.

The zeroth-order moment Q_{00} equals the total charge divided by $\sqrt{4\pi}$, since the corresponding spherical harmonic is $Y_{00} = 1/\sqrt{4\pi}$.

For $l = 1$ the three moments, corresponding to $m = -1, 0, 1$, are

$$\begin{aligned} Q_{10} &= p\, \sqrt{3/4\pi}\, \cos\Theta, \\ Q_{1,\pm 1} &= -p\, \sqrt{3/8\pi}\, \sin\Theta\, \exp(\mp \mathrm{i}\Phi), \end{aligned} \tag{4.46}$$

where p, Θ and Φ are the spherical components of the dipole moment $\boldsymbol{p}$. Introducing these moments in the multipole expansion and summing up the three terms with $l = 1$, the potential of a dipole, (4.33), is rederived. Higher order moments are obtained for values of $l > 1$.

For a bounded charge distribution axially symmetric around the z axis, (4.44) reduces to

$$\phi(\boldsymbol{r}) = \sum_{l=0}^{\infty} \sqrt{\frac{4\pi}{2l+1}} Q_{l0}\, P_l(\cos\theta)\, r^{-l-1}, \tag{4.47}$$

where $P_l \equiv P_l^0$ are the *Legendre polynomials* (see Appendix B). This expression allows us to determine the electrostatic potential over all space starting from the solution on the z axis. Indeed, expanding this solution in powers of $1/z$ we can identify the coefficients Q_{l0} of the expansion (4.47). On the z axis, where $\theta = 0$, the Legendre polynomials reduce to $P_l(1) = 1$, and $r \equiv |z|$.

In this way, for the charged disk studied in the previous section, it is possible to give the potential in all space starting from (4.29) as:

$$\phi(\theta, r) = \frac{Q}{r} P_0(\cos\theta) - \frac{a^2 Q}{4r^3} P_2(\cos\theta) + \frac{a^4 Q}{8r^5} P_4(\cos\theta) - \frac{5a^6 Q}{64r^7} P_6(\cos\theta) + \cdots . \tag{4.48}$$

where $Q = \pi a^2 \sigma$ is the total charge on the disk.

Let us finally mention that, due to the symmetrical form in which $\boldsymbol{r}$ and $\boldsymbol{r}'$ appear in the multipole expansion, it is possible to make a similar expansion for $r' > r$, in which both vectors exchange their roles. The resulting expansion is an asymptotic approximation to the potential produced near the origin by any charge distribution, with the condition that it does not contain point charges at the origin.

4.5 Electrostatic Energy

The fact that the electrostatic field can be derived from a potential makes it possible to define the potential energy of a test charge in the presence of a given charge distribution. This potential energy, called the *electrostatic energy,* equals the work required to bring the charge from infinity to its present position. The electric force on a charge q located at a point $\boldsymbol{r}$ is given by

$$\boldsymbol{F} = q\,\boldsymbol{E}(r) = -q\nabla\phi(\boldsymbol{r}), \tag{4.49}$$

where ϕ is the potential associated with the field $\boldsymbol{E}$. Since application of a force $-\boldsymbol{F}$ is needed to move the charge in the presence of the field, its potential energy is defined as $W_q = q\phi$. Here we have considered that the potential vanishes at infinity, as is in fact the case for a bounded source distribution.

The electrostatic energy of a system of point charges q_i – which coincides with the work necessary to build up the system bringing the charges from infinity – is given by the sum of the energy of each charge in the presence of the others:

$$W = \sum_i W_i = \frac{1}{2}\sum_{i,j}{}' \frac{q_i q_j}{|\boldsymbol{r}_i - \boldsymbol{r}_j|}, \tag{4.50}$$

where the primed sum excludes the term with $i = j$. The factor $1/2$ has been introduced to avoid double counting of each term in the sum.

Replacing the sums by integrals, this expression can be generalized to the case of a continuous charge distribution of density $\rho(\boldsymbol{r})$ generating a potential $\phi(\boldsymbol{r})$. If this distribution does not contain singularities due to point charges, the condition $i \neq j$ in the sum in (4.50) can be omitted in the continuum limit. The energy of the continuous system is

$$W = \frac{1}{2}\int_V \rho(\boldsymbol{r})\phi(\boldsymbol{r})\,\mathrm{d}^3r = -\frac{1}{8\pi}\int_V \phi(\boldsymbol{r})\nabla^2\phi(\boldsymbol{r})\,\mathrm{d}^3r, \tag{4.51}$$

where the integration volume V must contain the entire distribution. In the last expression the energy is given in terms of the potential using the Poisson equation (4.25). Equation (4.51) can be transformed into the form

$$W = -\frac{1}{8\pi}\int_S \phi\nabla\phi\cdot\mathrm{d}\boldsymbol{s} + \frac{1}{8\pi}\int_V |\nabla\phi|^2\mathrm{d}^3r, \tag{4.52}$$

where S is the surface of the volume V. If the charge distribution is bounded, the integration volume can be extended to infinity. In this limit the integral over S vanishes, since $\phi\nabla\phi$ falls off as R^{-3}, where R is a characteristic dimension of the volume V, and the surface grows as R^2. The energy turns out to be

$$W = \int \frac{|\nabla\phi|^2}{8\pi}\mathrm{d}^3r, \tag{4.53}$$

where the integral extends over all space.

Inspired by (4.53), J.C. Maxwell proposed considering the electric field as the source of the electrostatic energy. This idea can be qualitatively understood by means of a mechanical analogue: for two masses connected by a compressed spring, the potential energy of the system can be associated with the masses or with the spring. Maxwell's proposal corresponds to associate it with the spring.

According to this idea electrostatic energy is distributed in space, with a local density

$$w(\boldsymbol{r}) = \frac{|\nabla\phi(\boldsymbol{r})|^2}{8\pi} = \frac{|\boldsymbol{E}(\boldsymbol{r})|^2}{8\pi}. \tag{4.54}$$

The integral over space is identical to the potential energy of the charge distribution producing the field. Later on, we show that the density $w(\boldsymbol{r})$ appears in the energy conservation laws associated with the electric field.

Once the notion of energy density for the electrostatic field of a continuous charge distribution is introduced, it is reasonable to study the corresponding quantity for a system of point charges. The electric field generated by a system of charges is, according to the superposition principle, $\boldsymbol{E} = \sum_i \boldsymbol{E}_i$, where $\boldsymbol{E}_i$ is the field of each charge. Its modulus squared is given by $E^2 = \sum_i E_i^2 + \sum'_{i,j} \boldsymbol{E}_i \cdot \boldsymbol{E}_j$, so that the total energy "à la Maxwell" would be

$$W = \sum_i \frac{1}{8\pi}\int E_i^2\mathrm{d}^3r + {\sum_{i,j}}' \frac{1}{8\pi}\int \boldsymbol{E}_i\cdot\boldsymbol{E}_j\,\mathrm{d}^3r. \tag{4.55}$$

Using the identities $\boldsymbol{E}_i \cdot \boldsymbol{E}_j = \nabla\phi_i \cdot \boldsymbol{E}_j = \nabla(\boldsymbol{E}_j\phi_i) - \phi_i\nabla\cdot\boldsymbol{E}_j$ and $\nabla\cdot\boldsymbol{E}_j = 4\pi\rho_j(\boldsymbol{r}) = 4\pi q_j\delta(\boldsymbol{r}-\boldsymbol{r}_j)$, it is possible to show that the second term reduces to (4.50).

The first term in the right-hand side of (4.55), on the other hand, does not appear in the potential energy calculated above for a system of point charges.

This term is a superposition of contributions of each charge and, in fact, it differs from zero even when only one charge is present. The contribution of a given charge is called the *self-energy* of the charge, and is associated with the work needed to assemble that charge starting from elementary contributions brought from infinity. When considering point charges the problem arises that the self-energy of each one is infinite. Indeed, the quantity $E^2 \mathrm{d}^3 r$ diverges as r^{-2} for $r \to 0$.

The self-energy of a point charge is infinite because the work needed to bring elementary charges close together at infinitely small distances diverges. Thus, as for the calculation of energy, the concept of a point charge poses insurmountable difficulties. The correct calculation of the energy of a system of charges requires a more detailed model of the charge densities in the small volumes occupied by each particle. Note that these small scales are relevant to quantum phenomena, which are not covered by the theory that we study here. We shall come back to this problem in Chap. 9.

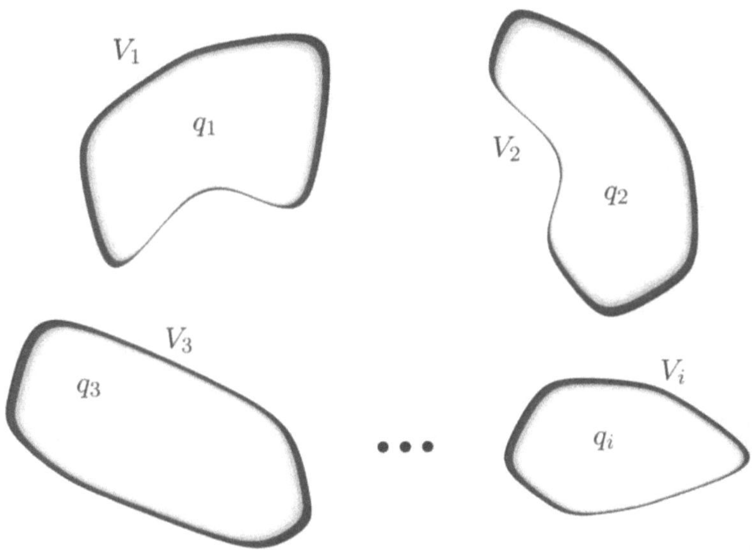

Fig. 4.5. System of bounded conductors

In order to complete the discussion of the energy in an electrostatic system, we now consider a group of closed conductors, with charges q_i and potentials V_i (Fig. 4.5). Since the field inside the conductors is zero it is sufficient to carry out the integral of the energy density in the space external to them:

$$W = \frac{1}{8\pi} \int |\nabla \phi|^2 \mathrm{d}^3 r = \frac{1}{8\pi} \int [\nabla \cdot (\phi \nabla \phi) - \phi \nabla^2 \phi] \, \mathrm{d}^3 r. \tag{4.56}$$

As in the space between conductors the potential satisfies the Laplace equation, the second term of this equation vanishes. The first term can be transformed into a surface integral by means of the divergence theorem, where the integration area consists of each conductor surface plus a surface surrounding the whole system. This surface can be expanded to infinity and, in the limit, the integral over it vanishes. The potential on each conductor surface is a constant V_i, so that we obtain

$$W = \frac{1}{8\pi} \sum_i V_i \int_{S_i} \nabla\phi \cdot \mathrm{d}\boldsymbol{s}_i = \frac{1}{8\pi} \sum_i V_i \int_{S_i} E_n \mathrm{d}s, \tag{4.57}$$

where E_n is the electric field component normal to the surface. This component is related to the surface charge density σ according to $E_n = 4\pi\sigma$, so that finally we get

$$W = \frac{1}{2} \sum_i V_i q_i. \tag{4.58}$$

The electrostatic energy of the system of conductors can be expressed in terms of only the charges q_i or the potentials V_i if a relationship between both quantities is found. In order to find this relation we assume first that only one of the conductors is charged, with charge q_j^0. Under these conditions, a solution to the Laplace equation exists such that the potential on each conductor i is constant and has the value V_i^j. Due to the linearity of the Laplace equation, if we multiply the charge q_j^0 by a constant λ the potential on each conductor is λV_i^j. A linear relationship exists thus between the potentials and the charge q_j, of the type $V_i^j = p_{ij} q_j$, where the quantities p_{ij}, called *potential coefficients*, depend only on the geometry of the system of conductors. In the case in which all conductors are charged, the superposition principle allows us to write

$$V_i = \sum_j p_{ij} q_j. \tag{4.59}$$

It can be proved that the potential coefficients are symmetrical, $p_{ij} = p_{ji}$, and positive. Also $p_{ii} - p_{ij} \geq 0$ for every i, j.

The energy of the system can thus be written in terms of the charges on each conductor as

$$W = \frac{1}{2} \sum_{i,j} p_{ij} q_i q_j. \tag{4.60}$$

Equation (4.59) can be inverted to express the charges as a function of the potentials, $q_i = \sum_j C_{ij} V_j$. The quantities C_{ij} are called *coefficients of capacity* if $i = j$, and *induction coefficients* if $i \neq j$. They satisfy $C_{ij} = C_{ji}$, $C_{ii} > 0$ and $C_{ij} \leq 0$. Using these coefficients, the energy of the system of conductors is

$$W = \frac{1}{2}\sum_{i,j} C_{ij}V_iV_j. \tag{4.61}$$

For a system of two conductors with opposite charges q and $-q$ which are at potentials V_1 and V_2 – a capacitor – the energy is $W = q(V_2 - V_1)/2 = q\Delta V/2 = C\Delta V^2/2$, where C is the capacity. This quantity is related to the coefficients C_{ij} and p_{ij} according to

$$C = \frac{C_{11}C_{22} - C_{12}^2}{C_{11} + C_{22} + 2C_{12}} = \frac{1}{p_{11} + p_{22} - 2p_{12}}. \tag{4.62}$$

In the cgs system of units the capacity is measured in cm, while in SI the corresponding unit is the farad (F) equal to 0.975×10^{12} cm.

Finally, we show that the electrostatic energy of the system of conductors allows us to calculate the force acting on one of them. Indeed, if we carry out a virtual displacement $\mathrm{d}\boldsymbol{l}$ of one of the conductors, keeping the charges of the assembly fixed, work $\mathrm{d}L = \boldsymbol{F}_Q \cdot \mathrm{d}\boldsymbol{l}$ is done. The label Q indicates that the process is performed keeping the charges constant. Since, during the virtual displacement the system stays isolated, the work carried out $\mathrm{d}W = -\mathrm{d}L$ equals the change in the electrostatic energy. From the relation $\mathrm{d}W = -\boldsymbol{F}_Q \cdot \mathrm{d}\boldsymbol{l}$ we obtain

$$\boldsymbol{F}_Q = -\left.\nabla W\right|_Q. \tag{4.63}$$

In the case where the virtual displacement is carried out keeping the potentials on each conductor fixed, the batteries must contribute charges and perform a work $\mathrm{d}W_\mathrm{b} = \sum_i V_i \mathrm{d}q_i$. The electrostatic energy of the system changes by the amount $\mathrm{d}W = \sum_i V_i \mathrm{d}q_i/2 = \mathrm{d}W_\mathrm{b}/2$. The conservation of the total energy is written as $\mathrm{d}W_\mathrm{b} = \mathrm{d}L + \mathrm{d}W$, where $\mathrm{d}L = \boldsymbol{F}_V \cdot \mathrm{d}\boldsymbol{l}$. From here it follows that

$$\boldsymbol{F}_V = \left.\nabla W\right|_V. \tag{4.64}$$

It is interesting to notice that the numerical values of the forces calculated at constant charge or at constant potential should be the same, since the displacement carried out is only virtual.

4.6 Variational Principle for Electrostatics

In several areas of physics, fundamental laws can be expressed so that the solution to a problem has the property of making a certain conveniently defined integral reach an extreme value. This formulation, called the *variational principle*, can replace framing the problem in terms of differential equations. In fact, the variational principle allows to find such equations. In that case these are called *Euler–Lagrange equations.* For instance, mechanics can be formulated in terms of the principle of least action (Chap. 3) and geometrical optics is derived from the Fermat principle of least time.

We show now that electrostatics can be formulated starting from a similar principle involving the energy. Assume that we have a distribution of charges $\rho(\boldsymbol{r})$ in a volume V bounded by a surface S. Let us assume further that this surface S is divided into two sectors S_1 and S_2 such that on S_2 there is a surface distribution of charge σ_0 and the potential satisfies the following boundary conditions

$$\begin{aligned} &\phi(\boldsymbol{r}) = \phi_0(\boldsymbol{r}) \text{ on } S_1 \quad \text{(Dirichlet boundary conditions)}, \\ &\nabla\phi \cdot \mathrm{d}\boldsymbol{s} = 4\pi\sigma_0 \mathrm{d}s \text{ on } S_2 \text{ (Neumann boundary conditions)}. \end{aligned} \tag{4.65}$$

It can be proved that the *functional*

$$W^*[\phi] = \int_V \left[\frac{1}{8\pi}|\nabla\phi|^2 - \rho\phi\right] \mathrm{d}^3r - \int_{S_2} \sigma_0\, \phi\, \mathrm{d}s, \tag{4.66}$$

is a minimum for the function ϕ satisfying Poisson's equation and the boundary conditions, and therefore it is the solution of the electrostatic problem.

To prove this property, let us calculate W^* for a function ϕ differing from the solution of Poisson's equation ϕ_0 by a function f, such that $\phi = \phi_0 + f$. The function f is chosen to be zero on the surface S_1 in such a way that ϕ satisfies the same boundary conditions as ϕ_0. From this calculation we obtain

$$W^*[\phi] = W^*[\phi_0] + \frac{1}{4\pi}\int \left[\nabla\phi_0 \cdot \nabla f - 4\pi\rho f + \frac{1}{2}|\nabla f|^2\right] \mathrm{d}^3r. \tag{4.67}$$

This quantity can be rewritten

$$\begin{aligned} W^*[\phi] = W^*[\phi_0] &- \frac{1}{4\pi}\int \left[\nabla^2\phi_0 + 4\pi\rho\right] f\, \mathrm{d}^3r + \frac{1}{4\pi}\int_{S_1} f\nabla\phi_0 \cdot \mathrm{d}\boldsymbol{s} \\ &+ \frac{1}{4\pi}\int_{S_2} f(\nabla\phi_0 - 4\pi\sigma) \cdot \mathrm{d}\boldsymbol{s} + \frac{1}{8\pi}\int |\nabla f|^2\, \mathrm{d}^3r. \end{aligned} \tag{4.68}$$

The first integral vanishes, since ϕ_0 satisfies Poisson's equation (4.25). The surface integral of f on S_1 is zero since this function vanishes on S_1. The surface integral on S_2 is also zero due to the boundary condition (4.65). On the other hand, if f is not constant, the last integral is a positive-definite quantity. As a consequence, $W^*[\phi] > W^*[\phi_0]$, and the solution of the electrostatic problem is a minimum of W^*.

Note that the quantity W^* is equal to the electrostatic energy of the charge distribution. In addition, as a corollary of the above proof, we note that if ϕ_0 is a solution of Poisson's equation, such a solution is unique, except for an arbitrary additive constant.

Besides its conceptual importance, the existence of a variational principle for electrostatics can be used as the basis for an approximate numerical or analytical calculation of the solution to the problem. Indeed, if through plausibility arguments it is possible to give an approximate solution depending on a group of parameters, these can be tuned in such a way as to minimize – within the functional space accessible through variation of the parameters –

the functional involved in the variational principle. In the following chapter we will show how this idea can be applied to the search for concrete solutions to electrostatic problems. In particular we will see how this procedure becomes the basis of the finite element method.

4.7 Classical Radius of the Electron

In the few years that elapsed between the development of special relativity in 1905 and the introduction of quantum mechanics in 1925, in an attempt to build a microscopic model of the electron, physicists postulated that all the inertia of this particle was associated with the energy of its field, through the relativistic relation $E = mc^2$. In this picture all of the electron mass would be "electromagnetic," and the rest mass would be equal to the energy of the electrostatic field. As we show next, this hypothesis allows us to estimate a size for the electron.

Assuming that the electron charge is evenly distributed over the surface of a sphere of radius r_0, the energy associated with its electrostatic field is

$$W = \frac{1}{8\pi}\int E^2\, \mathrm{d}^3r = \frac{1}{2}\int_{r_0}^{\infty} E^2 r^2 \mathrm{d}r = \frac{e^2}{2r_0}, \tag{4.69}$$

where e is the charge of the electron. Other hypotheses regarding the charge distribution inside the particle give energies of the same order. For example, for a charge uniformly distributed in the volume of a sphere of radius r_0, the result is $W = 3e^2/5r_0$. In order to obtain an order of magnitude result, independent of the model for the charge distribution, one usually takes $W = e^2/r_0$. Equating this quantity to the rest energy of the electron m_0c^2, we obtain the so-called *classical electron radius*

$$r_0 = \frac{e^2}{m_0c^2} \approx 3 \times 10^{-13}\,\mathrm{cm}. \tag{4.70}$$

Observe that the same calculation could be made for any charged particle. For the proton, which has the same charge as the electron, we would obtain a radius some two thousand times smaller than that of the electron.

The current evidence, based on experiments of inelastic collisions of electrons on protons, shows that while the electron is a point particle, the proton has an internal charge distribution. Experimental results indicate that the proton has a radius of the order of 10^{-13} cm and that the electric charge is concentrated on three small, point-like components [4.2]. These components of the proton are the already-mentioned *quarks*, which together with the *leptons* (electrons, muons and neutrinos) and the *intermediate bosons* (photons, W and Z and gluons) are the true elementary particles. Thus, assigning a radius to the electron turns out to be nothing but a historical curiosity, overtaken by experimental evidence and by modern theories of the structure of matter.

On the other hand, the expression e^2/m_0c^2 is the only combination of electromagnetic and relativistic (non-quantal) constants characteristic of the electron with dimensions of length. In fact, the quantity e^2/m_0c^2 frequently appears in some results of electromagnetic theory as a characteristic unit of distance. For example, the total cross section for dispersion of low frequency radiation by an electron (Thomson dispersion) is, except for a factor of $8\pi/3$, equal to the square of the classical electron radius. This result is consistent with assigning a radius to this particle.

4.8 Current Relevance of Electrostatics

In spite of being classical topics of electromagnetism, many of the problems treated in this chapter have been the object of study in recent years, in order to elucidate some of the fundamental questions in the theory. It is also important to point out that basic aspects mentioned in these chapters have different applications in present-day physics and technology. We now go on to discuss some of these aspects.

4.8.1 Electric Neutrality of Matter

Some authors have argued about the possibility that the charges of positrons and of antiprotons are not exactly equal to those of electrons and protons. Starting from quantum electrodynamics, the possible limits to the inequality of these charges have been calculated, a higher limit of the order of 10^{-21} being established for the possible relative difference between the charges of the electron and of the proton, and of the order of 10^{-18} for the equivalent difference between electrons and positrons. Experimental results confirm these values. In connection with this problem, we recommend reading the paper by B. Müller and M.H. Thoma, *Vacuum Polarization and the Electric Charge of the Positron* [4.3].

4.8.2 Validity of Coulomb's Law

The more or less strict validity of Coulomb's law is related to the experimental limits of the mass of the photon, which – according to quantum electrodynamics – is the light quantum responsible for the interaction of charged particles. According to this theory, the interaction between charged particles is due to the exchange of photons emitted and absorbed by these particles. The range of the resulting force is directly related to the mass of the exchanged particles. The Coulomb force, inversely proportional to the square of the distance between charges, appears only when the mass of these particles is zero. To clarify this question experiments of different types have been carried out. On the one hand, delicate electrostatic experiments help to clarify such a basic problem as that of the mass of the photon. On the other hand, experiments involving elementary particles give as a result a confirmation of Coulomb's law, established with a torsion balance for the first time in 1785. This relationship between the mass of the photon and the electrostatic field is discussed by R. Shaw, in his paper *Symmetry, Uniqueness and the Coulomb Law of Force* [4.4].

4.8.3 Control of Environmental Pollution

Many industrial processes require the combustion of different fuels, with a rising production of smoke including not only gases but also a certain amount of solid residues that contribute to environmental pollution. One of the methods used to control the emission of solid residues into the atmosphere is based on electrostatic phenomena. By means of a corona discharge, electric charge is transmitted to the particles circulating through a chimney. Next, by application of an electric field, these particles are extracted from the upward current of gases of the chimney and are eliminated from the residues emitted into the atmosphere.

4.8.4 Optimization of Industrial Painting

The application of paints on an industrial scale can imply an enormous waste of material if it is not carried out efficiently, that is to say, if an important fraction of the paint used does not reach the object to be covered. A way of optimizing this process consists in electrostatically charging the paint particles at the exit of the torch, by friction or by means of a corona discharge. If the object to be colored is also electrified, electric field lines guide the trajectory of the paint toward the desired target.

4.8.5 Photocopying

The photocopying process makes use of static electricity to record the images to be copied. An essential part of a photocopying machine is a metallic cylinder covered with a layer of a photoconductive material, i.e. a material whose electric conduction properties change when exposed to light. The parts exposed to light become conducting while those in the dark remain insulating. In a first step of the process, the photoconductor is charged in darkness, through a corona discharge. In this way, charges of opposite sign are evenly distributed on each side of the photoconductor. In a second step, the photoconductor is illuminated with the image to be copied. The most brightly exposed part becomes conducting so that charges in that area escape toward the metallic cylinder.

The number of charges escaping is proportional to the intensity of light received, and a latent electrostatic image is thus formed. At this point carbon particles (toner) are dispersed over the cylinder. These particles are more strongly attracted towards the most intensely charged areas. Finally a sheet of positively charged paper is placed near the cylinder, so that the carbon particles are attracted towards it and the image copied to the cylinder is transferred to the paper sheet. The image is fixed by a process of coalescence of the carbon particles induced by heat, while the cylinder is cleaned and recharged to restart the process.

Problems

4.1 Fifteen equal point charges q are placed at the vertices of a regular 16-sided polygon. Determine the electric field at the center of the polygon, and the electrostatic energy of the system.

Hint: What would be the force at the origin if all the vertices were occupied? Use the superposition principle.

4.2 A linear uniform charge distribution of length L is situated along the x axis and centered at the origin. Calculate the electric field at the point $(0, 0, z)$. Analyze the limits $z >> L$ and $z << L$. Use the previous result to calculate the field on the axis of a uniformly charged regular polygon.

4.3 Consider a sphere of radius R uniformly charged with density ρ. (a) Find the electric field inside and outside the sphere. (b) Consider two spheres of radii R with uniform charge density ρ and $-\rho$ which partially intersect. Show that the electric field in the common region is constant.

Hint: Use the superposition principle both in (a) and in (b).

4.4 Using Gauss's theorem (a) solve problem 4.3a; (b) calculate the field produced by an infinite line of charge and compare it with the result of problem 4.2; (c) calculate the field produced by two infinite parallel plates, with surface charge densities σ and $-\sigma$.

4.5 Use Gauss's theorem to demonstrate the following statements: (a) Any charge excess in a conductor lies entirely on its surface (in a conductor, by definition, the charges move freely under the action of electric fields); (b) a closed hollow conductor shields the volume inside it from the fields due to external charges, but it does not shield the external space from the fields produced by charges placed inside; (c) the electric field next to the surface of a conductor is normal to its surface and its modulus is $4\pi\sigma$, σ being the surface charge density.

4.6 A linear distribution of charge is placed on the x axis. The distribution is close to the origin of coordinates, it is bounded, and has density $\lambda(x)$. Show that the potential produced by these charges at a point on the x axis sufficiently far away from the origin can be written as

$$\phi(x) = \sum_{n=0}^{\infty} \frac{\mu_n}{x^{n+1}}.$$

Find the value of μ_n, and interpret the first terms of the series. Making use of the expansion of the potential in Legendre polynomials, find ϕ over all space.

Hint: Compare with the multipole expansion.

4.7 Determine the Cartesian quadrupole coefficients $\bar{q}_{ij}$ for a uniformly charged rod and for a uniformly charged disk. Establish the correlation between the sign of the quadrupole moments and the departure of each of the distributions from a sphere.

4.8 Consider an electric dipole as defined in the text. (a) Calculate the energy of an electric dipole placed into an electric field. (b) Find the equilibrium orientations of the dipole in the field, and analyze its stability. (c) Calculate the force on the dipole and show that it equals the negative gradient of the energy obtained in (a). (d) An induced dipole is a dipole whose moment is proportional to the electric field in which it is placed, $\boldsymbol{p} = \alpha \boldsymbol{E}$. Calculate the force and the energy of an induced dipole in an electric field, and show that the force is not given by the negative gradient of the energy. Interpret this result.

Hint: (a) Calculate the energy of a pair of opposite charges and take the limit as in the text. (c) What is the energy used to induce the dipole?

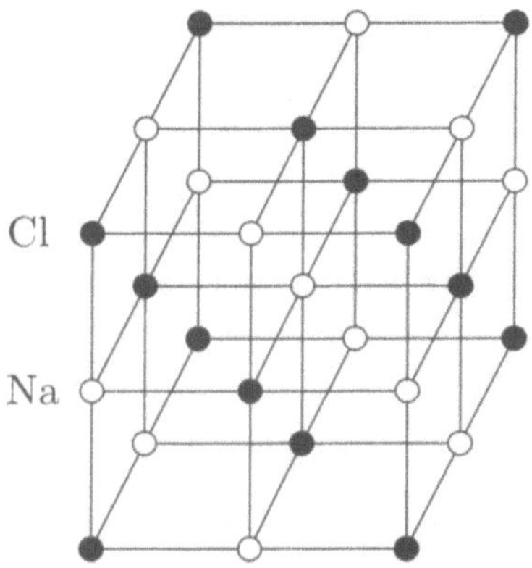

Fig. 4.6. A sodium chloride crystal

4.9 A sodium chloride (NaCl) crystal is a three-dimensional cubic array of positive and negative ions that can be considered, to a first approximation, as point charges (Fig. 4.6). The spacing between ions determined by X-ray diffraction is 2.81 Å. According to this model, the crystal stays bonded by electrostatic forces. If the model were correct, the electrostatic energy of the crystal should be equal to the heat of vaporization of NaCl plus the energy of dissociation of the molecules. Experimentally, it is known that the total energy of dissociation of NaCl is 7.92 eV per molecule. Calculate the electrostatic energy per molecule of the crystal. Is the model realistic?

Hint: To calculate the energy of an ion, first consider the contribution of a line containing that ion; then, the contribution of the four lines next to the first one, and so forth, terminating the series when it seems convenient.

4.10 Consider a capacitor formed by a conductor of square cross section, 1 cm per side, kept at a potential of 1 statvolt, surrounded by a hollow conductor at zero potential, also of square cross section, coaxial and parallel to the previous one, 2 cm per side. Choosing a test function appropriate to the boundary conditions, determine the electrostatic potential in the space between the two conductors. The energy per unit length evaluated by the finite element method is $w = 0.407\,\mathrm{erg/cm}$.

Solution: Given the symmetry of the problem, we can limit our analysis to the region $x > 0$ enclosed between the straight lines $y = \pm x$ passing through the vertices of both conductors, and discuss an approximation to the function $\phi(x, y)$ in that region. The total energy per unit length normal to the (x, y) plane is four times the value of the energy in the region under consideration. The sharp tips of the central electrode produce a singularity in the electric field that manifests itself as an infinite slope in the potential. This is not easy to describe by approximate solutions which all satisfy the boundary conditions. For this reason we limit ourselves to functions $\phi_n(x, y)$ depending on only one of the coordinates $\phi_n(x, y) \approx \phi_n(x)$. Let us consider as test functions a combination of polynomial functions satisfying the boundary conditions, of the type $\phi_1(x) = 2-x$, and $\phi_2(x) = (2-x)^2$ such that $\phi_n(1) = 1$ and $\phi_n(2) = 0$, so that the potential is expressed as

$$\phi(x, y) = V_0 \left[(1 - c)\,\phi_1(x) + c\phi_2(x)\right].$$

The energy functional per unit length perpendicular to the (x, y) plane is

$$\begin{aligned} W^*[\phi] &= \frac{1}{8\pi} \int \left[\left(\frac{\partial \phi}{\partial x}\right)^2 + \left(\frac{\partial \phi}{\partial y}\right)^2\right] \mathrm{d}x\, \mathrm{d}y \\ &= \frac{1}{2\pi} \int_1^2 \mathrm{d}x \int_{-x}^{x} \left(\frac{\partial \phi}{\partial x}\right)^2 \mathrm{d}y = \frac{1}{\pi} \int_1^2 x \left(\frac{\partial \phi}{\partial x}\right)^2 \mathrm{d}x. \end{aligned}$$

Therefore, W^* reduces to a function of the parameter c and the problem reduces to the determination of the minimum of the function

$$\begin{aligned} W^*(c) &= \frac{V_0^2}{\pi} \int_1^2 x \left\{\frac{\mathrm{d}}{\mathrm{d}x}\left[(1 - c)(2 - x) + c(2 - x)^2\right]\right\}^2 \mathrm{d}x \\ &= \frac{V_0^2}{\pi} \left(\frac{1}{2}c^2 - \frac{1}{3}c + \frac{3}{2}\right). \end{aligned}$$

The condition for a minimum is $\mathrm{d}W^*/\mathrm{d}c = 0$, which implies $c = 1/3$. This gives, for the variational value of the energy, $W_{\mathrm{var}} = 13V_0^2/9\pi$. For

$V_0 = 1$ statvolt, it is approximately $W_{\text{var}} = 0.45978$ erg/cm per unit length perpendicular to the plane. This result lies above the value $W = 0.407$ erg/cm obtained by the finite element method, with a precision of the order of 10^{-3} and therefore very close to the exact value. An improvement of this approximation could be obtained by taking more terms in the series of functions, but the presence of the *tip effect* at the angles of the central electrode makes the convergence very slow. One more term of the same type – adding the function $(2-x)^3$ – gives the value 0.45924 erg/cm for the energy. The reader may try other functions giving a better approximation.

References

4.1 R.A. Millikan: *On the Elementary Electric Charge and the Avogadro Constant*, Phys. Rev. **32**, 349 (1911).
4.2 H. Kendall and W.K.H. Panofsky: *The Structure of the Proton and the Neutron*, Sci. Am. **224**, 60 (1971).
4.3 B. Müller and M.H. Thoma: *Vacuum Polarization and the Electric Charge of the Positron*, Phys. Rev. Lett. **69**, 3432 (1992).
4.4 R. Shaw: *Symmetry, Uniqueness and the Coulomb Law of Force*, Am. J. Phys. **33**, 300 (1965).

Further Reading

R.P. Feynman, R.B. Leighton and M. Sands: *The Feynman Lectures on Physics*, Vol. II (Addison–Wesley, Reading, 1977).
J.D. Jackson: *Classical Electrodynamics* (Wiley, New York, 1966) Chap. 1.
A. D. Moore: *Electrostatics*, Sci. Am. **226**, 47 (1972).
W.K.H. Panofsky and M. Phillips: *Classical Electricity and Magnetism* (Addison–Wesley, Reading, 1955) Chap. 1.
J.R. Reitz and F.J. Milford: *Foundations of Electromagnetic Theory* (Addison–Wesley, Reading, 1960) Chap. 3.

5. The Poisson and Laplace Equations

As we have shown in the previous chapter, the Poisson and Laplace equations govern the space dependence of the electrostatic potential. The general form of Poisson's equation for a field $\psi(\boldsymbol{r})$ is

$$\nabla^2\psi(\boldsymbol{r}) = f(\boldsymbol{r}), \tag{5.1}$$

where the inhomogeneity $f(\boldsymbol{r})$ represents a density of field sources. When this density vanishes, the Poisson equation transforms into Laplace's equation

$$\nabla^2\psi(\boldsymbol{r}) = 0. \tag{5.2}$$

From the mathematical point of view, both (5.1) and (5.2) are linear partial differential equations to be solved specifying adequate boundary conditions.

In the next section we show that these equations are used to describe a wide variety of physical phenomena. As a consequence, it is also convenient to study general mathematical methods of solution. At the present time the solution by means of appropriate computational algorithms has been highly developed and, in practice, it is the most widely used method. However, in many cases relevant information about the physical systems under study, such as their symmetry properties or their detailed behavior in limiting cases of interest, is provided by analytic solutions only. For this reason in the following sections we describe some analytical methods. Of the numerical methods, the finite element method is presented in the last section, because it has been highly developed in recent years.

5.1 The Poisson and Laplace Equations in Other Branches of Physics

Let us consider, in the first place, the phenomenon of matter diffusion. This is a transport process characterized by a specific relationship between the mass current density $\boldsymbol{j}$ and the density $n(\boldsymbol{r}, t)$, i.e.

$$\boldsymbol{j} = -D\,\nabla n, \tag{5.3}$$

where D is a positive factor, called the diffusion coefficient or diffusivity. It may depend in general on the temperature and the concentration as well as on position and time. The negative sign in (5.3) implies that, as observed in practice, mass flows from regions of higher density towards those of lower density. Equation (5.3), known as Fick's law, describes the transport process exhibited for example by a system of particles immersed in a fluid whose molecules have much smaller mass, as for example ink in water. Diffusion processes are also observed in more complex systems, for instance in the dynamics of populations of some living organisms. In classical physics, independently of the details of the transport mechanism involved – diffusion, convection, etc. – the law of mass conservation demands that an equation of continuity for the mass density similar to (4.6) be satisfied

$$\partial_t n + \nabla \cdot \boldsymbol{j} = f(\boldsymbol{r}, t). \tag{5.4}$$

The term $f(\boldsymbol{r}, t)$ describes the mass per unit volume and unit time that sources exchange with the system. The equation obtained when substituting Fick's law into the continuity equation for mass (5.4) is the diffusion equation. When diffusivity is independent of position, this equation has the form

$$\partial_t n - D\nabla^2 n = f(\boldsymbol{r}, t). \tag{5.5}$$

For a time-independent problem we obtain

$$\nabla^2 n = -\frac{1}{D} f(\boldsymbol{r}). \tag{5.6}$$

This equation has exactly the form of Poisson's equation. The problem of stationary diffusion without sources, on the other hand, is described by the Laplace equation. As a consequence, stationary diffusion and electrostatics are equivalent problems from the mathematical point of view.

Heat conduction in a solid material medium or a convection-free medium, which is a phenomenon closely related to the diffusion problem, is also described by a continuity equation derived from the energy conservation law. This is written

$$\partial_t w + \nabla \cdot \boldsymbol{q} = f(\boldsymbol{r}, t), \tag{5.7}$$

where $w(\boldsymbol{r}, t)$ is the energy density and $\boldsymbol{q}(\boldsymbol{r}, t)$ is the density of heat flow.

Both quantities can be related to the temperature field $T(\boldsymbol{r}, t)$. In fact, temperature differences are proportional to differences in energy density, $\Delta T = c_V \Delta w$, where c_V is the heat capacity at constant volume of the material. Also, when thermal gradients between the different parts of the material are not too large, heat flow can be assumed to be proportional to the temperature gradient $\boldsymbol{q} = -\kappa \nabla T$, where κ is the thermal conductivity of the material. Here, the negative sign indicates that heat flows from the warmer regions to the colder regions. These constitutive relations imply that the temperature field satisfies

$$\partial_t T - \frac{\kappa}{c_V} \nabla^2 T = \frac{1}{c_V} f(\boldsymbol{r}, t), \tag{5.8}$$

where $f(\boldsymbol{r},t)$ represents the heat sources. Such an equation is totally similar to (5.5), implying that the stationary heat conduction problem is also mathematically equivalent to electrostatics. The theory of thermal conduction was developed in 1811 by Jean Baptiste Fourier (1768–1830) who in order to solve the Laplace equation introduced the series expansion method that bears his name. Some decades before, Pierre Simon de Laplace had studied this equation for the first time in connection with gravitational systems.

As we pointed out above, the mass continuity equation (5.4) is applied within the framework of classical mechanics to many phenomena of matter transport. In particular, for an incompressible fluid the density is a constant independent of position and time. The current density $\boldsymbol{j}$ can be factored, in general, as $\boldsymbol{j} = n(\boldsymbol{r},t)\boldsymbol{v}(r,t)$, where $\boldsymbol{v}(\boldsymbol{r},t)$ is the velocity field of the mass distribution. As a consequence, if there are no matter sources, the equation of continuity reduces to an equation for $\boldsymbol{v}(\boldsymbol{r},t)$:

$$\nabla \cdot \boldsymbol{v} = 0. \tag{5.9}$$

If the flow is irrotational – free of vortices – the curl of the velocity field is zero. This implies that this field can be derived from a potential:

$$\boldsymbol{v} = -\nabla\phi. \tag{5.10}$$

According to (5.9), ϕ satisfies Laplace's equation. The description of an irrotational and incompressible fluid is therefore another problem which from the mathematical point of view, is analogous to electrostatics.

Let us finally mention that the Schrödinger equation – which governs the dynamics of quantum systems – also involves the Laplacian operator ∇^2. Indeed, the stationary Schrödinger equation for the wave function $\psi(\boldsymbol{r})$ of a particle of mass m and energy E moving in a potential $V(\boldsymbol{r})$ is

$$-\frac{\hbar^2}{2m}\nabla^2\psi + V\psi = E\psi. \tag{5.11}$$

Analytical solutions of this equation are important because they reveal basic information about the quantization of the physical quantities involved.

In the following sections we study analytical methods of solution of the Poisson and Laplace equations and in Sect. 5.5 we describe the finite element method.

5.2 Solution of Poisson's Equation. Green's Function

Poisson's equation

$$\nabla^2\phi = -4\pi\rho, \tag{5.12}$$

has a unique solution in a closed space domain, when the value of $\rho(\boldsymbol{r})$ and appropriate boundary conditions are specified. The latter can be of two types, namely, Dirichlet boundary conditions when the value of ϕ on the boundary is specified, or Neumann boundary conditions when we specify the value of the normal derivative of ϕ on the boundary. Poisson's equation also allows mixed conditions or linear combinations of Dirichlet and Neumann conditions.

Under appropriate circumstances the domain of uniqueness of the solution can be extended up to infinity. This requires the boundary conditions at infinity to be consistent. Typically, when the sources of electric potential are located in a bounded region of space, the reference potential at infinity is taken to be zero: $\phi = 0$. Here we will not extend the discussion on conditions of existence and uniqueness of the solution of Poisson's equation, since in most interesting situations both properties are guaranteed.

In the presence of boundary conditions, (4.22) is not the solution to the electrostatic problem. In order to link the solution to the boundary conditions, we observe firstly that, given two scalar functions ϕ and ψ, continuous and with continuous derivatives in a regular domain of volume V, the divergence theorem implies

$$\int_V \nabla \cdot (\psi \nabla \phi)\ \mathrm{d}^3 r = \int_S \psi \nabla \phi \cdot \mathrm{d}\boldsymbol{s}, \tag{5.13}$$

where S is the surface of the domain and $d\boldsymbol{s}$ is an element of area whose normal is drawn outward with respect to S. From (5.13) it follows immediately that

$$\int_V (\nabla \psi \cdot \nabla \phi + \psi \nabla^2 \phi)\ \mathrm{d}^3 r = \int_S \psi\ \partial_n \phi\ \mathrm{d}s. \tag{5.14}$$

Here, ∂_n is the derivative in the direction normal to the surface. Exchanging the roles of ϕ and ψ in this relation and subtracting the ensuing equations, we obtain

$$\int_V (\psi \nabla^2 \phi - \phi \nabla^2 \psi)\ \mathrm{d}^3 r = \int_S (\psi \partial_n \phi - \phi \partial_n \psi)\ \mathrm{d}s. \tag{5.15}$$

This identity is known as the second Green's theorem. We can apply it to the solution of electrostatic problems if we identify the function $\phi(\boldsymbol{r})$ with the electrostatic potential and take $\psi(\boldsymbol{r}) = |\boldsymbol{r} - \boldsymbol{r}'|^{-1}$. Keeping in mind that

$$\begin{aligned} \nabla^2 \phi &= -4\pi\rho, \\ \nabla^2 \psi &= -4\pi\delta(\boldsymbol{r} - \boldsymbol{r}') \end{aligned} \tag{5.16}$$

and using Green's theorem in the variable $\boldsymbol{r}'$, we find

$$\begin{aligned} \phi(\boldsymbol{r}) = & \int_V \frac{\rho(\boldsymbol{r}')}{|\boldsymbol{r} - \boldsymbol{r}'|}\ \mathrm{d}^3 r' \\ & + \frac{1}{4\pi} \int_S \left[\frac{1}{|\boldsymbol{r} - \boldsymbol{r}'|}\ \partial_{n'} \phi - \phi \partial_{n'} \frac{1}{|\boldsymbol{r} - \boldsymbol{r}'|} \right] \mathrm{d}s'. \end{aligned} \tag{5.17}$$

This relationship links the potential at point $\boldsymbol{r}$ to the charge density contained in the volume V and to the values of the potential and of its normal derivative on the surface S, that is to say, to the boundary conditions. Thus, (5.17) would be the general solution to the electrostatic problem.

However, as we pointed out previously, the potential is determined by its value on the surface or the value of its derivative on S. Therefore, (5.17) involves too much information regarding the potential on the right-hand side. If we were to provide such information the problem would be overdetermined.

The English mathematician George Green (1793–1841) introduced in the first decades of the ninteenth century a general method to deal with this problem. The method can be used with great generality since it can be extended to a large class of ordinary and partial linear differential equations.

In electrostatic problems, we call *Green's function* $G(\boldsymbol{r},\boldsymbol{r}')$ the solution of Poisson's equation with a point source of unit intensity, that is to say:

$$\nabla^2 G(\boldsymbol{r},\boldsymbol{r}') = -4\pi\, \delta(\boldsymbol{r}-\boldsymbol{r}'). \tag{5.18}$$

In free space the solution of this equation is given by $G(\boldsymbol{r},\boldsymbol{r}') = |\boldsymbol{r}-\boldsymbol{r}'|^{-1}$. In order to satisfy the boundary conditions the general solution can be given as

$$G(\boldsymbol{r},\boldsymbol{r}') = \frac{1}{|\boldsymbol{r}-\boldsymbol{r}'|} + F(\boldsymbol{r},\boldsymbol{r}'), \tag{5.19}$$

where $F(\boldsymbol{r},\boldsymbol{r}')$ should satisfy the Laplace equation, $\nabla^2 F = 0$. The boundary conditions for G, to be determined below, translate into boundary conditions for F.

Let us insert into the second Green identity (5.15) $\psi = G$ and let us again identify ϕ with the electrostatic potential. From these identifications we find

$$\begin{aligned}\phi(\boldsymbol{r}) = & \int_V G(\boldsymbol{r},\boldsymbol{r}')\rho(\boldsymbol{r}')\ \mathrm{d}^3r' \\ & + \frac{1}{4\pi}\int_S \left[G(\boldsymbol{r},\boldsymbol{r}')\ \partial_{n'}\phi - \phi\partial_{n'}G(\boldsymbol{r},\boldsymbol{r}')\right]\ \mathrm{d}s'.\end{aligned} \tag{5.20}$$

We can specify the boundary conditions to be satisfied by G so that, starting from (5.20), this function can be used to solve different types of problem. In this way, imposing $G(\boldsymbol{r},\boldsymbol{r}') \equiv 0$ for $\boldsymbol{r}$ and $\boldsymbol{r}'$ in S, this expression gives us the potential over all space if the value of $\phi(\boldsymbol{r})$ is specified on the surface:

$$\phi(\boldsymbol{r}) = \int_V G_{\mathrm{D}}(\boldsymbol{r},\boldsymbol{r}')\rho(\boldsymbol{r}')\ \mathrm{d}^3r' - \frac{1}{4\pi}\int_S \phi\partial_{n'}G_{\mathrm{D}}(\boldsymbol{r},\boldsymbol{r}')\ \mathrm{d}s', \tag{5.21}$$

where G_{D} indicates the Green's function for the Dirichlet problem.

The case in which $\partial_n\phi$ is specified on S is equivalent to the case in which the value of the normal component of the electric field or the surface charge density is fixed. Choosing $\partial_n G \equiv -4\pi/a$, where a is the area of the surface S, we obtain

$$\begin{aligned}\phi(\boldsymbol{r}) = & \int_V G_{\mathrm{N}}(\boldsymbol{r},\boldsymbol{r}')\rho(\boldsymbol{r}')\ \mathrm{d}^3r' + \frac{1}{4\pi}\int_S G_{\mathrm{N}}(\boldsymbol{r},\boldsymbol{r}')\ \partial_{n'}\phi\ \mathrm{d}s' \\ & + \frac{1}{a}\int_S \phi\ \mathrm{d}s,\end{aligned} \tag{5.22}$$

where the label N indicates Neumann boundary conditions. The last term in this expression is the average value of ϕ on the surface. This is an arbitrary constant since it sets the reference level for the potential. Observe that for these boundary conditions we cannot impose $\partial_n G \equiv 0$, as for the Dirichlet problem, because this choice would violate Gauss's theorem.

The problem of finding a solution of Poisson's equation is in this way transformed into that of obtaining Green's function. This is equivalent to determining the electrostatic potential generated by a point charge in the presence of certain boundary conditions. In general, this latter problem might not be trivial but in some cases can be simpler than the original problem.

As an example we treat here the solution to a Dirichlet problem when the potential $\phi(x, y, 0)$ is specified on the infinite plane $z = 0$. For this we need to determine Green's function $G_{\mathrm{D}}(\boldsymbol{r}, \boldsymbol{r}')$ for the plane. This coincides with the electrostatic potential generated at $\boldsymbol{r}$ by an unitary point charge located at $\boldsymbol{r}'$, when we impose the condition that the potential vanishes on the plane (Fig. 5.1).

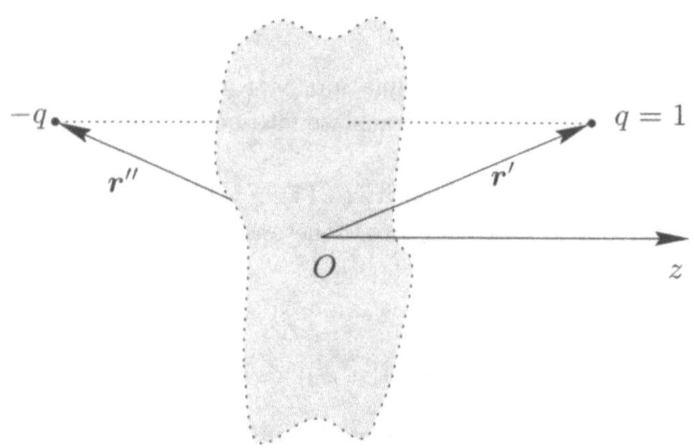

Fig. 5.1. Green's function for Dirichlet conditions on an infinite plane

The solution of this problem can be obtained by the *method of images.* Indeed, the solution in the halfspace $z > 0$ – in which the unit charge is placed – is equivalent to the potential produced in free space by a system of two charges: the original charge at $\boldsymbol{r}' \equiv (x', y', z')$ plus an image charge of the same value and opposite sign placed in the halfspace $z < 0$ at the point $\boldsymbol{r}'' = (x', y', -z')$. The Green's function is then

$$G_{\mathrm{D}}(\boldsymbol{r}, \boldsymbol{r}') = \frac{1}{\sqrt{(x - x')^2 + (y - y')^2 + (z - z')^2}} - \frac{1}{\sqrt{(x - x')^2 + (y - y')^2 + (z + z')^2}}. \tag{5.23}$$

This function vanishes for $z = 0$, so that it satisfies the required boundary condition. Let us emphasize that this solution is valid in the region $z > 0$ only, where the source of the potential is placed.

Having obtained the Green's function of our problem, we can now face the calculation of the electrostatic potential in the whole halfspace $z > 0$, assuming that the value of the potential on the plane is given. According to (5.21), we need to know the normal derivative of Green's function. From (5.23) we obtain

$$\partial_{n'} G_{\mathrm{D}} = -\partial_{z'} G_{\mathrm{D}}|_{z'=0} = -\frac{2z}{[(x-x')^2 + (y-y')^2 + z^2]^{3/2}}. \tag{5.24}$$

Note that this derivative is proportional to the charge density induced in a conducting plane at zero potential by a charge placed in front of it. Assuming that there are no charges for $z > 0$, and that the potential on the plane is specified by the function $\phi(x, y, 0)$, the solution of Poisson's equation for this problem is

$$\begin{aligned} \phi(x, y, z) = & \frac{1}{2\pi} \int_{-\infty}^{+\infty} \int_{-\infty}^{+\infty} \phi(x', y', 0) \\ & \times \frac{z}{[(x-x')^2 + (y-y')^2 + z^2]^{3/2}} \, \mathrm{d}x' \mathrm{d}y'. \end{aligned} \tag{5.25}$$

If we consider now the particular situation in which $\phi(x, y, 0)$ does not depend on y, $\phi(x, y, 0) = f(x)$, the potential is

$$\phi(x, y, z) = \frac{1}{\pi} \int_{-\infty}^{+\infty} f(x') \, \frac{z}{(x-x')^2 + z^2} \, \mathrm{d}x'. \tag{5.26}$$

The function $L_z(x - x') = z/\pi[(x - x')^2 + z^2]$ is called a Lorentzian. It is bell-shaped, and its width is proportional to z. The Lorentzian is an *approximant* to the Dirac delta function $\delta(x - x')$ for $z \to 0$ (Appendix A). This implies that, as expected, very near the plane $\phi(x, y, z) \approx f(x)$.

Consider now the special case in which $f(x) = V_0 + A \sin kx$. On the plane, then, the potential varies periodically around V_0. We can calculate $\phi(x, y, z)$ in the rest of the halfspace from (5.26) by means of an integration by residues. The result is

$$\phi(x, y, z) = V_0 + A \, \exp(-kx) \, \sin kx. \tag{5.27}$$

The potential thus decays exponentially as z grows, with a decay length proportional to k^{-1}. Therefore, the smaller the wave length $2\pi/k$ of the oscillations at $z = 0$ the faster the potential decreases as z grows. This explains the efficiency of a reticulated structure as electrostatic shielding, in place of a metallic sheet that would give total shielding at a higher cost.

This result can also be applied to thermal conduction problems, given the mathematical analogy discussed in the first section of this chapter. Indeed, the same reasoning explains why a few meters underground the temperature is practically uniform in spite of the thermal variations that take place from one point to another of the Earth's surface. By virtue of the solution to the diffusion equation, something similar happens with the hourly or seasonal temperature variations. These do not propagate downward into the ground and this is the reason why a few meters below the surface the temperature is practically constant all year round.

In subsequent chapters we show that the time variation of the electrostatic potential is different from that of diffusion fields, since it is governed by the wave equation. Contrary to the diffusion equation, the wave equation contains second derivatives with respect to time with opposite sign to that of the space derivatives. From the mathematical point of view, the diffusion equation and the wave equation belong to different categories: the first one is of *parabolic* type while the second is of *hyperbolic* type. Laplace's equation, on the other hand, is of *elliptic* type.

5.3 Separation of Variables

As we mentioned in the previous section, the Green's function for a given problem can be written, in general, as $G = |\boldsymbol{r} - \boldsymbol{r}'|^{-1} + F$, where $F(\boldsymbol{r}, \boldsymbol{r}')$ should satisfy Laplace's equation and adequate boundary conditions. Therefore, it is important to have a method of solving Laplace's equation in more or less general situations. Moreover, the problem of solving this equation is of interest in itself, since it corresponds to finding the electrostatic potential in a charge free space.

The method that we describe here is called the *separation of variables.* In general terms it consists of assuming that the solution to the equation can be written as a product of functions, each factor depending on only one of the coordinates. The set of these separate solutions is generally a complete system of solutions of Laplace's equation, so that the general solution can be given as a linear combination of them. The appropriate linear combination is determined by the boundary conditions of the problem.

There are only eleven systems of coordinates in which the Laplace equation is separable. Among them, fortunately, we find the more common systems of coordinates, such as Cartesian, spherical, cylindrical and elliptic. The system of coordinates to be used in a given problem is suggested by its symmetries. Here we only discuss separation of variables in Cartesian coordinates and in spherical coordinates with azimuthal symmetry. In order to simplify the presentation, we proceed to analyze examples.

5.3.1 Cartesian Coordinates

In Cartesian coordinates, Laplace's equation is

$$\partial^2_{xx}\phi + \partial^2_{yy}\phi + \partial^2_{zz}\phi = 0. \tag{5.28}$$

Substituting a solution of the type $\phi(x, y, z) = X(x)Y(y)Z(z)$, in the above equation and dividing throughout by ϕ, we obtain

$$\frac{X''}{X} + \frac{Y''}{Y} + \frac{Z''}{Z} = 0. \tag{5.29}$$

Here the primed functions indicate derivatives with respect to the corresponding variables. In this expression each one of the terms depends on a different variable, and the sum must be independent of any of the variables. This implies that each term must be a constant

$$\frac{X''}{X} = \alpha, \quad \frac{Y''}{Y} = \beta, \quad \frac{Z''}{Z} = \gamma. \tag{5.30}$$

The separation constants α, β and γ should satisfy $\alpha + \beta + \gamma = 0$. These three equations are equivalent, so that it is enough to analyze one of them. The first one, for example, has two independent solutions

$$X(x) = \exp(\pm\sqrt{\alpha}\ x), \tag{5.31}$$

and any linear combination of them is also a solution. When $\alpha > 0$, $X(x)$ grows or decays exponentially. When $\alpha < 0$, the real solutions for $X(x)$ are harmonic functions, such as $\sin\sqrt{|\alpha|}x$ and $\cos\sqrt{|\alpha|}x$. The solutions for $Y(y)$ and $Z(z)$ are also real or complex exponentials. Let us note that at least one of the three separation constants must be negative.

For each value of the separation constants, there is a solution in separate variables,

$$\begin{aligned} \phi_{\alpha\beta} = \quad & \left(A_1 e^{\sqrt{\alpha}x} + A_2 e^{-\sqrt{\alpha}x}\right)\left(B_1 e^{\sqrt{\beta}y} + B_2 e^{-\sqrt{\beta}y}\right) \\ & \times \left(C_1 e^{\sqrt{\gamma}z} + C_2 e^{-\sqrt{\gamma}z}\right), \end{aligned} \tag{5.32}$$

with $\gamma = -\alpha - \beta$. It can be shown that these solutions form a complete set of functions, so that the general solution of Laplace's equation in Cartesian coordinates is a linear combination of the separate solutions:

$$\phi(x, y, z) = \sum_{\alpha}\sum_{\beta} a_{\alpha\beta}\phi_{\alpha\beta}. \tag{5.33}$$

The boundary conditions determine the coefficients $a_{\alpha\beta}$ of this combination and the appropriate values of the separation constants. Let us show how to proceed with regard to this point by means of a particular example.

Consider a cube of side a whose lateral faces are at zero potential, while the lower and upper faces are at constant potentials V_1 and V_2, respectively (Fig. 5.2). The problem consists of finding the potential inside the cube. The potential has the form of (5.33), where the values of α, β and $a_{\alpha\beta}$ should be such that the specified boundary conditions are satisfied.

In this case, simple inspection shows that $\phi_{\alpha\beta}$, and consequently ϕ, vanishes on the faces at zero potential if both $X(x)$ and $Y(y)$ are sinusoidal periodic functions with halfperiod a:

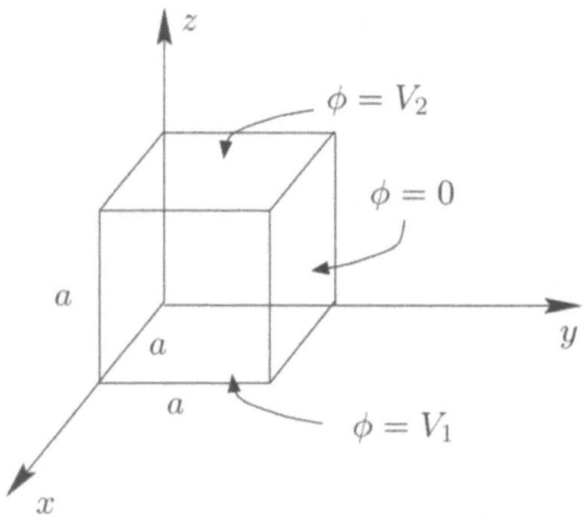

Fig. 5.2. An electrostatic problem in Cartesian coordinates

$$X(x) = \sin(n\pi x/a), \quad Y(y) = \sin(m\pi y/a). \tag{5.34}$$

Here n and m are integers corresponding to $\alpha = -n^2\pi^2/a^2$ and $\beta = -m^2\pi^2/a^2$. This fixes $\gamma = (n^2 + m^2)\pi^2/a^2 > 0$, so that $Z(z)$ is a combination of growing and falling exponentials. The solution has the form

$$\begin{aligned}\phi(x,y,z) = & \sum_n \sum_m \sin(n\pi x/a)\ \sin(m\pi y/a) \\ & \times \left[c_{nm} e^{\sqrt{\gamma} z} + d_{nm} e^{-\sqrt{\gamma} z}\right],\end{aligned} \tag{5.35}$$

that indeed satisfies the boundary conditions where the potential vanishes. Observe that due to the parity of the sinusoidal functions, it is enough to sum over the positive integer values of n and m. The contributions of the negative values can be taken into account by properly defining the coefficients c_{nm} and d_{nm}.

The values of the coefficients c_{nm} and d_{nm} must be determined by the boundary conditions $\phi(x,y,0) = V_1$ and $\phi(x,y,a) = V_2$. To find them, we use the orthogonality relations

$$\int_0^a \sin(n\pi x/a)\ \sin(n'\pi x/a)\ \mathrm{d}x = \frac{a}{2}\,\delta_{nn'} \tag{5.36}$$

of the sinusoidal functions involved in the expansion. Taking $z = 0$ and $z = a$ in (5.35), multiplying both sides by $\sin(n'\pi x/a)$ and $\sin(m'\pi y/a)$, and integrating this identity in the intervals $0 < x < a$ and $0 < y < a$, we obtain respectively

$$\begin{cases} c_{nm} + d_{nm} = \dfrac{4V_1}{nm\pi^2}[1 - \cos(n\pi)][1 - \cos(m\pi)], \\ c_{nm} e^{\sqrt{\gamma} a} + d_{nm} e^{-\sqrt{\gamma} a} = \dfrac{4V_2}{nm\pi^2}[1 - \cos(n\pi)][1 - \cos(m\pi)]. \end{cases} \tag{5.37}$$

This is a system of linear equations for c_{nm} and d_{nm} whose solution completes the solution of the problem. In Fig. 5.3 the resulting equipotentials, for $a = 1$ and $V_1 = V_2$ in the plane $y = 1/2$ are plotted. The irregularity in some of the equipotential lines near $z = 0$ and $z = a$ is due to the fact that the plot was obtained by means of an approximate calculation, considering the first 256 terms of the sum. We leave it as an exercise for the reader to show that this sum can be evaluated explicitly in analytical form.

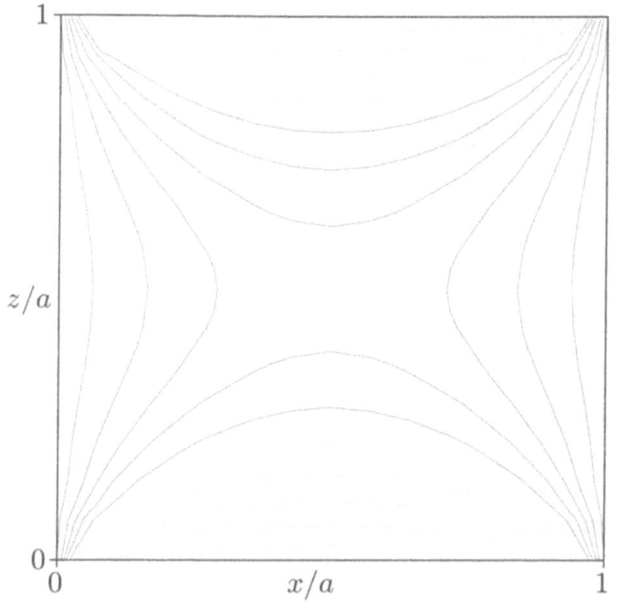

Fig. 5.3. Equipotentials for the cube problem, with $y = a/2$

The problem analyzed above through separation of variables in Cartesian coordinates shows the typical rules for dealing with this type of problem. Successful application of these rules to any problem is essentially a matter of practice.

5.3.2 Spherical Coordinates

Let us consider now problems in spherical coordinates with azimuthal symmetry. This implies that the solution of Laplace's equation does not depend on the azimuthal angle φ, and the relevant variables are only r and θ. Laplace's equation in spherical coordinates with azimuthal symmetry ($\partial_\varphi \equiv 0$) is

$$\frac{1}{r}\,\partial^2_{rr}(r\phi) + \frac{1}{r^2\sin\theta}\partial_\theta(\sin\theta\;\partial_\theta\phi) = 0. \tag{5.38}$$

Assuming for the potential the separable form $\phi(r,\theta) = R(r)P(\theta)$, Laplace's equation transforms into

$$\frac{r}{R}\,(rR)'' + \frac{1}{P\sin\theta}\,(\sin\theta\; P')' = 0, \tag{5.39}$$

where the primed functions are derivatives with respect to the corresponding variables. Separation of variables implies

$$\frac{r}{R}\,(rR)'' = -\frac{1}{P\sin\theta}\,(\sin\theta\; P')' = \alpha, \tag{5.40}$$

where α is, in this case, the only independent separation constant.

The equation for $R(r)$ can be solved immediately. Indeed, it can be proved that its solution is given in powers of r

$$R(r) = r^{\lambda}, \tag{5.41}$$

where $\lambda(\lambda+1) = \alpha$. For each value of α there are then two values of λ. Calling these values λ_1 and λ_2, the general solution to the equation for the radial coordinate is

$$R = Ar^{\lambda_1} + Br^{\lambda_2}, \tag{5.42}$$

where A and B are arbitrary constants.

The equation for $P(\theta)$ is more complicated. Defining a new variable $\mu = \cos\theta$ we get

$$\frac{\mathrm{d}}{\mathrm{d}\mu}\left[(1-\mu^2)\frac{\mathrm{d}P}{\mathrm{d}\mu}\right] + \alpha P = 0. \tag{5.43}$$

This equation is known as the Legendre equation. Its solutions can be obtained assuming an expansion in powers of μ. It can be shown that these are singular at $\mu = \pm 1$ unless the constant α satisfies $\alpha = l(l+1)$ with l a non-negative integer. Since in a given electrostatic problem we should expect a regular solution for $\cos\theta = \pm 1$, the only acceptable solutions to the Legendre equation are those obtained for $\alpha = l(l+1)$ with $l = 0, 1, 2, \ldots$. Then, the regularity condition imposed determines the possible values of the separation constant α. For $\alpha = l(l+1)$, the solution to the Legendre equation is a polynomial of degree l in the variable μ, called the Legendre polynomial of degree (or order) l and denoted by $P_l(\mu)$. The properties of these polynomials are studied in Appendix B.

Since the possible values of α have been fixed, we can now determine the powers of r in the solution to the equation for $R(r)$. The relation $\lambda(\lambda+1) = \alpha = l(l+1)$ should be satisfied, which implies $\lambda_1 = l$ and $\lambda_2 = -l-1$. The general solution of Laplace's equation for azimuthal symmetry is then a combination of the solutions obtained for each value of l

$$\phi(r,\theta) = \sum_{l=0}^{\infty}\left(A_l r^l + B_l r^{-l-1}\right) P_l(\cos\theta). \tag{5.44}$$

As above, the coefficients of the combination A_l and B_l are determined by the boundary conditions. Let us consider, as a particular example, the electrostatic potential generated over all space by a spherical surface of radius

a, on which the potential varies according to $\phi(a,\theta) = f(\cos\theta)$, where f is any bounded function (Fig. 5.4).

Due to the absence of charges inside the sphere the potential should be a regular function there (region I). This implies that all negative powers in (5.44) should vanish, since otherwise ϕ would diverge at the origin. This condition imposes $B_l = 0$ for all l in that region. The potential is then

$$\phi_{\mathrm{I}}(r,\theta) = \sum_{l=0}^{\infty} A_l r^l P_l(\cos\theta). \tag{5.45}$$

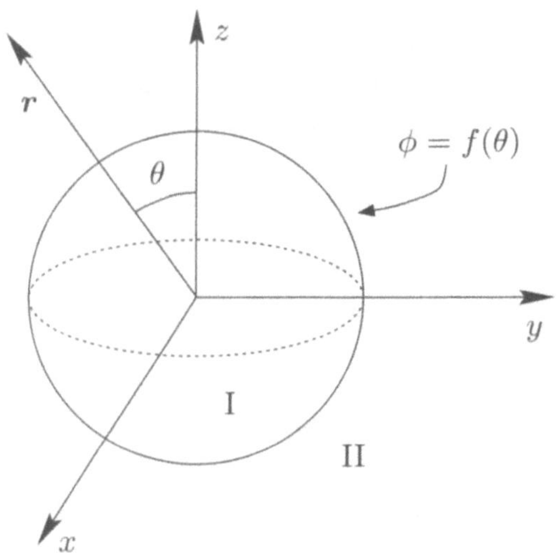

Fig. 5.4. An electrostatic problem in spherical coordinates with azimuthal symmetry

For $r = a$, ϕ_{I} should coincide with the potential on the spherical surface. This implies

$$f(\cos\theta) = \sum_{l=0}^{\infty} A_l a^l P_l(\cos\theta). \tag{5.46}$$

To find the values of A_l, we use the orthogonality relation of the Legendre polynomials (Appendix B),

$$\int_{-1}^{1} P_l(\mu)\; P_{l'}(\mu)\; \mathrm{d}\mu = \frac{2}{2l+1}\;\delta_{ll'}. \tag{5.47}$$

Integrating (5.46) multiplied on each side by $\sin\theta\; P_{l'}(\cos\theta)$ we obtain

$$A_l = \frac{2l+1}{2}\; a^{-l} \int_{-1}^{1} f(\mu)\; \mathrm{d}\mu, \tag{5.48}$$

which completes the solution of the problem inside the sphere.

For region II, with a similar reasoning, we eliminate the positive powers in (5.44). Indeed, the potential cannot diverge for $r \to \infty$, since there are no sources at infinity. Therefore, all A_l should be zero, except possibly A_0. But since $P_0(\mu) = 1$, the term with $l = 0$ is an additive constant that, in fact, coincides with the potential at infinity. Taking for convention $\phi_{\mathrm{II}}(r \to \infty) = 0$, we fix $A_0 = 0$. The values of B_l are obtained in a way similar to the one used in region I to find A_l. The result is

$$B_l = \frac{2l+1}{2}\, a^{l+1} \int_{-1}^{1} f(\mu)\, \mathrm{d}\mu. \tag{5.49}$$

We note that (5.44) generalizes the multipole expansion analyzed in the previous chapter. In this way, if it is possible to find the electrostatic potential along the axis of azimuthal symmetry and to expand it in powers of z, multiplying each term by the corresponding Legendre polynomial gives the potential in all space.

The solution of problems in spherical coordinates without azimuthal symmetry requires consideration of the angular variable φ and, therefore, of a new separation constant. The functions that solve the angular part of the Laplace equation are the spherical harmonics already discussed in the previous chapter (see also Appendix B). The general form of the solution is

$$\phi(r,\theta,\varphi) = \sum_{l=0}^{\infty} \sum_{m=-l}^{l} \left(A_{lm} r^l + B_{lm} r^{-l-1}\right) Y_{lm}(\theta,\varphi). \tag{5.50}$$

Application of this expansion to the integration of differential equations involving the Laplace operator is usually studied in detail in the courses on quantum mechanics, in connection with the Schrödinger equation.

5.4 The Finite Element Method

Numerical methods of solution of differential equations are used to solve problems in situations where there are no symmetry conditions allowing the application of some of the procedures described above. A method extensively used in certain situations is the so-called finite difference method, which provides an approximation to the solution starting from a discretization of the domain of interest. This transforms the differential equations into algebraic equations that can be solved by iterative methods. The solution can be improved by increasing the density of points used in the discretization. In spite of its usefulness, this method has limitations when there are irregular geometries or when the boundary conditions are not specified in a conventional way.

As mentioned in the previous chapter, the fact that the electrostatic energy is a minimum for the potential distribution satisfying Laplace's equation and the boundary conditions allows us to design methods of solution based on the choice of test functions depending on a certain number of parameters, which make the energy functional a function of these parameters.

This idea is the basis of the so called Ritz method, which constitutes the starting point for the finite element method. For this reason, we briefly present it here. We select a series of test functions $\phi_1(\boldsymbol{r}), \phi_2(\boldsymbol{r}), \ldots$ satisfying the boundary conditions. These allow us to build an approximate solution of the form

$$\phi(\boldsymbol{r}) = c_1\phi_1(\boldsymbol{r}) + c_2\phi_2(\boldsymbol{r}) + \ldots \tag{5.51}$$

With this, the functional $W^*[\phi]$ becomes a function $W^*(c_1, c_2, \ldots)$ of the parameters $c_1, c_2, \ldots$ The best solution within the family of linear combinations of the type (5.51) is that which minimizes $W^*(c_1, c_2, \ldots)$.

In more than one space dimension it is not easy to find a group of functions $\phi_i(\boldsymbol{r})$ which satisfy the boundary conditions and allow us to proceed with this program. Another limitation of this method is that it does not allow control of the degree of approximation in a given region of the domain of interest. This can be of importance in cases where a higher precision is required in a given region of the solution domain. These inconveniences are overcome by the finite element method, described next.

Firstly, we divide the multidimensional domain where the solution is sought by means of hyperplanes defining the boundaries of the *elements*. In a two-dimensional problem, these hyperplanes are lines determining a network of points in the domain of interest. This divides the domain into a series of triangles. In some cases, the elements can also take other forms more appropriate to the symmetry of the domain. This partition of the domain is called the *mesh*. One advantage of the method lies in the fact that the meshes can be chosen with a higher density of points in those regions where a more abrupt variation of the potential is expected, whereas a less dense mesh can be taken in regions where the solution is expected to have a smoother variation. This procedure can save calculation effort and time.

Once the mesh has been defined, the potential $\phi(\boldsymbol{r})$ in each one of the elements is approximated by continuous interpolation functions, expressed in terms of the nodal values of the potential and, in some cases, of their derivatives. The group of interpolation functions in the whole solution domain provides a polygonal approximation to the potential.

Assume we are looking for the solution of Poisson's equation in the two-dimensional domain illustrated in Fig. 5.5. A possible subdivision in triangles is also indicated in the figure. With this discretization we can assume the potential to vary linearly inside each triangular element. If x_i and y_i are the coordinates of the vertex i and ϕ_i is the value of the potential at that point, the function

$$\phi^{(e)}(x,y) = \sum_i \frac{a_i + b_i x + c_i y}{2\Delta} \phi_i, \tag{5.52}$$

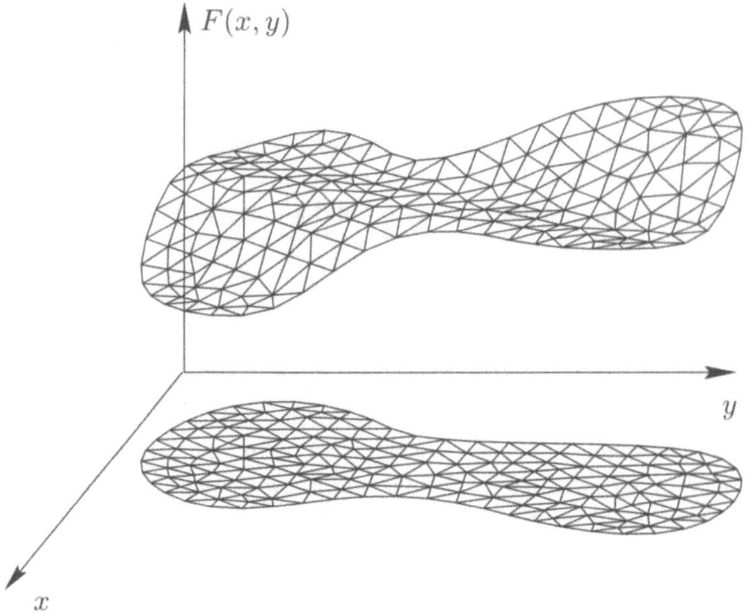

Fig. 5.5. Polygonal approximation to the potential by the finite element method

where

$$2\Delta = \begin{vmatrix} 1 & x_i & y_i \\ 1 & x_j & y_j \\ 1 & x_k & y_k \end{vmatrix} \tag{5.53}$$

is twice the area of the triangle with vertices (i, j, k), gives a linear approximation to ϕ in the triangle or element e. The coefficients a_i, b_i, c_i are determined by

$$a_i = x_j x_k - x_k x_j, \quad b_i = y_i - y_k, \quad c_i = x_k - x_j, \tag{5.54}$$

where (i, j, k) form a cyclic permutation. The expression for $\phi^{(e)}(x, y)$ can be written in a simpler way if we define

$$N_l^{(e)} = \frac{a_l + b_l + c_l}{2\Delta}, \quad l = i, j, k. \tag{5.55}$$

Indeed,

$$\phi^{(e)}(x, y) = N_i \phi_i + N_j \phi_j + N_k \phi_k. \tag{5.56}$$

The combination of all values of $\phi^{(e)}$ for all the elements provides a faceted representation of the potential having no discontinuities at the boundaries between elements, because the value of $\phi(\boldsymbol{r})$ at those boundaries is determined by the common nodal values. For the elements at the boundary of the domain it is necessary to consider the boundary conditions. In general more complete interpolation forms than (5.52) can be used, including quadratic or cubic terms. In all cases the form of equations (5.56) is the same, although increasing the degree of the approximation causes the dimension of the matrices involved to grow accordingly.

The solution of the problem by the finite element method consists of choosing the nodal values of ϕ such as to make the functional $W^*[\phi]$ stationary. This implies

$$\delta W^*[\phi] = \sum_{i=1}^{n} \frac{\partial W^*}{\partial \phi_i} \delta\phi_i = 0. \tag{5.57}$$

Here n is the total number of discrete values of ϕ. As $\delta\phi_i$ are independent for each i, the solution to this equation requires

$$\frac{\partial W^*}{\partial \phi_i} = 0; \quad i = 1, 2, 3, ..., n. \tag{5.58}$$

If the interpolation functions obey certain rules of continuity and compatibility, the functional $W^*[\phi]$ can be written as a sum of individual functions defined for each one of the elements of the group, that is to say

$$W^*[\phi] = \sum_{e=1}^{M} W^{*(e)} \left[\phi^{(e)}\right]. \tag{5.59}$$

The condition for an extremum is written then

$$\delta W^* = \sum_{e=1}^{M} \delta W^{*(e)} = 0, \tag{5.60}$$

which leads to

$$\frac{\partial W^{*(e)}}{\partial \phi_j} = 0; \quad j = 1, 2, 3, ..., r. \tag{5.61}$$

Here r is the number of nodes of the element e, equal to three in the case of a partition into triangles.

Handling the equations obtained from (5.61) must necessarily be done with a computer. To apply the method in practice one can use software to generate the meshes and number the nodes and in this way automate the whole operation. In the references indicated below the reader will find a complete presentation of the method and examples of application to specific cases.

Appendix E contains a description of the operation of the PhysicSolver software provided with this book which can be used to solve electrostatic and magnetostatic problems. In the exercises given below those problems which can be solved with PhysicSolver are indicated explicitly. The reader should be able to imagine many more applications for the software.

Problems

5.1 Using the method of images, find the electrostatic potential in the following configurations: (a) a point charge, in the presence of a metallic sphere at zero potential; (b) a point charge, between two parallel planes at zero potential; (c) an infinite line of charge placed inside a cylinder at zero potential, parallel to the axis.

Using appropriate approximations for the ideal configurations given above, apply PhysicSolver to obtain solutions for these cases.

Hint: (a) Try with charge inside the sphere. (b) May require inifinitely many images!

5.2 A square of side a has three of its sides at zero potential. On the fourth side, the potential varies sinusoidally, so that it is zero at its ends. Find the potential inside the square. Solve this problem through separation of variables and compare the solution with that obtained with PhysicSolver.

5.3 A conducting hollow cube of side a has its bases at constant potential $\phi = V$. The other four faces are grounded. (a) Find the potential $\phi(x, y, z)$ inside the cube. (b) Find the surface charge density on the bases.

Hint: (b) Calculate first the electric field near the bases.

5.4 Calculate, by the method of separation of variables, the potential inside a semi-infinite rectangular metallic tube, of sides a and b. The walls are at potential $\phi = 0$ and the cover at $\phi = V$.

Hint: How is the potential expected to decay along the tube?

5.5 Consider a square domain of side a such that three of its sides are at zero potential and one of them is at potential V. Schematize the solution to the Laplace equation by the finite element method, taking a partition with sixteen nodes as the one indicated in Fig. 5.6, and compare the precision of this method with the solution obtained by separation of variables, with respect to the number of iterations in each case.

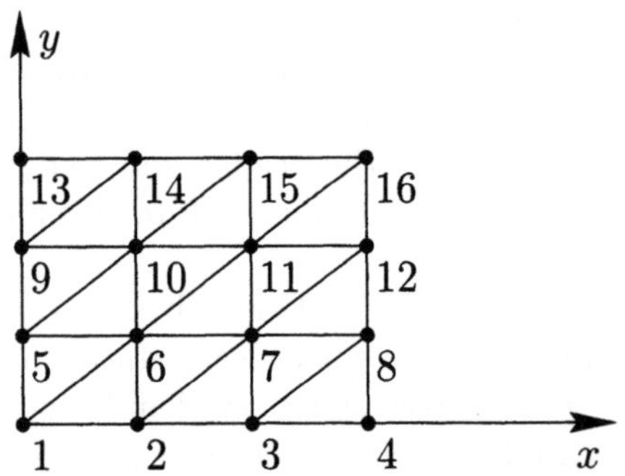

Fig. 5.6.

5.6 An uncharged conducting sphere of radius R is placed within a uniform electric field $\boldsymbol{E} = E_0\hat{\boldsymbol{z}}$. Calculate the potential inside and outside the sphere, and the density of induced surface charge. Consider what happens if the sphere has a net charge Q.

Hint: How does the net charge in the sphere affect the boundary conditions?

5.7 Two concentric spheres are divided into two hemispheres by the same horizontal plane. The upper hemisphere of the interior sphere and the lower hemisphere of the external one are at potential V. The other two hemispheres are at potential $-V$. Determine the potential over all space. Analyze the limit in which the radius of the smaller sphere tends to zero. Solve the problem by separation of variables and using PhysicSolver.

5.8 A conducting disk of radius R is placed on the (x, y) plane with its center at the origin, and kept at a potential V. The charge density on the disk is known to be proportional to $(R^2 - \rho^2)^{-1/2}$, with ρ the distance to the center. (a) Calculate the potential on the axis of the disk. (b) Show that the potential in any point of space is, for $r > R$,

$$\phi(r, \theta) = \frac{2V}{\pi}\frac{R}{r}\sum_{l=0}^{\infty}\frac{(-1)^l}{2l+1}\left(\frac{R}{r}\right)^{2l} P_{2l}(\cos\theta).$$

(c) Find the potential for $r < R$. (d) Solve this problem using PhysicSolver.

Hint: After having found the potential on the axis, recall the multipole expansion.

5.9 A square of side a has its four sides at zero potential, except for a small area of length ε centered in one of the sides, where the potential is V. Find the potential inside the square, in the limit $\varepsilon \to 0$ and $V \to \infty$, such that $\varepsilon V \to K$. Draw schematically the equipotential lines.

5.10 Solve Problem 4.10 using the finite element method explicitly, and applying PhysicSolver.

5.11 Calculate the potential produced over all space by a charged hollow metallic sphere, in the center of which an electric dipole is placed. Discuss the distribution of charges on the sphere.

Hint: How does the potential behave as $r \to 0$? Is the charge distribution on the inner surface of the sphere equal to that on the outer surface?

5.12 In scanning tunnel microscopy, a region of the surface of a material is scanned by a fine metallic tip at distances of the order of a few angstroms from the surface. This allows electrons of the material to be collected by the tip, thanks to the *tunnel effect.* This means that the electrons cross a classically impenetrable potential barrier. The circulating electric current depends on the capacitance inserted in the circuit by this system. Determine the potential distribution generated by the tip at potential V, assuming the surface of the material to be a grounded metallic plane and approximating the test tip by a sphere of radius a – in practice, one takes for the radius of the sphere $a = 100$ Å [5.1]. Compare the result given below for the exact solution with that obtained using PhysicSolver.

Solution: This problem can be solved by means of the method of images. To do this we first consider the problem of a charge q placed on the z axis, a distance $d > a$ away from the center O of a sphere of radius a, at potential $\phi = 0$. If we consider an image charge, inside the sphere, of value $q' = -aq/d$ at a distance a^2/d from O, the total potential on the sphere is $\phi = 0$. Indeed the potential generated by the two charges

$$\phi(\boldsymbol{r}) = q\left(\frac{1}{|\boldsymbol{r} - d\hat{\boldsymbol{z}}|} - \frac{a}{d}\frac{1}{|\boldsymbol{r} - a^2\hat{\boldsymbol{z}}/d|}\right)$$

vanishes on the surface of the sphere $r = a$.

Let us consider now the sphere of radius a held at a potential V with its center O separated by a distance $d > a$ to the left of a plane – perpendicular to the z-axis – and at zero potential. The system of image charges giving rise to the desired configuration can be built up step by step in the following way. We first increase the potential of the sphere to V_1 by putting a charge $q_1 = a$ at the center of the sphere. This increases the potential of the plane to a value different from zero. The potential of the plane can be recovered by placing a charge $p_1 = -q_1$ a distance $b_1 = d$ to the right of the plane. Due to this second charge, the potential of the sphere differs from the value V_1. This can be restored by placing a third charge $q_2 = -p_1 a/2d$ at a point a distance $a_2 = a^2/2d$ to the right of O. To reestablish the plane to zero potential, we must locate an image $p_2 = -q_2$ at the point $b_2 = d - a^2/2d$. To return to V_1 the potential of the sphere, we locate a charge $q_3 = -p_2 a/(2d - a^2/2d)$ at the point $a_3 = a^2/(2d - a^2/2d)$.

As each one of the successive images is smaller than the previous one, we can expect that successive iterations of this process converge to a solution. We call S_n the inverse of the successive charges located inside the sphere and T_n the inverse of those located to the right of the plane. With these variables, the relationships we have just obtained can be written in the form:

$$a_{n+1} = a\frac{S_n}{S_{n+1}},$$

$$S_{n+1} = \frac{2d - A_n}{a} S_n.$$

We also have $b_n = d - a_n$, $T_n = -S_n$. From these equations it is possible to obtain the following relationship between S_n, S_{n+1} and S_{n-1}:

$$S_{n+1} + S_{n-1} = \frac{2d}{a} S_n.$$

We can try to solve this equation assuming $S_n = u^n$. It becomes an equation of the second degree for u, whose solutions are

$$u = \frac{d}{a} \pm \sqrt{\frac{d^2}{a^2} - 1}.$$

This expression can be simplified by taking $\cosh\alpha = d/a$. With this we obtain $u = \cosh\alpha \pm \sinh\alpha$. The general solution for S_n has the form $S_n = M\cosh n\alpha + N\sinh n\alpha$.

The constants M and N can be determined by fixing the end values of S_n for $n = 1$ and for $n = 2$ expressed in terms of the charges calculated above. This gives $M = 0,\ N = (a\sinh\alpha)^{-1}$.

The value of the image charges inside the sphere and their positions are given by

$$q_n = \frac{a\sinh\alpha}{\sinh n\alpha}, \quad z_n = -d + a_n = -d + a\frac{\sinh(n-1)\alpha}{\sinh n\alpha},$$

while the charges $p_n = -a\sinh\alpha/\sinh n\alpha$ located to the right of the plane, outside the domain of interest, because they are images, are at $z'_n = b_n = d - a_n$.

The physically relevant potential in the halfspace of interest is:

$$\phi(\boldsymbol{r}) = V\sum_{n=1}^{\infty} q_n \left[\frac{1}{|\boldsymbol{r} - (-d + a_n)\hat{\boldsymbol{z}}|} - \frac{1}{|\boldsymbol{r} - (d - a_n)\hat{\boldsymbol{z}}|}\right].$$

The total charge on the sphere, Q_{sph}, is given by the sum of the charges q_n and is

$$Q_{\mathrm{sph}} = \sum_{n=1}^{\infty} q_n = V\sum_{n=1}^{\infty}\frac{a\sinh\alpha}{\sinh n\alpha}.$$

Therefore, the coefficient of capacity C_{11} for the sphere in front of the plane is

$$C_{11} = \sum_{n=1}^{\infty}\frac{a\sinh\alpha}{\sinh n\alpha}.$$

According to (5.24), we can now calculate the normal field on the plane, and from there the charge induced on the plane. According to Gauss's theorem, the total charge induced Q_{pl} should be identical to the sum of the image charges located behind the plane:

$$Q_{\mathrm{pl}} = \sum_{n=1}^{\infty} p_n = -\sum_{n=1}^{\infty} q_n = C_{12}V.$$

We conclude from this that the coefficients of capacity satisfy in this case the relationship $C_{11} = C_{12}$.

References

5.1 J. M. Pitarke, F. Flores and P. M. Echenique: *Tunneling Spectroscopy: Surface Geometry and Interface Potential Effects*, Surface Science **234**, 1 (1990).

Further Reading

R.P. Feynman, R.B. Leighton and M. Sands: *The Feynman Lectures on Physics, Vol. II* (Addison–Wesley, New York, 1964) Chap. 12.

K.H. Huebner, E.A. Thorton and T.G. Byrom: *The Finite Element Method for Engineers*, (John Wiley & Sons, New York, 3rd edition, 1995).

R.L. Huston and C.E. Passerello: *Finite Element Methods, An Introduction* (Marcel Dekker Inc., New York 1984).

W. Magnus and F. Oberhettinger: *Special Functions of Mathematical Physics* (Chelsea, New York, 1949).

P.M. Morse and H. Feshbach: *Methods of Theoretical Physics* (McGraw–Hill, New York, 1955).

6. Magnetic Field

We have indicated in Chap. 1 that magnetic phenomena were originally observed in certain minerals – the magnets – capable of attracting each other and some metals. Terrestrial magnetism and the possibility of inducing magnetic properties in metals through rubbing were well known many centuries before the basic phenomena were understood. The first quantitative studies of magnetic force, in particular, those due to Coulomb, led to the introduction of the notion of magnetic poles, in analogy with electric charges in electrostatics.

The phenomenology of magnetism began to be understood around 1820, when Oersted detected magnetic effects originating from electric currents. The experiments of Biot, Savart, Arago and Ampère quantifying such effects were also very important. In fact, today we know that magnetic phenomena are the manifestation of the interaction of moving charges. Magnetism in material media is a consequence of the existence of currents and of elementary dipoles at atomic level. In fact, magnetism in matter is one of the most complex manifestations of elementary magnetic phenomena. Its description requires a microscopic model of the distribution and of the dynamics of charges in a solid, with the necessary quantum ingredients for such a description.

Although the microscopic foundation of magnetism in material media is today well understood, during its historical development concepts were often introduced which led to confusion concerning the correct identification of the fields relevant to the description of the phenomena. In this book we assign a fundamental role to the magnetic field $\boldsymbol{B}$.

It is an experimental fact that between moving electric charges there exists a force additional to that of the electrostatic interaction studied in Chap. 4. On a test charge q moving with velocity $\boldsymbol{v}$, a total force $\boldsymbol{F}$ given by

$$\boldsymbol{F} = q\left(\boldsymbol{E} + \frac{1}{c}\boldsymbol{v} \times \boldsymbol{B}\right) \tag{6.1}$$

acts. As we have mentioned in Chap. 3, this is called the Lorentz force. It has two components: the first, independent of the velocity of the particle, is proportional to the electric field created by the sources, and has the same form as the electrostatic force. The second component corresponds to a new type of interaction, produced by moving charges or by magnets and proportional to the field $\boldsymbol{B}$, called the *magnetic field*.

From (6.1) we see that in the cgs system the magnetic field has the same units as $\boldsymbol{E}$, namely, gauss (G). Although the SI unit for the magnetic field $\boldsymbol{B}$ is the tesla (T), equal to 10^4 gauss, in many practical situations it is common to use the gauss. To have an idea of the magnitude of these units let us mention that the Earth's magnetic field is of the order of 0.5 gauss.

In this chapter we study the properties of the magnetic field in stationary conditions, that is to say, independent of time. When the field is produced by moving charges a stationary state corresponds to electric currents which do not vary in time. This property defines the area of electromagnetism known as *magnetostatics.*

6.1 Ampère's Law

Given a stationary current I circulating in a conductor, it is found experimentally that the magnetic field produced satisfies Ampère's law:

$$\oint_C \boldsymbol{B} \cdot \mathrm{d}\boldsymbol{l} = \frac{4\pi}{c} I. \tag{6.2}$$

Here C is a closed circuit around the conductor through which the current circulates. More generally, Ampère's law (6.2) is valid for any closed circuit C, where I is the (stationary) current crossing a surface determined by C (Fig. 6.1).

Transforming the line integral into a surface integral, and introducing the current density $\boldsymbol{j}$ associated with I, (6.2) can be written

$$\int_S (\nabla \times \boldsymbol{B}) \cdot \mathrm{d}\boldsymbol{s} = \frac{4\pi}{c} \int_S \boldsymbol{j} \cdot \mathrm{d}\boldsymbol{s}, \tag{6.3}$$

where S is any surface bounded by C. Since the integration is carried out on an arbitrary surface, the last identity should be valid in local form

$$\nabla \times \boldsymbol{B} = \frac{4\pi}{c} \boldsymbol{j}. \tag{6.4}$$

This is the differential form of Ampère's law.

Another fundamental property defining the magnetic field is the experimental evidence that no free magnetic charges or monopoles exist. Indeed – in contrast to the case of electrostatics, where the field lines begin or end at charges – the lines of magnetic field are closed. Consequently, for an arbitrary volume V, the total flux of the magnetic field through the surface S enclosing V is zero,

$$\int_S \boldsymbol{B} \cdot \mathrm{d}\boldsymbol{s} = \int_V \nabla \cdot \boldsymbol{B}\, \mathrm{d}^3 r = 0, \tag{6.5}$$

or, in differential form,

$$\nabla \cdot \boldsymbol{B} = 0. \tag{6.6}$$

Together with Ampère's law (6.4), (6.6) determines the magnetostatic field from the density of stationary current $\boldsymbol{j}$. It is important to notice the analogy between these two equations and those satisfied by the electrostatic field: $\nabla \cdot \boldsymbol{E} = 4\pi\rho$ and $\nabla \times \boldsymbol{E} = 0$. In fact, we should consider Ampère's law as the magnetic analog of Gauss's law, since both relate the fields to the sources.

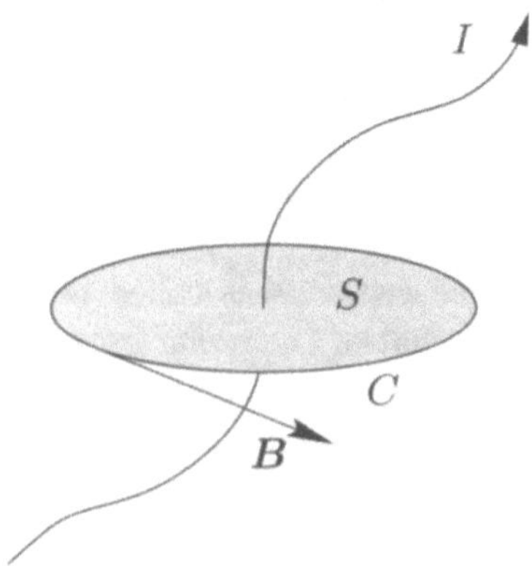

Fig. 6.1. Illustration of Ampère's law

On the other hand, comparing (6.4) and (6.6) with Maxwell equations (4.1), we notice that (6.4) is not one of the fundamental equations of electromagnetism, while (6.6) is. Indeed, under nonstatic conditions – that is to say, when the currents depend on time – Ampère's law must be modified. The relation $\nabla \cdot \boldsymbol{B} = 0$ on the other hand, is a law associated with a fundamental property of the magnetic field and is valid in any situation, as discussed above.

Note that in current-free regions $\boldsymbol{j}$ vanishes and therefore $\nabla \times \boldsymbol{B} = 0$ is satisfied. This implies that the magnetic field can be written as $\boldsymbol{B} = -\nabla\phi_{\mathrm{M}}$, where ϕ_{M} is called the *magnetic scalar potential.* According to (6.6), in current-free regions this potential satisfies Laplace's equation. Consequently, in this case the problem of finding the magnetic field can be reduced to a problem similar to electrostatics. However, it is necessary to keep in mind that Ampère's law (6.2) implies that the magnetic scalar potential is a multi-valued function. In fact, the line integral of $\boldsymbol{B}$ along a closed curve enclosing currents is non-zero. Thus, in the definition of ϕ_{M} one should introduce cuts similar to the Riemann sheets. The scalar magnetic potential – a useful tool in the solution of problems – is not a fundamental element of electromagnetic theory.

6.2 Vector Potential. The Biot–Savart Law

We have already shown that the introduction of a potential to analyze electrostatic phenomena implies a simplification of the problem. On the other hand, at the end of the previous section, we mentioned the disadvantages of defining a magnetic scalar potential. We show now that a coherent description of magnetic phenomena by means of a potential requires the use of a vector quantity.

Equation (6.6) implies that the magnetic field can be written as the curl of a vector

$$\boldsymbol{B} = \nabla \times \boldsymbol{A}, \tag{6.7}$$

since the divergence of a curl is always zero. The vector $\boldsymbol{A}$ is called *vector potential*, and has in magnetism a role similar to that of the scalar potential $\phi(\boldsymbol{r})$ of electrostatics. In Chap. 3 we have mentioned that the vector potential can be interpreted as a momentum reservoir for the dynamics of particles in a magnetic field.

In magnetostatics, starting from Ampère's law (6.4) we obtain an equation for the vector potential, given by

$$\nabla \times \nabla \times \boldsymbol{A} = \nabla(\nabla \cdot \boldsymbol{A}) - \nabla^2 \boldsymbol{A} = \frac{4\pi}{c}\boldsymbol{j}, \tag{6.8}$$

where we have used the definition of the Laplacian of a vector (Appendix D), $\nabla^2 \boldsymbol{A} = \nabla(\nabla \cdot \boldsymbol{A}) - \nabla \times \nabla \times \boldsymbol{A}$.

In the same way as the electrostatic potential is defined to within a constant, an arbitrary term with zero curl can be added to the vector potential without changing the value of the magnetic field. This freedom in defining $\boldsymbol{A}$ implies that its divergence can be arbitrarily chosen. The choice of the divergence of the vector potential, called *gauge* selection, will be more extensively discussed in Chap. 7, when we study the formulation of general electrodynamic problems in terms of potentials.

Equation (6.8) is simplified if we choose $\nabla \cdot \boldsymbol{A} = 0$, and leads to

$$\nabla^2 \boldsymbol{A} = -\frac{4\pi}{c}\boldsymbol{j}. \tag{6.9}$$

Notice the formal parallelism between (6.9) and Poisson's equation. However, attention must be paid to the mathematical character of the equation for the vector potential. It reduces to three Poisson's equations for the components of $\boldsymbol{A}$ only in Cartesian coordinates. Indeed, only in this system of coordinates can we write $\nabla^2 \boldsymbol{A} = (\nabla^2 A_x, \nabla^2 A_y, \nabla^2 A_z)$; in other systems of coordinates it is necessary to recall the definition of the Laplacian of a vector indicated above (Appendix D).

In general, to obtain the solution to (6.9) we can appeal to the so-called vector Green's theorem, prescribing the necessary and sufficient boundary conditions required to completely specify the solution in a given domain. In this way, it is shown that the solution is totally determined if the tangential and normal components of the vector potential $\boldsymbol{A}$ are given on the surface limiting the volume of interest.

For a bounded system of currents and with the condition that $\boldsymbol{A}(\boldsymbol{r})$ tends to zero for $r \to \infty$, the solution to (6.9) is

$$\boldsymbol{A}(\boldsymbol{r}) = \frac{1}{c} \int \frac{\boldsymbol{j}(\boldsymbol{r}')}{|\boldsymbol{r} - \boldsymbol{r}'|} \, \mathrm{d}^3 r'. \tag{6.10}$$

It can be shown that this solution satisfies $\nabla \cdot \boldsymbol{A} = 0$.

In many applications the current I associated with the density $\boldsymbol{j}$ circulates through a circuit of small and constant section s, such that $I = js$. In this case, the volume integral is limited to the volume of the circuit and can be written

$$\boldsymbol{A}(\boldsymbol{r}) = \frac{1}{c} \oint js \frac{\mathrm{d}\boldsymbol{l}'}{|\boldsymbol{r} - \boldsymbol{r}'|} = \frac{I}{c} \oint \frac{\mathrm{d}\boldsymbol{l}'}{|\boldsymbol{r} - \boldsymbol{r}'|}, \tag{6.11}$$

where $\mathrm{d}\boldsymbol{l}'$ is the line element along the circuit. This expression allows us to calculate the vector potential starting from a given current distribution. Both (6.10) and (6.11) show that the vector potential is in general parallel to the currents generating it.

Taking the curl of (6.11) it is possible to find the magnetic field generated by a stationary current. The resulting equation is called the Biot–Savart law, which explicitly links the field to the current from which it originates:

$$\boldsymbol{B}(r) = \nabla \times \boldsymbol{A}(\boldsymbol{r}) = \nabla \times \frac{I}{c} \oint \frac{\mathrm{d}\boldsymbol{l}'}{|\boldsymbol{r} - \boldsymbol{r}'|} = \frac{I}{c} \oint \frac{\mathrm{d}\boldsymbol{l}' \times (\boldsymbol{r} - \boldsymbol{r}')}{|\boldsymbol{r} - \boldsymbol{r}'|^3}. \tag{6.12}$$

If we consider a surface S bounded by a closed curve C, the *magnetic flux* Φ associated with the field $\boldsymbol{B}$ is defined as

$$\Phi = \int_S \boldsymbol{B} \cdot \mathrm{d}\boldsymbol{s}. \tag{6.13}$$

The condition (6.7) allows us to associate the vector potential in direct form to the magnetic flux Φ. Indeed, the Stokes' theorem implies

$$\Phi = \int_S \boldsymbol{B} \cdot \mathrm{d}\boldsymbol{s} = \oint_C \boldsymbol{A} \cdot \mathrm{d}\boldsymbol{l}. \tag{6.14}$$

This means that circulation of the vector potential along the curve C is equal to the magnetic flux through the surface S. In the cgs system the magnetic flux is measured in $\mathrm{G\,cm^2}$ and in the SI system the corresponding unit is the weber (Wb) equivalent to $10^8\,\mathrm{G\,cm^2}$.

6.3 Examples of Potentials and Fields. Multipole Expansion

As an application of the concepts introduced in the previous sections, we shall present here some examples of the explicit calculation of fields and magnetic potentials.

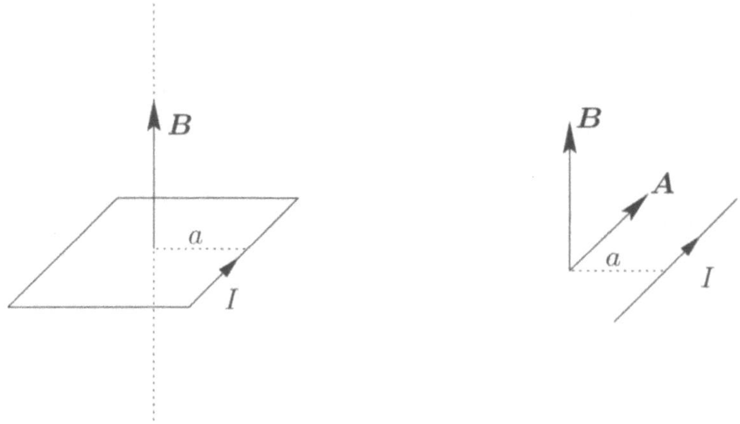

Fig. 6.2. Margnetic field of a current polygon and contribution of each side

Let us consider, in the first place, a regular polygon of n sides, such that the distance from its center to the middle point of any of the sides is a (Fig. 6.2) and through which a current I circulates. We want to find the magnetic field at the center of the polygon. According to the Biot–Savart law this field is given by the superposition of identical contributions from each side of the polygon.

We can evaluate the field by first calculating the vector potential generated by a segment of current of length $2L$ at distance a of the center (Fig. 6.2). The vector potential $\boldsymbol{A}$ has non-zero component in the direction of the current only. If this direction is z, we have

$$A_z = \frac{I}{c}\int_{-L}^{L} \frac{\mathrm{d}z'}{\sqrt{z'^2 + a^2}} = \frac{I}{c}\ln\frac{\sqrt{L^2+a^2}+L}{\sqrt{L^2+a^2}-L}. \tag{6.15}$$

The magnetic field $\nabla \times \boldsymbol{A}$ has azimuthal component B_φ only,

$$B_\varphi = -\partial_a A_z = \frac{2I}{ac}\frac{L}{\sqrt{L^2+a^2}}. \tag{6.16}$$

At the center of the polygon, $\boldsymbol{B}$ is therefore normal to the plane of the polygon. Since the half-length of each side is $L = a\tan(\pi/n)$, the total magnetic field is given by

$$B_\varphi = \frac{2\pi I}{ac}\frac{\sin(\pi/n)}{\pi/n}. \tag{6.17}$$

As $n \to \infty$, the polygon becomes a circle of radius a, and the factor $\sin(\pi/n)/(\pi/n)$ in (6.17) tends to unity. The factor $2\pi I/ac$ corresponds thus to the field produced by a circular loop of radius a carrying a current I. This limit can be proved by calculating $B_z(0,0,0)$ explicitly for a circular loop by means of the Biot–Savart law (6.12). This law makes it possible to calculate the magnetic field at an arbitrary point on the axis of the circular loop. This turns out to be

$$\begin{aligned} B_z(0,0,z) &= \frac{I}{c}\oint \frac{\mathrm{d}\boldsymbol{l} \times (\boldsymbol{a} - \boldsymbol{r})|_z}{|\boldsymbol{a} - \boldsymbol{r}'|^3} = \frac{2\pi I a^2}{\left(a^2 + z^2\right)^{3/2}} \\ &= \frac{2\pi I}{a} \sin^3 \alpha, \end{aligned} \tag{6.18}$$

where α is the angle subtended by the loop from the point where the field is calculated.

In the second place, we study the magnetic field produced by a straight wire of infinite length carrying a current I and placed a distance r from the point of observation. In this situation it is not possible to use the integral (6.11) because we are not dealing with a bounded system of currents. On the other hand, it is possible to use the Ampère law to calculate the magnetic field directly from (6.2). Assuming that the magnetic field has only an azimuthal component around the wire, and since – from symmetry arguments – its modulus should only depend on the distance r, we have

$$\oint_C \boldsymbol{B} \cdot \mathrm{d}\boldsymbol{l} = 2\pi r|\boldsymbol{B}| = \frac{4\pi}{c} I. \tag{6.19}$$

Here C is a circumference of radius r, centered at the wire and placed perpendicular to it. From this we obtain

$$\boldsymbol{B} = \frac{2I}{cR}\, \hat{\boldsymbol{\varphi}}, \tag{6.20}$$

where $\hat{\boldsymbol{\varphi}}$ is the unit vector in cylindrical coordinates in the azimuthal direction. This magnetic field can be derived from a vector potential parallel to the current given by

$$\boldsymbol{A} = -\frac{2I}{c} \ln R \hat{\boldsymbol{z}}. \tag{6.21}$$

It is interesting to compare this form of the vector potential produced by an infinite line of current with the result obtained for a finite segment of length $2L$, given in (6.15), in the limit $L \to \infty$. This result, when $L \gg R \equiv a$, reduces to

$$A_z = \frac{2I}{c}(\ln 2L - \ln R). \tag{6.22}$$

This diverges for $L \to \infty$. The divergence is a consequence of the fact that (6.15) was obtained starting from (6.11) which is valid only for bounded current distributions. However, we observe that A_z depends on L only through a term independent of r. As this term has no relevance in the calculation of the magnetic field it can be eliminated, and we obtain again the result given in (6.21).

Finally, we consider the magnetic equivalent of the electric dipole studied in Chap. 4. Since, as we discussed above, magnetic monopoles do not exist in nature, the magnetic dipole is the fundamental source of the field $\boldsymbol{B}$. Its relevance in the study of magnetic phenomena lies in that, at the atomic level, the electric currents inside atoms and the spin of the subatomic particles can be modeled as magnetic dipole moments.

A magnetic dipole can be represented by a circular loop of radius ε, carrying a stationary current I, in the limit $\varepsilon \to 0$ and $I \to \infty$, such that $\varepsilon^2 I$ is constant. To simplify, we assume the loop to be placed on the xy plane and centered at the origin (Fig. 6.3).

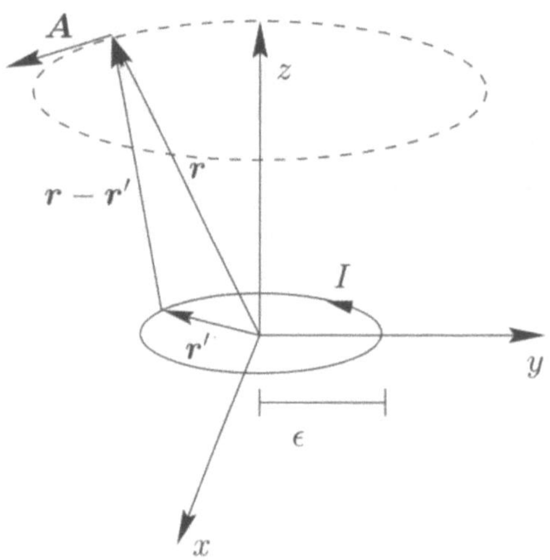

Fig. 6.3. Magnetic dipole as limit of a current loop

The vector potential can be calculated as

$$\boldsymbol{A}(\boldsymbol{r}) = \frac{I}{c} \oint \frac{\mathrm{d}\boldsymbol{l}'}{|\boldsymbol{r} - \boldsymbol{r}'|}, \tag{6.23}$$

where the integral is carried out along the loop, and $\boldsymbol{r}' = \varepsilon(\cos\varphi, \sin\varphi, 0)$. Expanding $|\boldsymbol{r} - \boldsymbol{r}'|^{-1}$ to first order in ε, we obtain

$$\frac{1}{|\boldsymbol{r} - \boldsymbol{r}'|} \approx \frac{1}{r} + \frac{1}{r^3}(x\cos\varphi + y\ \sin\varphi)\ \varepsilon. \tag{6.24}$$

Introducing this expansion in the expression for $\boldsymbol{A}$ the first term, which is analogous to the monopolar term in the electrostatic case, gives a vanishing contribution proportional to the circulation integral of $\mathrm{d}\boldsymbol{l}$ along a closed circuit. The first non-zero contribution is calculated keeping in mind that $\mathrm{d}\boldsymbol{l}' = \varepsilon(-\sin\varphi, \cos\varphi, 0)\mathrm{d}\varphi$, giving

$$\boldsymbol{A}(\boldsymbol{r}) = \frac{\pi\varepsilon^2 I}{r^3 c}(-y, x, 0). \tag{6.25}$$

In the limit indicated above, $\pi\varepsilon^2 I/c \to m$ is a characteristic constant of the loop proportional to the area $\pi\varepsilon^2$ and to the current I.

We observe that the vector potential can be expressed in a form independent of the orientation of the loop, as

$$\boldsymbol{A}(\boldsymbol{r}) = \frac{\boldsymbol{m} \times \boldsymbol{r}}{r^3}, \tag{6.26}$$

where $\boldsymbol{m}$ is a vector of modulus m, perpendicular to the plane of the loop and oriented, in connection with the sense of circulation of the current, in the positive direction according to the "right-hand rule." This vector is called the *magnetic dipole moment* of the loop. Equation (6.26) is the general expression for the potential created at a point $\boldsymbol{r}$ in space by a magnetic dipole placed at the origin. The resulting magnetic field is

$$\boldsymbol{B} = \frac{3\boldsymbol{r}(\boldsymbol{r} \cdot \boldsymbol{m}) - r^2 \boldsymbol{m}}{r^5}. \tag{6.27}$$

Notice that this magnetic field has the same behavior as a function of distance as the dipole electric field (4.34). Unlike the electric dipole moment, $\boldsymbol{m}$ is a pseudovector since it depends on the chosen orientation for the loop surface. This is compatible with the fact that $\boldsymbol{B}$ is also a pseudovector. The potential $\boldsymbol{A}$, on the other hand, is an ordinary – axial – vector (see Chap. 7).

The result obtained for the magnetic dipole suggests the possibility of carrying out a multipole expansion of the vector potential generated by an arbitrary and bounded distribution of stationary currents. Expanding $|\boldsymbol{r} - \boldsymbol{r}'|^{-1}$ in (6.10) in a series of spherical harmonics – just as we did in the electrostatic case – we find

$$\boldsymbol{A}(\boldsymbol{r}) = \frac{4\pi}{c} \sum_{l=0}^{\infty} \sum_{m=-l}^{l} \frac{1}{2l+1} \, r^{-l-1} \, \boldsymbol{a}_{lm} \, Y_{lm}(\theta, \varphi), \tag{6.28}$$

for $r > r'$. Here

$$\boldsymbol{a}_{lm} = \int r'^l \boldsymbol{j} \, Y^*_{lm}(\theta', \varphi') \, \mathrm{d}^3 r' \tag{6.29}$$

are the multipolar moments of the distribution of currents with density $\boldsymbol{j}$. It can be shown that the zeroth order moment,

$$\boldsymbol{a}_{00} = \frac{1}{\sqrt{4\pi}} \int \boldsymbol{j} \, \mathrm{d}^3 r, \tag{6.30}$$

vanishes. The first non-vanishing contribution corresponds to the term with $l = 1$ and has the form of the vector potential for the magnetic dipole, with

$$\boldsymbol{m} = \frac{1}{2c} \int \boldsymbol{r}' \times \boldsymbol{j}(\boldsymbol{r}')\, \mathrm{d}^3 r'. \tag{6.31}$$

We leave the reader the problem of finding the relationship between $\boldsymbol{m}$ and the multipole moments of order $l = 1$. Higher order terms are interpreted in a way similar to the electrostatic case.

6.4 Magnetic Energy. Variational Principle

In this section, we determine the energy associated with a group of circuits through which stationary currents circulate. This system is the magnetic equivalent of the group of charges or conductors studied in electrostatics. In fact, the aim of this analysis is to find an expression for the energy density associated with the magnetic field, similar to the one found for the electric field in Chap. 4.

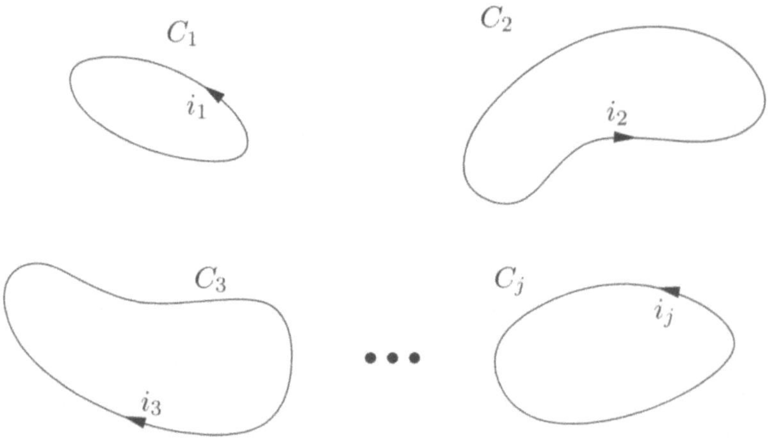

Fig. 6.4. System of currents

Let us consider a group of closed circuits C_j carrying currents i_j (Fig. 6.4). The magnetic field $\boldsymbol{B}_j$ generated by the circuit C_j is proportional to the current i_j. Therefore, the flux produced by that field through the circuit C_k can be written $\Phi_{kj} = cL_{kj}i_j$, where c is the speed of light and L_{kj} is a constant depending on the geometry of both circuits, called the *mutual inductance*. This quantity, in cgs units, is measured in s^2/cm; in the SI system the unit is the henry (H) equal to $10^9\,\mathrm{s}^2/\mathrm{cm}$. The total magnetic flux through the circuit C_k is the sum of contributions of all circuits and is given by

$$\Phi_k = c \sum_j L_{kj} i_j. \tag{6.32}$$

To reach a stationary situation in which through each circuit C_k a constant current I_k circulates, we imagine that the configuration of the circuits stays fixed, and the currents i_k grow from zero at $t = 0$ to I_k at time t. During the growth of the currents, an electromotive force is generated in the circuit C_k given by (see Chap. 7)

$$\mathcal{E}_k = -\frac{1}{c}\frac{\mathrm{d}\Phi_k}{\mathrm{d}t} = -\sum_j L_{kj}\frac{\mathrm{d}i_j}{\mathrm{d}t}. \tag{6.33}$$

This electromotive force perfoms work on the charges in C_k, in such a way that to keep the current i_k in that circuit it is necessary to compensate the energy exchange. The power delivered to circuit C_k under the application of the electromotive force $\mathcal{E}_k$ is $\mathrm{d}W_k/\mathrm{d}t = -i_k\mathcal{E}_k$. The total work carried out in an infinitesimal interval dt to maintain the system of currents in the presence of the electromotive forces induced by the variations of magnetic field is thus

$$\mathrm{d}W = \sum_k \mathrm{d}W_k = -\mathrm{d}t \sum_k i_k \mathcal{E}_k. \tag{6.34}$$

We associate this work with the energy accumulated in the system of currents.

To simplify the calculation of the total energy, we assume that all currents i_j vary in time in the same proportion,

$$i_j = \alpha(t) I_j, \tag{6.35}$$

where $\alpha(0) = 0$ and $\alpha(t) = 1$. The total work to be delivered to the system while the currents increase to their final values is

$$\begin{aligned} W &= \int_0^W \mathrm{d}W = -\int_0^t \sum_k i_k \mathcal{E}_k \, \mathrm{d}t' \\ &= \int_0^W \sum_k \sum_j i_k L_{kj} \frac{\mathrm{d}i_j}{\mathrm{d}t'} \, \mathrm{d}t' = \frac{1}{2} \sum_{k,j} L_{kj} I_k I_j. \end{aligned} \tag{6.36}$$

The energy can be rewritten as a function of the fluxes in each circuit as

$$W = \frac{1}{2c} \sum_k \Phi_k I_k. \tag{6.37}$$

Inverting the matrix of inductances L_{kj} it is possible to express the currents in terms of the fluxes, in a way similar to (6.32). Therefore, the energy can be given in terms of the fluxes as $W = \sum_{k,j} N_{kj}\Phi_k\Phi_j$. Note the analogy between these relationships and those corresponding to the electrostatic case in which the energy could be expressed alternatively in terms of the charges or of the potentials.

On the other hand, each flux is given by a surface integral of the magnetic field, that is to say, of the curl of the vector potential $\boldsymbol{A}$. From (6.37) we obtain

$$W = \frac{1}{2c} \sum_k I_k \int_{S_k} (\nabla \times \boldsymbol{A}) \cdot \mathrm{d}\boldsymbol{s} = \frac{1}{2c} \sum_k I_k \int_{C_k} \boldsymbol{A} \cdot \mathrm{d}\boldsymbol{l}$$
$$= \frac{1}{2c} \int \boldsymbol{A} \cdot \boldsymbol{j}\, \mathrm{d}^3 r. \tag{6.38}$$

The last integral is carried out over the volume of the circuits, and $\boldsymbol{j}$ is the associated current density. This expression can be transformed in terms of the currents using (6.10):

$$W = \frac{1}{2c^2} \int \mathrm{d}^3 r \int \mathrm{d}^3 r' \frac{\boldsymbol{j}(\boldsymbol{r}) \cdot \boldsymbol{j}(\boldsymbol{r}')}{|\boldsymbol{r} - \boldsymbol{r}'|} = \sum_{k,j} \frac{I_k I_j}{2c^2} \int_{C_k} \int_{C_j} \frac{\mathrm{d}\boldsymbol{l} \cdot \mathrm{d}\boldsymbol{l}'}{|\boldsymbol{r} - \boldsymbol{r}'|}. \tag{6.39}$$

From here we can obtain an expression for the induction coefficients – the Müller formula – which allows evaluation of the induction coefficients in specific cases:

$$L_{kj} = \frac{1}{c^2} \int_{C_k} \int_{C_j} \frac{\mathrm{d}\boldsymbol{l} \cdot \mathrm{d}\boldsymbol{l}'}{|\boldsymbol{r} - \boldsymbol{r}'|}. \tag{6.40}$$

It is clear from this that $L_{kj} = L_{jk}$.

Keeping in mind that outside of the circuits there are no currents, and if the system of circuits is bounded, the integration volume in (6.38) can be extended to all space. Replacing $\boldsymbol{j} = c\nabla \times \boldsymbol{B}/4\pi$ and using vector identities we obtain the following expression for the magnetic energy

$$W = \frac{1}{8\pi} \int \boldsymbol{B} \cdot (\nabla \times \boldsymbol{A})\, \mathrm{d}^3 r = \frac{1}{8\pi} \int B^2 \mathrm{d}^3 r. \tag{6.41}$$

By analogy with the electrostatic case, we can associate the energy of a system of stationary currents with its magnetic field, introducing the energy density of the field as

$$w(\boldsymbol{r}) = \frac{|\boldsymbol{B}|^2}{8\pi}. \tag{6.42}$$

This result should be independent of the way in which the system is built up. Moreover its local character and its association with the magnetic field makes it independent of the particular structure of the system of currents which are sources of the field. In the next chapter we show that this density of magnetic energy together with that introduced in Chap. 4 for the electric field, are the relevant magnitudes involved in the conservation laws of electromagnetism.

To illustrate the order of magnitude of the energy stored in an accessible magnetic field let us evaluate expression (6.41) for the case of a nuclear magnetic imaging device. In these devices a field of about 10^4 gauss is generated in a cylindrical volume some 2 m long and 60 cm diameter by means of a superconducting magnet. The total magnetic energy is about 2×10^{12} erg, or about 60 W h. If an accident produces the extinction of the superconducting current in the magnet this energy must be dissipated. Assuming that the dissipation takes place in a period of one second, the mean power released would be 216 kW. This power would cause the coolant (liquid helium) refrigerating the superconductor to abruptly evaporate, with catastrophic consequences. Superconducting magnets are designed to minimize this effect.

The possibility of defining an energy density for the magnetic field allows us, as for electrostatics, to formulate a variational principle for the calculation of $\boldsymbol{B}$. The functional that reaches an extremum in magnetostatics has the form

$$W^*[\boldsymbol{A}] = \int \left(\frac{1}{8\pi} |\nabla \times \boldsymbol{A}|^2 - \frac{1}{c} \boldsymbol{j} \cdot \boldsymbol{A} \right) \, \mathrm{d}^3 r. \tag{6.43}$$

This variational principle makes it possible to apply the method of finite elements already described in Chap. 4.

The magnetic energy defined above allows us also to calculate the force on one of the circuits, if it is assumed to be rigid. To do this, we carry out a virtual displacement $\mathrm{d}\boldsymbol{l}$ in one circuit, keeping the currents of the system constant. This produces a variation of the fluxes Φ_k in each circuit giving rise to a change in the total magnetic energy $\mathrm{d}W = \sum_k I_k \mathrm{d}\Phi_k / 2c$. Also, according to (6.33) and (6.34), the batteries maintaining the currents in the circuits should carry out a work $\mathrm{d}W_\mathrm{b} = \sum_k I_k \mathrm{d}\Phi_k / c = 2\ \mathrm{d}W$. The total energy balance implies that $\mathrm{d}W_\mathrm{b} = \mathrm{d}L + \mathrm{d}W$, where $\mathrm{d}L = \boldsymbol{F} \cdot \mathrm{d}\boldsymbol{l}$ is the mechanical work carried out when displacing the circuit. This work is $\mathrm{d}L = \mathrm{d}W_\mathrm{b} - \mathrm{d}W = \mathrm{d}W$, from which we obtain

$$\boldsymbol{F} = \nabla W|_I \, . \tag{6.44}$$

The label I indicates a process at constant current. Notice that this expression for the force is similar to the one obtained in electrostatics for the case when the potentials stay constant in the virtual displacement. The alternative process in which the fluxes stay constant is difficult to carry out in practice since both a battery or a current source connected to a normal conductor tend to keep the current in the circuit constant. A possible way of carrying out such a process would consist of working with superconducting circuits that have the property of maintaining the flux constant (Chap. 13).

6.5 Induction Coefficients

In agreement with (6.32) the induction coefficient of a circuit C_k with respect to another circuit C_l is defined as the flux induced in C_k when the current i_l circulates in C_l. This definition includes the case of self-induction, when the flux in a circuit is generated by the current circulating in it. We show in a specific case how such coefficients are calculated.

Since the induction coefficients determine both the magnetic flux (6.32) and the energy of the current system (6.37), these coefficients can be determined by evaluating any of these two quantities. The most convenient method in each case depends on the particular problem.

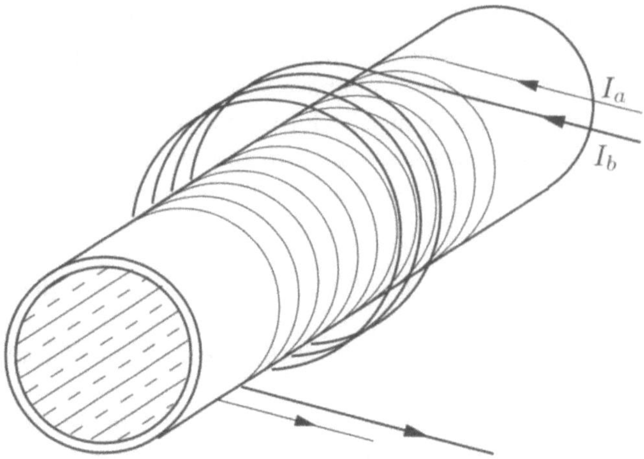

Fig. 6.5. Calculation of the mutual induction coefficient between a long and a short winding

As an example of a case in which the coefficients of mutual induction can be evaluated by calculating the magnetic flux, let us consider the circuits a and b of Fig. 6.5. If the winding a contains n very tight turns per unit length, the magnetic field produced is confined inside the long solenoid and has modulus

$$B = \frac{4\pi}{c} In. \tag{6.45}$$

Moreover, if the winding b has $m \ll n$ turns per unit length, a total length l_b and radius r_b the flux linked by this circuit is

$$\Phi_b = \pi r_b^2 m l_b B = \frac{4\pi^2}{c} r_b^2 m n l_b I = c L_{ab} I. \tag{6.46}$$

We conclude that

$$L_{ab} = \frac{4\pi^2}{c^2} r_b^2 m n l_b. \tag{6.47}$$

The evaluation of the self-induction coefficient for each winding requires keeping in mind the contribution of the field contained inside the wire. Generally, a winding is built with thin wires. For this reason it is useful to have an expression for the coefficient of internal self-induction of a thin wire that can be applied in different cases. The calculation of the magnetic energy of the system yields the result looked for. We assume a long rectilinear wire of radius r_a through which a current flows, evenly distributed through the section. The magnetic field inside the wire is directed in the azimuthal direction and its modulus is

$$B = \frac{2I}{c r_a^2} \rho, \tag{6.48}$$

where ρ is the radial coordinate normal to the wire. The magnetic energy per unit length is

$$W = \frac{1}{8\pi}\int B^2 \mathrm{d}^2 r = \frac{I^2}{4c^2} = \frac{1}{2}L_{aa}I^2. \tag{6.49}$$

It follows that the coefficient of internal self-induction per unit length of a thin wire is

$$L_{aa} = \frac{1}{2c^2}. \tag{6.50}$$

This result can still be used as a good approximation even in cases in which the wire is not rectilinear, whenever the diameter is much smaller than the length and than the radius of curvature of the winding.

Induction coefficients can also be evaluated using the magnetic flux. In this case it is not always necessary to have an explicit expression for the field. Indeed, (6.14) implies that the flux can be determined directly knowing the vector potential. As an example, we consider a current loop of radius r_a, as shown in Fig. 6.3, and calculate the mutual induction coefficient with a second loop of radius r_b placed on the same axis and in a parallel plane, a distance d from the first one. To obtain $\boldsymbol{A}$ it is convenient to write the current in the loop in spherical coordinates, in the form:

$$j_\varphi = I\delta(\cos\theta)\frac{\delta(r' - r_a)}{r_a}. \tag{6.51}$$

It is possible to find an expression for the vector potential in spherical coordinates noting that, due to the symmetry of the problem, we can choose as the only non-zero component of $\boldsymbol{A}$ the component $A_\varphi(r,\theta)$. On the other hand, at the point with coordinates $(x, 0, z)$, $A_\varphi(r,\theta)$ coincides with the component $A_y(x, 0, z)$ of $\boldsymbol{A}$ in Cartesian coordinates. In this case, its expression is given by (6.10). It follows that

$$A_\varphi(r,\theta) = \frac{Ir_a}{c}\int_0^{2\pi} \frac{\cos\varphi' \mathrm{d}\varphi'}{\sqrt{r_a^2 + r^2 - 2r_a r\sin\theta\cos\varphi'}}. \tag{6.52}$$

This integral can be expressed in terms of Legendre polynomials in the form

$$A_\varphi(r,\theta) = \frac{\pi I}{c}\sum_{n=0}^{\infty}(-1)^n\frac{(2n-1)!!}{2^n(n+1)!}P^1_{2n+1}(\cos\theta)\rho^{2n+2}, \tag{6.53}$$

with $\rho = r_a/r$ for $r > r_a$ and $\rho = r/r_a$ for $r < r_a$.

Due to the factor $\cos\varphi'$ in (6.52) this multipolar expansion contains the associated Legendre polynomials $P^1_n(\cos\theta)$. The term $n = 0$ of (6.53) gives the dipole approximation for the vector potential:

$$A_\varphi(r,\theta) = \frac{\pi r_a^2 I}{c}\frac{\sin\theta}{r^2}. \tag{6.54}$$

where $\pi r_a^2 I$ is the dipole moment of the loop.

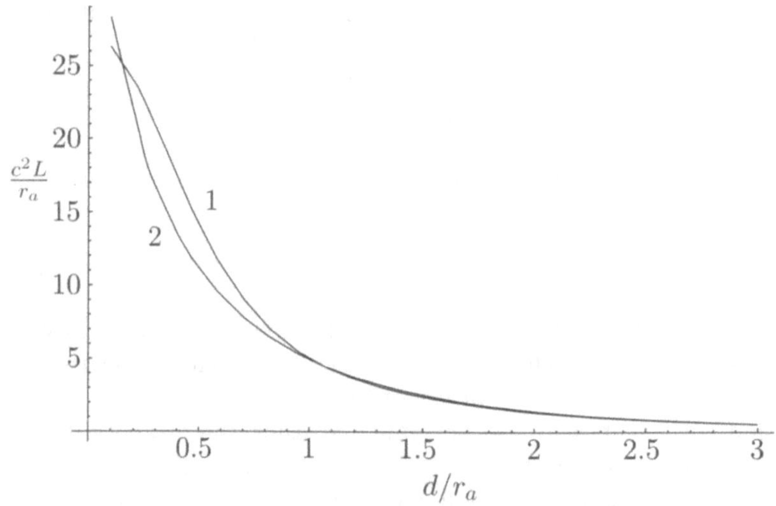

Fig. 6.6. Coefficient of mutual induction for two loops of radius r_a, placed on parallel planes and over a common axis, as a function of the distance d/r_a between them: (*1*) dipole approximation; (*2*) 20-term approximation

These expressions for the vector potential can be used to evaluate with different degrees of approximation the coefficient of mutual induction of two coaxial loops of radius r_a and r_b, placed on the same axis and in parallel planes a distance d apart. Figure 6.6 shows the coefficient of mutual induction obtained from (6.53) and (6.54) as a function of the distance d between the loops.

6.6 Symmetry Between Electrostatics and Magnetostatics

Before discussing time dependent electromagnetic phenomena, it is instructive to make a comparative review of the laws of electrostatics and magnetostatics. In particular, we want to highlight the symmetry between both parts of electromagnetism. This symmetry relates to the covariance underlying the theory. It extends also to time dependent phenomena, and shows up more clearly when expressing the laws of electromagnetism in covariant form, consistently with the theory of special relativity. Here we limit ourselves to listing the more relevant expressions obtained in Chaps. 4 and 6. Simple inspection of Table 6.1 shows the parallelism we wish to focus on.

Table 6.1. Symmetry between electrostatics and magnetostatics

Electrostatics	Magnetostatics
$\nabla \cdot \boldsymbol{E} = 4\pi\rho$	$\nabla \times \boldsymbol{B} = \frac{4\pi}{c}\boldsymbol{j}$
$\nabla \times \boldsymbol{E} = 0$	$\nabla \cdot \boldsymbol{B} = 0$
$\boldsymbol{E} = -\nabla\phi$	$\boldsymbol{B} = \nabla \times \boldsymbol{A}$
$\nabla \cdot \nabla\phi = -4\pi\rho$	$\nabla \times \nabla \times \boldsymbol{A} = \frac{4\pi}{c}\boldsymbol{j}$
$\phi(\boldsymbol{r}) = \int \rho(\boldsymbol{r}')\frac{\mathrm{d}^3r'}{\lvert\boldsymbol{r}-\boldsymbol{r}'\rvert}$	$\boldsymbol{A}(\boldsymbol{r}) = \int \boldsymbol{j}(\boldsymbol{r}')\frac{\mathrm{d}^3r'}{\lvert\boldsymbol{r}-\boldsymbol{r}'\rvert}$
$\boldsymbol{E}(\boldsymbol{r}) = \int \rho(\boldsymbol{r}')\frac{\mathrm{d}^3r'}{\lvert\boldsymbol{r}-\boldsymbol{r}'\rvert^3}$	$\boldsymbol{B}(\boldsymbol{r}) = \int \boldsymbol{j}(\boldsymbol{r}')\frac{\mathrm{d}^3r'}{\lvert\boldsymbol{r}-\boldsymbol{r}'\rvert^3}$
$\phi_{\mathrm{dip}} = \frac{\boldsymbol{p}\cdot\boldsymbol{r}}{r^3}$	$\boldsymbol{A}_{\mathrm{dip}} = \frac{\boldsymbol{m}\times\boldsymbol{r}}{r^3}$
$\boldsymbol{E}_{\mathrm{dip}} = \frac{3\boldsymbol{r}(\boldsymbol{r}\cdot\boldsymbol{p}) - r^2\boldsymbol{p}}{r^5}$	$\boldsymbol{B}_{\mathrm{dip}} = \frac{3\boldsymbol{r}(\boldsymbol{r}\cdot\boldsymbol{m}) - r^2\boldsymbol{m}}{r^5}$
$\boldsymbol{p} = \int \boldsymbol{r}'\rho(\boldsymbol{r}')\mathrm{d}^3r'$	$\boldsymbol{m} = \frac{1}{2c}\int \boldsymbol{r}' \times \boldsymbol{j}(\boldsymbol{r}')\,\mathrm{d}^3r'$
$Q_{lm} = \int r'^l\rho(r')Y^*_{lm}(\theta',\varphi')\,\mathrm{d}^3r'$	$\boldsymbol{a}_{lm} = \int r'^l\boldsymbol{j}(r')Y^*_{lm}(\theta',\varphi')\,\mathrm{d}^3r'$
$W = \frac{1}{2}\int \rho\phi\,\mathrm{d}^3r$	$W = \frac{1}{2}\int \boldsymbol{j}\cdot\boldsymbol{A}\,\mathrm{d}^3r$
$W = \frac{1}{2}\sum_{i,j} C_{ij}V_iV_j$	$W = \frac{1}{2}\sum_{i,j} L_{ij}\,I_i\,I_j$
$w(\boldsymbol{r}) = \frac{1}{8\pi}\lvert\boldsymbol{E}(\boldsymbol{r})\rvert^2$	$w(\boldsymbol{r}) = \frac{1}{8\pi}\lvert\boldsymbol{B}(\boldsymbol{r})\rvert^2$

6.7 Problems of Current Interest in Magnetism

As in the case of electrostatics, some of the problems treated in this chapter have given rise in recent years both to research in fundamental aspects of physics and to applications to current technology. We describe some of these topics below.

6.7.1 Earth's Magnetic Field

Geophysical determinations indicate that the Earth's magnetic field is approximately dipolar. Geological history shows that this magnetic field suffers inversions approximately every million years, a phenomenon of great importance for the development of our planet. The more accepted theories on the structure of the Earth's interior claim that the central nucleus of the planet has a radius of approximately 1200 km and that – in spite of being at temperatures higher than 3000 K – it is solid, due to the high pressure to which it is subjected. This central nucleus is surrounded by an external liquid nucleus, made up mainly of iron and nickel, and in smaller proportion by lighter elements such as copper, sulfur and oxygen. The principle called the *self-sustained dynamo* by which the magnetic field is generated has been understood since the 1950s. Nevertheless, a satisfactory model explaining in detail how the convective currents generating the field originate and how they are maintained does not exist. The origin of the periodic inversions is also not well understood. The principle of the dynamo can be explained by means of the so-called Faraday disk consisting of a metallic disk rotating in the presence of a bar magnet. Under these conditions an electric current is generated that produces losses through the Joule effect. This eventually leads to the detention of the disk. If the magnet is replaced by a small solenoid fed by a current, the same effect occurs. In fact, if the current in the solenoid is allowed to close the circuit through the conducting disk – and, at the same time, the losses are compensated giving enough energy to maintain the rotation – the field is self sustained. In relation with this topic, we recommend reading the papers by C.R. Carrigan and D. Gubbins [6.1], and by G. Glatzmaier and P.H. Roberts [6.2].

6.7.2 Isotope Separation

Separation of the isotopes of a chemical element is an important achievement of modern physics and chemistry. Different methods have been developed to attain this objective, taking advantage of the mass differences among different isotopes. One of these methods uses the different trajectories that ions of the same charge but with slightly different masses follow in a magnetic field. The lightest isotope suffers a larger deviation in the field, and this allows collection in different places of atoms or ions of each type. The design of magnets giving maximum efficiency in this separation process is a technological challenge.

6.7.3 Particle Acceleration

The development of physics requires constant advances in technology leading to the development of new tools for fundamental research. Examples are the large particle accelerators of modern research centres dedicated to the study of the structure of matter. Some of these centres are CERN (European Center for Nuclear Research) in Geneva, FERMILAB (National Laboratory of USA) operated by the University of Chicago and SLAC (Stanford Linear Accelerator) at the National Laboratory of USA and operated by Stanford University. These machines use accurately designed electromagnets to focus particles moving at speeds close to that of light. The magnets used are normally dipoles or quadrupoles, according to the type of correction one wants to introduce in the trajectories of the particles.

6.7.4 Motors

The design of electric motors has advanced considerably in recent years. Efficiency has improved in a substantial way: one example is the decrease in size observed if we compare modern motors with those performing the same function some years ago. Many applied researchers and engineers are working on building a motor for an electric automobile able to compete successfully with internal combustion engines. Dramatic progress has been made in recent years in this direction.

At present, ordinary cars already have a large number of small electric motors operating different kinds of mechanisms: windshield wipers, nozzles for automatic admission, etc. In each of these motors magnet design leading to the maximum efficiency is sought.

6.7.5 The Existence of Monopoles

The possibility that magnetic monopoles may exist in the Universe has long intrigued both theorists and experimentalists. Dirac first showed that magnetic monopoles could be accomodated within electromagnetic theory if their charge, g, is given by an integer multiple of hc/2e.

Indeed, the English physicist Paul Adrien Maurice Dirac (1902–1984) proposed in 1948 a theory allowing the possibility that magnetic monopoles can exist under these conditions. Since then there have been many attempts to detect particles of this type, but without success up to now. In this respect we recommend reading the paper by F.C. Adams *et al.*, *Extension of the Parker Bound on the Flux of Magnetic Monopoles* [6.3], from which the citation at the beginning was extracted.

Problems

6.1 Calculate the magnetic field on the axis perpendicular to the plane of a regular polygon of n sides and apothem R, through which a current I circulates. The axis passes through the center of the polygon. Take the limit $n \to \infty$ to obtain the field generated by a circular current loop.

6.2 Show that the solution to the Poisson equation for the vector potential

$$\boldsymbol{A}(\boldsymbol{r}) = \frac{1}{c} \int \frac{\boldsymbol{j}(\boldsymbol{r})}{|\boldsymbol{r} - \boldsymbol{r}'|} \, \mathrm{d}^3 r',$$

satisfies $\nabla \cdot \boldsymbol{A} = 0$ and $\nabla \times (\nabla \times \boldsymbol{A}) = 4\pi \boldsymbol{j}/c$.

Hint: Apply carefully the vector operators to $\boldsymbol{A}$, taking into account that they act on the non-primed variables. Then make use of Maxwell equations and the continuity equation.

6.3 Use the Ampère law to determine the magnetic field generated by (a) an infinite line of current; (b) an infinite solenoid with N tight turns per unit length; (c) a wire rolled tightly on a torus.

6.4 Consider a circular loop carrying a constant current. (a) Calculate the field produced by the loop on its axis. (b) Obtain the scalar potential on the axis. (c) Using the result of (b), obtain the scalar potential over all space. (d) Calculate $\boldsymbol{B}$ everywhere in space. (e) Use PhysicSolver to obtain the field over all space and compare the results obtained.

Hint: (c) Recall the multipole expansion.

6.5 A very small loop which carries a current is placed in the vicinity of an infinite line of current. Calculate the energy of the small loop in the field produced by the line. Plot it as function of the loop orientation. Which are the equilibrium positions of the loop? Which of these positions are stable?

Hint: Consider the force on the small loop. How is it derived from the energy of the loop?

6.6 Determine the field in all space produced by two paralell circular coils placed a distance d apart. Determine the conditions for having an approximately uniform field in a certain volume around the middle point between both coils. Use PhysicSolver and compare the results.

Hint: Use the result of Problem 6.4 and the superposition principle.

6.7 Show that the magnetic energy stored in a system of circuits can be written as function of the linked fluxes, as $W = \sum_{i,j} N_{ij}\Phi_i\Phi_j$. Relate the coefficients N_{ij} to the matrix of inductances.

6.8 Determine the coefficient of self-induction of a solenoid assuming that its diameter is much smaller than its length.

6.9 Calculate the coefficient of mutual induction between two current loops, assuming they are placed in arbitrary positions, with the only restriction that the axes of both loops intersect.

6.10 Determine the magnetic energy stored in a solenoid of radius a and length $L \gg a$, carrying a current I, assuming that the magnetic field is uniform in each section of the solenoid, but keeping in mind the variation along the axis due to the different contributions of each loop. Use PhysicSolver and compare the results.

Hint: Calculate first the magnetic field along the axis of the solenoid.

6.11 On the surface of a long cylinder of radius a there is a coil such that the current circulating in each generatrix returns by the one which is diametrically opposite, as indicated in Fig. 6.7. Study the magnetic field produced by this configuration. Assume the current intensity to vary with the polar angle according to $I = I_{0a} \sin\varphi$. Consider the effect on a second winding of radius $b < a$ similar to the first and placed inside it, when in the second winding a current $I = I_{0b} \sin(\varphi - \varphi_0)$ circulates. This configuration is the basis of the operation of an induction motor.

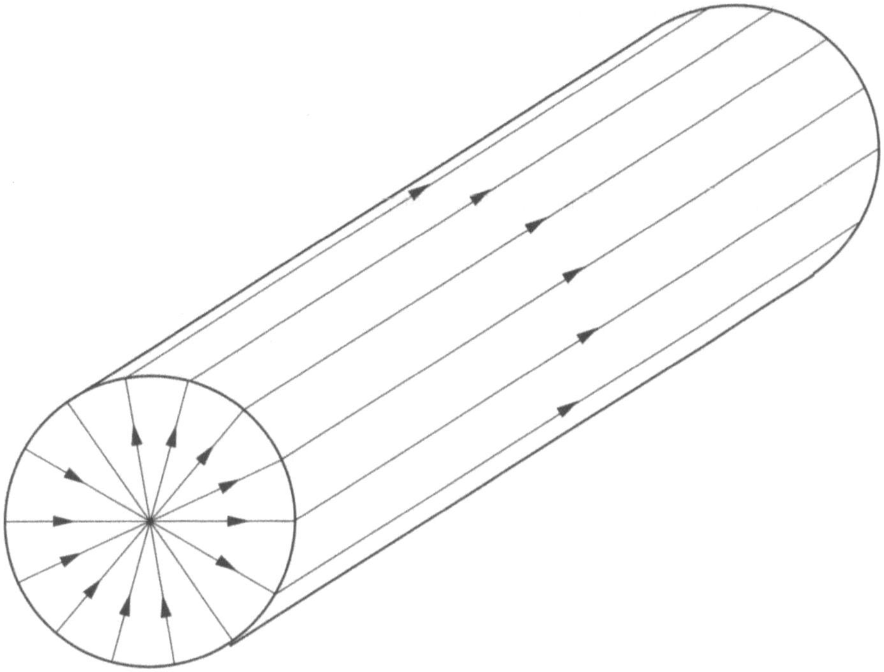

Fig. 6.7. Model winding for an induction motor

Solution: In order to solve this problem we consider in the first place the vector potential (6.21) generated at point $\boldsymbol{r}'$ by a single wire located at point $\boldsymbol{r}$ and we express the distance $R = |\boldsymbol{r} - \boldsymbol{r}'|$ in terms of polar coordinates in the plane perpendicular to the axis of the cylinder. We have

$$\begin{aligned}\ln R &= \ln \sqrt{r^2 + r'^2 - 2rr' \cos(\varphi - \varphi')} \\ &= \frac{1}{2} \left\{ 2 \ln r' + \ln[1 + (r/r')^2 - 2(r/r') \cos(\varphi - \varphi')] \right\} ,\end{aligned}$$

where φ and φ' are the polar angles. When $r < r'$ this expression can be expanded in powers of r/r' in the form

$$\ln R = \ln r' - \sum_{m=1}^{\infty} \frac{1}{m} \left(\frac{r}{r'} \right)^m (\cos m\varphi \cos m\varphi' + \sin \varphi \sin m\varphi')$$

For $r > r'$ an equivalent expansion in powers of r'/r can be given. This expression allows us to determine the vector potential inside and outside the cylinder of Fig. 6.7. Indeed, using the superposition principle, the potential inside the cylinder can be expressed in the form:

$$A_z^{<}(\rho, \varphi) = -\frac{1}{2\pi} \int_0^{2\pi} \mathrm{d}\varphi' \frac{2I(\varphi')}{c} \ln R = \frac{I_{0a}}{c} \frac{\rho}{a} \sin \varphi .$$

Similarly, using the expansion valid when $r > a$ we obtain for the potential outside the cylinder

$$A_z^{>}(\rho, \varphi) = \frac{I_{0a}}{c} \frac{a}{\rho} \sin \varphi.$$

This potential is continuous at the surface of the cylinder and gives rise to a uniform magnetic field inside and to a dipole field outside (see Fig. 6.8) given by

$$\begin{cases} B_x^{<}(\rho, \varphi) = I_{0a}(1/ac), \\ B_y^{<}(\rho, \varphi) = 0, \end{cases} \qquad \text{field inside}$$

$$\begin{cases} B_x^{>}(\rho, \varphi) = I_{0a}(a/c\rho^2) \cos 2\varphi, \\ B_y^{>}(\rho, \varphi) = I_{0a}(a/c\rho^2) \sin 2\varphi, \end{cases} \qquad \text{field outside.}$$

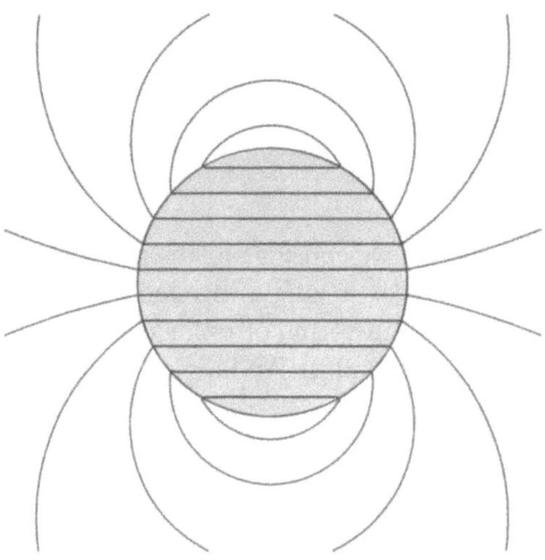

Fig. 6.8. Lines of magnetic field in an induction motor

The second winding, of diameter $b < a$, placed inside the first one is in the presence of a uniform field. The torque acting on this winding can be evaluated by first considering the force acting on one of the wires, parallel to the z axis and carrying a current I in presence of a uniform magnetic field in the x direction. Using the expression for the Lorentz force we find

$$\boldsymbol{F} = \frac{I}{c} \hat{\boldsymbol{z}} \times \boldsymbol{B} = \frac{1}{c} I B \hat{\boldsymbol{y}}$$

The torque $\boldsymbol{t}(b, \varphi)$ produced by this force is given by

$$t(b,\varphi) = \boldsymbol{r} \times \boldsymbol{F} = Fx\hat{\boldsymbol{z}} = \frac{IBx}{c}\hat{\boldsymbol{z}} = \frac{jBb\cos\varphi}{c}\hat{\boldsymbol{z}},$$

where $x = b\cos\varphi$ and $y = b\sin\varphi$ are the Cartesian coordinates of the wire in the perpendicular plane. To determine the total torque on the winding of radius b we must add the torques on each wire. We obtain

$$\boldsymbol{T} = \int_0^{2\pi} \mathrm{d}\varphi\, \boldsymbol{t}(b,\varphi) = \frac{bB}{c}\hat{\boldsymbol{z}} \int_0^{2\pi} \mathrm{d}\varphi\, j(\varphi)\cos\varphi.$$

If we take for the current in the internal cylinder a law of the type $I = I_{0b}\sin(\varphi - \varphi_0)$, the total torque $\boldsymbol{T}$ is given by

$$\boldsymbol{T} = -\pi \frac{I_{0a} I_{0b}}{c^2} \frac{b}{a} \sin\varphi_0 \hat{\boldsymbol{z}}.$$

This torque is a maximum for $\varphi_0 = -\pi/2$. This means that the optimum torque is obtained when the current in the internal winding is $\pi/2$ out of phase relative to the current in the external winding. In Chap. 7 we shall show that a current of this type can be generated by induction, if the original current in the external winding varies sinusoidally in time.

References

6.1 C.R. Carrigan and D. Gubbins: *The Source of the Earth Magnetic Field*, Sci. Am. 777 (1988).

6.2 G. Glatzmaier and P.H. Roberts: *A Three-Dimensional Self-Consistent Computer Simulation of a Geomagnetic Field Reversal*, Nature **377**, 203 (1995).

6.3 F.C. Adams, M. Fatuzzo, K. Freese, G. Tarlé and R. Watkins: *Extension of the Parker Bound on the Flux of Magnetic Monopoles*, Phys. Rev. Lett. **70** , 2511 (1993).

Further Reading

R.P. Feynman, R.B. Leighton and M. Sands: *The Feynman Lectures on Physics, Vol. II* (Addison–Wesley, New York, 1964).

J.D. Jackson: *Classical Electrodynamics* (John Wiley & Sons, Inc., New York, 1975).

P.M. Morse and H. Feshbach: *Methods of Theoretical Physics* (McGraw–Hill, New York, 1955).

W.K.H. Panofsky and M. Phillips: *Classical Electricity and Magnetism* (Addison–Wesley, Reading, 1955) Chaps. 7 and 8.

7. Maxwell's Equations

We have analyzed, up to this point, the origin and the consequences of four physical laws associated with static electromagnetic phenomena. The first of them, Gauss's law, relates the electric field to its sources. In differential form it is written as

$$\nabla \cdot \boldsymbol{E} = 4\pi\rho. \tag{7.1}$$

Within the area of application of electromagnetism, this law is of general validity.

In static situations, the electric field also satisfies

$$\nabla \times \boldsymbol{E} = 0. \tag{7.2}$$

As we show in the next section, this equation must be modified in the presence of time dependent magnetic fields, giving rise to Faraday's law, which describes induction phenomena.

The experimental observation of the non-existence of magnetic monopoles establishes that the magnetic field satisfies

$$\nabla \cdot \boldsymbol{B} = 0. \tag{7.3}$$

This relationship, like Gauss's law, has general validity.

Finally, Ampère's law links the magnetostatic field to the currents originating it, according to

$$\nabla \times \boldsymbol{B} = \frac{4\pi}{c}\, \boldsymbol{j}. \tag{7.4}$$

We shall however show that Ampère's law violates the conservation of charge, and that (7.4) has to be modified to preserve this important physical principle. This correction was introduced by J.C. Maxwell who in such a way unified in a single consistent theory the description of electromagnetic phenomena.

7.1 Time-Dependent Magnetic Fields. Faraday's Law

In 1831 Michael Faraday and Joseph Henry (1797–1878) almost simultaneously discovered a new phenomenon associated with electric and magnetic fields. This discovery was to be of fundamental importance in technological applications. This is the phenomenon of *magnetic induction*, which implies that the time variation of a magnetic field gives rise to electric fields.

More precisely, Faraday's law indicates that for a closed circuit C in the presence of a varying magnetic field an electromotive force $\mathcal{E}$ is induced (Fig. 7.1). This electromotive force is proportional to the time derivative of the magnetic flux Φ through the circuit:

$$\mathcal{E} = -\frac{1}{c}\frac{\mathrm{d}\Phi}{\mathrm{d}t}. \tag{7.5}$$

Here Φ is the magnetic flux introduced in the previous chapter. The unit for the electromotive force in the cgs system is the statvolt, the same as for the potential ϕ. In the SI system the unit is the volt.

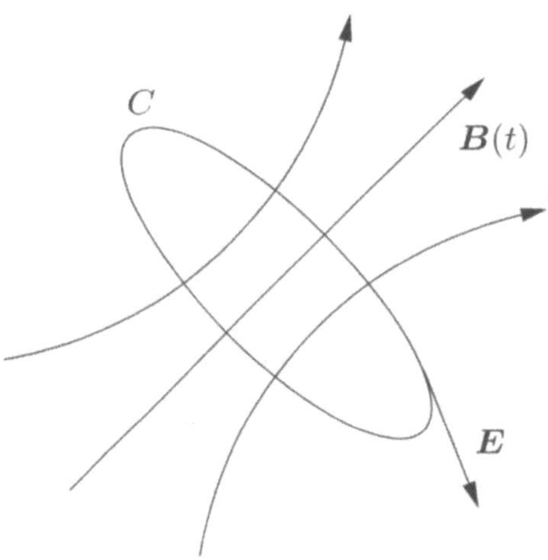

Fig. 7.1. Illustration of Faraday's law

The electromotive force generated in the circuit can be written as the line integral of the electric field along C. In integral form, Faraday's law is then

$$\oint_C \boldsymbol{E}\cdot \mathrm{d}\boldsymbol{l} = -\frac{1}{c}\int_S \partial_t \boldsymbol{B}\cdot \mathrm{d}\boldsymbol{s}, \tag{7.6}$$

where in order to introduce the time derivative into the surface integral we have assumed that the curve C does not vary in time.

Since Faraday's law is valid for any closed curve – whether it coincides with a material circuit or not – it is possible to find a differential form for it by converting the line integral of $\boldsymbol{E}$ into a surface integral of its curl. The differential version of Faraday's law is

$$\nabla \times \boldsymbol{E} + \frac{1}{c}\partial_t \boldsymbol{B} = 0. \tag{7.7}$$

Comparing with Maxwell's equations (4.1), we notice that Faraday's law is one of the fundamental laws of electromagnetism. In fact, it modifies the electrostatic equation $\nabla \times \boldsymbol{E} = 0$, indicating that the rotational component of the electric field has a contribution due to time variations of the magnetic field. In particular, we note that in non-static situations it is not possible to derive the electric field from a potential.

7.2 Displacement Current and Maxwell's Equations

To show the above-mentioned inconsistency between Ampère's law and charge conservation, we take the divergence of both sides of (7.4). Since $\nabla \cdot (\nabla \times \boldsymbol{B}) = 0$, Ampère's law implies $\nabla \cdot \boldsymbol{j} = 0$. On the other hand, the equation of continuity for the charge (4.6) – which is a direct consequence of the corresponding conservation principle – establishes that $\partial_t \rho + \nabla \cdot \boldsymbol{j} = 0$. Therefore $\nabla \cdot \boldsymbol{j}$ vanishes only in static situations, where $\partial_t \rho \equiv 0$. As a consequence, Ampère's law should be modified to be extended to time-dependent problems.

Maxwell proposed to rewrite Ampère's law as

$$\nabla \times \boldsymbol{B} = \frac{4\pi}{c}\,(\boldsymbol{j} + \boldsymbol{j}_{\mathrm{D}}), \tag{7.8}$$

where the additional term $\boldsymbol{j}_{\mathrm{D}}$, called the *displacement current*, must be determined in such a way as to make this equation compatible with the conservation of the electric charge. Taking the divergence of (7.8) we obtain $\nabla \cdot \boldsymbol{j} + \nabla \cdot \boldsymbol{j}_{\mathrm{D}} = 0$, which is compatible with the continuity equation if

$$\partial_t \rho = \nabla \cdot \boldsymbol{j}_{\mathrm{D}}. \tag{7.9}$$

On the other hand, the time evolution of the charge density ρ can be linked to the time variation of the electric field using Gauss's law,

$$\partial_t \rho = \frac{1}{4\pi}\,\nabla \cdot \partial_t \boldsymbol{E}. \tag{7.10}$$

Thus, a possible identification is

$$\boldsymbol{j}_{\mathrm{D}} = \frac{1}{4\pi}\,\partial_t \boldsymbol{E}. \tag{7.11}$$

Replacing in (7.8) we get

$$\nabla \times \boldsymbol{B} - \frac{1}{c}\,\partial_t \boldsymbol{E} = \frac{4\pi}{c}\,\boldsymbol{j}. \tag{7.12}$$

This is the modification to Ampère's law proposed by Maxwell in order to extend it to dynamical situations. It is interesting to note that, in contrast to the rest of Maxwell's equations – which were obtained directly from experiments– (7.12) was derived from theoretical arguments. It was experimentally verified indirectly in 1888, when H. Hertz showed the existence of the electromagnetic waves predicted by Maxwell.

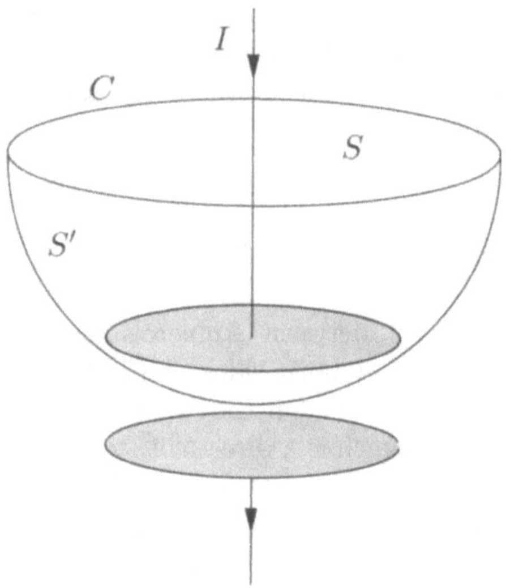

Fig. 7.2. A charging parallel-plate capacitor

We can check the need to introduce the displacement current by analyzing the following experiment. Let us consider the process by which a parallel-plate capacitor is charged (Fig. 7.2).

Ampère's law implies that the circulation of magnetic field along the curve C is proportional to the total current flowing through the surface S. Now, according to Stokes' theorem, such a relationship should be valid for any surface bounded by C, in particular for the surface S' passing between the capacitor plates without intercepting any conductor. On S', however, the flow of conduction current is zero, and Ampère's law would be violated.

The concept of the displacement current avoids this contradiction. Indeed, in the space between the plates of the capacitor there is a time-dependent electric field. According to the modification proposed by Maxwell the time variation of the field compensates the current flow though the surface S with a flow of displacement current through S'.

If we assume the diameter of the plates to be much larger than their separation, we can neglect boundary effects and write the electric field as $E = 4\pi\sigma = 4\pi Q/A$. Here σ is the surface charge density, Q is the total charge on each plate, and A is the area. In the charging process

$$\partial_t E = \frac{4\pi}{A}\,\partial_t Q = \frac{4\pi}{A}\,I. \tag{7.13}$$

According to (7.12), the circulation of $\boldsymbol{B}$ along the curve C can be written as

$$\oint \boldsymbol{B}\cdot \mathrm{d}\boldsymbol{l} = \begin{cases} \frac{4\pi}{c}\int_S \boldsymbol{j}\cdot \mathrm{d}\boldsymbol{s} \\ \\ \frac{1}{c}\int_{S'} \partial_t \boldsymbol{E}\cdot \mathrm{d}\boldsymbol{s} \end{cases} = \frac{4\pi}{c} I. \tag{7.14}$$

Indeed, the introduction of the displacement current extends the validity of Ampère's law to dynamical situations.

Returning to the general formulation of electromagnetism, (7.12) completes the group of four laws governing the dynamics of electromagnetic fields, called Maxwell's equations. These are

$$\begin{cases} \nabla \cdot \boldsymbol{E} = 4\pi\rho & \text{Gauss's law,} \\ \nabla \times \boldsymbol{B} - \frac{1}{c}\partial_t \boldsymbol{E} = \frac{4\pi}{c}\boldsymbol{j} & \text{Modified Ampère's law,} \\ \nabla \cdot \boldsymbol{B} = 0 & \text{Non-existence of monopoles,} \\ \nabla \times \boldsymbol{E} + \frac{1}{c}\partial_t \boldsymbol{B} = 0 & \text{Faraday's law.} \end{cases} \tag{7.15}$$

From here on, we analyze the properties and the solutions of these equations in some situations of particular interest. It should be pointed out that the description of systems formed by electromagnetic fields and charged particles requires Maxwell's equations to be supplemented by the laws of relativistic dynamics and by the Lorentz force (6.1) already introduced in Chap. 3.

7.3 Symmetries of Maxwell's Equations

The intuitive concept of symmetry is linked in our daily life both to natural forms and to forms created by humans. We say that a tree or a building are symmetric if, when we change our point of view, they look the same. In this section we show that the *symmetries* of Maxwell's equations and of the laws of mechanics help us reach a deeper understanding of them. In certain cases new solutions can be obtained by starting from known ones through a symmetry analysis.

We consider the symmetry properties of Maxwell's laws under geometrical transformations, such as rotations and reflections, as well as under the sign inversion of charges and of time. Each one of these transformations is linked to a property of space-time and to the properties of the physical laws. The theory of relativity presented in Chaps. 2 and 3 shows us another type of symmetry of physical laws, i.e. invariance under the change in the state of motion of the observer. In Chap. 9 we will show how Maxwell's theory satisfies the symmetries required by the theory of relativity.

In problems involving the dynamics of a system of charged particles, Maxwell's equations (7.15) should be supplemented with the equation of motion for each particle, on which the Lorentz force acts. As we have already shown, the Lorentz force acting on the particle j is

$$\boldsymbol{F}_j = q_j \left(\boldsymbol{E} + \frac{1}{c}\boldsymbol{v}_j \times \boldsymbol{B} \right), \tag{7.16}$$

and, therefore, the law of motion for that particle is

$$\frac{\mathrm{d}\boldsymbol{p}_j}{\mathrm{d}t} = \boldsymbol{F}_j . \tag{7.17}$$

The solution of these equations gives us the momentum $\boldsymbol{p}_j(t)$ and the position $\boldsymbol{r}_j(t)$ of each particle as functions of time. The system of equations (7.15), (7.16) and (7.17) summarize electromagnetism and mechanics. They therefore describe the properties of fields and the dynamics of charged particles in classical physics. In what follows we simply call them the *classical laws* and analyze their symmetry properties.

7.3.1 Rotations

Performing a rotation of the coordinate axes, we replace the coordinate system (x, y, z) by a system (x', y', z'), related to the previous one through

$$\begin{pmatrix} x' \\ y' \\ z' \end{pmatrix} = \mathcal{R} \begin{pmatrix} x \\ y \\ z \end{pmatrix} , \quad \mathcal{R} = \begin{pmatrix} a_x^x & a_y^x & a_z^x \\ a_x^y & a_y^y & a_z^y \\ a_x^z & a_y^z & a_z^z \end{pmatrix} \tag{7.18}$$

where the elements a_j^i are determined by the Euler angles of rotation of the coordinate axes. Both Maxwell's equations and the laws of mechanics are written in vector notation so that their form does not change under this rotation. These equations, therefore, are written in the same way in all reference systems differing from the original one by a rotation of the coordinate axes. A rotation of this type is usually called a *passive* rotation, because the axes but not the physical system are rotated.

As we have shown in Chap. 3, Lorentz transformations written in covariant form have the form (7.18), generalized to four dimensional space–time. The covariance of mechanics discussed in that chapter – and of electromagnetism, to be discussed in Chap. 9 – refers to transformations of the same type including space rotations in a given reference system of the type (7.18).

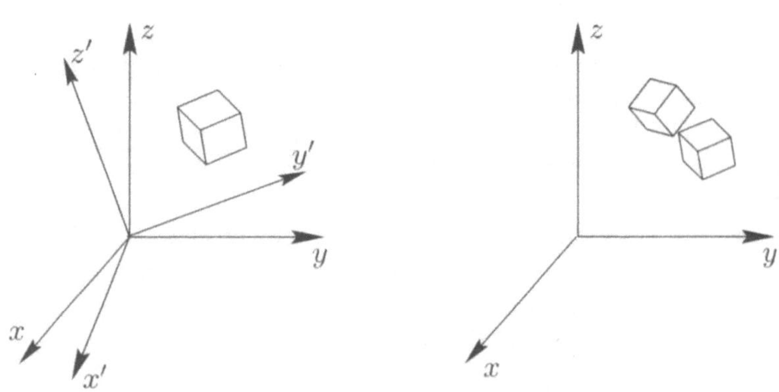

Fig. 7.3. Passive rotation and active rotation

The operation defined by (7.18) can also be interpreted as a real rotation of the physical system in space, while the coordinate axes remain fixed. If our physical system can be described by the classical laws, this real rotation also leaves the equations invariant since space is isotropic. As a consequence, if we have a solution of the classical equations given by the group of quantities

$$\{\rho(\boldsymbol{r},t),\ \boldsymbol{j}(\boldsymbol{r},t),\ \boldsymbol{E}(\boldsymbol{r},t),\ \boldsymbol{B}(\boldsymbol{r},t),\ \boldsymbol{p}_j(\boldsymbol{r},t),\ \boldsymbol{v}_j(\boldsymbol{r},t),\ \boldsymbol{r}_j(t)\}, \tag{7.19}$$

the new group of quantities obtained from the original ones by application of the rotation matrix

$$\rho'(\boldsymbol{r},t) = \rho(\boldsymbol{r},t) \tag{7.20}$$

and

$$\left\{\begin{matrix} \boldsymbol{j}'(\boldsymbol{r},t) \\ \boldsymbol{E}'(\boldsymbol{r},t) \\ \boldsymbol{B}'(\boldsymbol{r},t) \\ \boldsymbol{F}'_j(t) \\ \boldsymbol{p}'_j(t) \\ \boldsymbol{v}'_j(t) \\ \boldsymbol{r}'_j(t) \end{matrix}\right\} = \mathcal{R} \left\{\begin{matrix} \boldsymbol{j}(\boldsymbol{r},t) \\ \boldsymbol{E}(\boldsymbol{r},t) \\ \boldsymbol{B}(\boldsymbol{r},t) \\ \boldsymbol{F}_j(t) \\ \boldsymbol{p}_j(t) \\ \boldsymbol{v}_j(t) \\ \boldsymbol{r}_j(t) \end{matrix}\right\} \tag{7.21}$$

is also a solution. The quantities that depend on position must be expressed as functions of $\boldsymbol{r}'$.

Consider for example the case of N particles of charge $-q$ and mass m moving in the presence of a fixed charge of value Nq. This is a classical model for the atom, equivalent to the Kepler problem. The property we have just mentioned implies that knowing the trajectories of the N particles of mass m, all other trajectories obtained from the original one through rotation around the center where the charge Nq is placed are also solutions. In Problems 7.4 and 7.5 we invite the reader to apply the concepts of symmetry to the solution of specific problems.

7.3.2 Space Reflection

We can imagine a space-reflection operation on an arbitrary plane in the classical equations as being of active or passive character, as in the case of rotations. In an active reflection, we consider the dynamics of a real physical system which is the specular image of the original system.

We indicate by $\mathcal{R}_{\text{sp}}$ the operation of specular reflection, defined as the operation generating, at the image point, a reflected vector as indicated in Fig. 7.4. Note that **the magnetic field in the mirror is opposite to the reflected vector**. In fact, "behind the mirror" the right hand rule is valid both to determine $\boldsymbol{B}$ and vector products. The following relations summarize the effect of an active reflection on each one of the classical quantities

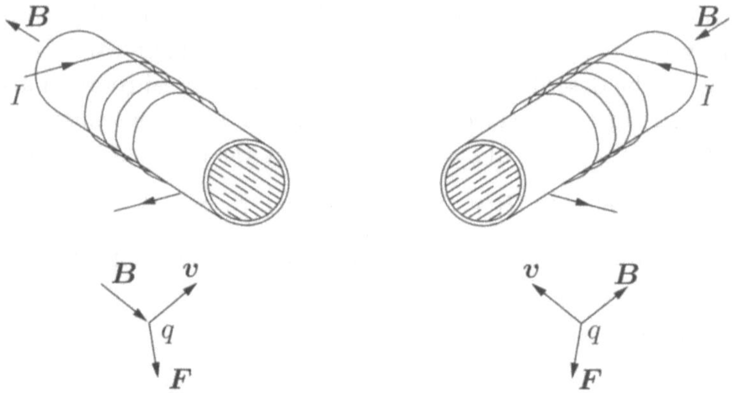

Fig. 7.4. Active reflection. The left half of the figure represents a physical system obtained from the original one through specular reflection

$$\left\{\begin{array}{c} \rho'(\boldsymbol{r},t) \\ \boldsymbol{j}'(\boldsymbol{r},t) \\ \boldsymbol{E}'(\boldsymbol{r},t) \\ \boldsymbol{B}'(\boldsymbol{r},t) \\ \boldsymbol{F}'_j(t) \\ \boldsymbol{p}'_j(t) \\ \boldsymbol{v}'_j(t) \\ \boldsymbol{r}'_j(t) \end{array}\right\} = \mathcal{R}_{\mathrm{sp}} \left\{\begin{array}{c} \rho(\boldsymbol{r},t) \\ \boldsymbol{j}(\boldsymbol{r},t) \\ \boldsymbol{E}(\boldsymbol{r},t) \\ -\boldsymbol{B}(\boldsymbol{r},t) \\ \boldsymbol{F}_j(t) \\ \boldsymbol{p}_j(t) \\ \boldsymbol{v}_j(t) \\ \boldsymbol{r}_j(t) \end{array}\right\}. \tag{7.22}$$

On the other hand, a *passive reflection* is a transformation of coordinate axes of the type $(x, y, z) \to (-x, y, z)$. Physical laws should not change under this transformation since the choice of axes is completely arbitrary. Indeed, the equations can be written in the same way in both systems of coordinates if we keep in mind that the new coordinate system is a left-handed system. Therefore:

- such vectors as the electric field, the current density, velocity and force remain invariant under this transformation. For this reason its components change in the form $(V'_x, V'_y, V'_z) = (-V_x, V_y, V_z)$. This produces the *reflected* vector. Vectors of this first type are called *polar.*
- the components of the magnetic field change as $(B'_x, B'_y, B'_z) = (B_x, -B_y, -B_z)$, implying that the magnetic field $\boldsymbol{B}$ changes to $-\boldsymbol{B}$. Vectors whose definition, as in the case of the magnetic field, depend on a convention relative to the orientation of the coordinate axes used, are called *axial vectors* or *pseudovectors.*

The following relations summarize the behavior of the basic quantities under a passive reflection

$$\left\{\begin{array}{c} \rho'(\boldsymbol{r},t) \\ \boldsymbol{j}'(\boldsymbol{r},t) \\ \boldsymbol{E}'(\boldsymbol{r},t) \\ \boldsymbol{B}'(\boldsymbol{r},t) \\ \boldsymbol{F}'_j(t) \\ \boldsymbol{p}'_j(t) \\ \boldsymbol{v}'_j(t) \\ \boldsymbol{r}'_j(t) \end{array}\right\} = \left\{\begin{array}{c} \rho(\boldsymbol{r},t) \\ \boldsymbol{j}(\boldsymbol{r},t) \\ \boldsymbol{E}(\boldsymbol{r},t) \\ -\boldsymbol{B}(\boldsymbol{r},t) \\ \boldsymbol{F}_j(t) \\ \boldsymbol{p}_j(t) \\ \boldsymbol{v}_j(t) \\ \boldsymbol{r}_j(t) \end{array}\right\}. \tag{7.23}$$

A force is a polar vector and both terms in the Lorentz force have such a characteristic. The pseudovector nature of $\boldsymbol{B}$ is compensated by the vector product, whose definition also depends on the type of coordinate system used. In Chap. 9 we will show that an axial vector is in fact a second-rank antisymmetric tensor.

The interpretation of (7.23) indicates that if a group of quantities such as that in the right-hand column is a solution of the classical equations for the original coordinates, the group in the left-hand column is a solution for the reflected coordinates.

7.3.3 Charge Inversion

Let us now study the effect on the classical laws of a change of sign of **all** intervening charges. This transformation affects the definition of all electromagnetic quantities. The following changes are induced in the system variables:

$$\left\{\begin{array}{c} \rho'(\boldsymbol{r},t) \\ \boldsymbol{j}'(\boldsymbol{r},t) \\ \boldsymbol{E}'(\boldsymbol{r},t) \\ \boldsymbol{B}'(\boldsymbol{r},t) \\ \boldsymbol{F}'_j(t) \\ \boldsymbol{p}'_j(t) \\ \boldsymbol{v}'_j(t) \\ \boldsymbol{r}'_j(t) \end{array}\right\} = \left\{\begin{array}{c} -\rho(r,t) \\ -\boldsymbol{j}(\boldsymbol{r},t) \\ -\boldsymbol{E}(\boldsymbol{r},t) \\ -\boldsymbol{B}(\boldsymbol{r},t) \\ \boldsymbol{F}_j(t) \\ \boldsymbol{p}_j(t) \\ \boldsymbol{v}_j(t) \\ \boldsymbol{r}_j(t) \end{array}\right\}. \tag{7.24}$$

A charge distribution of opposite sign to the original one produces fields which are the negatives of those originally obtained. However, as the sign of all charges has been inverted the dynamics remains unaffected. This means that knowing only the dynamics of the system we cannot decide on the sign of the charges a given system is made of.

A different situation arises when considering the charges as divided into two groups, as they appear in many situations of interest. On one hand we have the *sources*, i.e. charges that create the fields and whose dynamics is predetermined. These are, for example, the charges moving in a conductor connected to a current source, determining the external magnetic field. On the other hand, there are the *test charges*, whose magnitude is in principle much smaller and whose fields do not alter those produced by the sources. It is the motion of these particles that we want to study.

We can, in the first place, consider the sources as fixed and study the effect of inverting the sign of the test charges. In that case, we have the following relations for the transformation of the relevant quantities:

$$\left\{\begin{array}{l} \rho'(\boldsymbol{r},t) \\ \boldsymbol{j}'(\boldsymbol{r},t) \\ \boldsymbol{E}'(\boldsymbol{r},t) \\ \boldsymbol{B}'(\boldsymbol{r},t) \end{array}\right\} = \left\{\begin{array}{c} \rho(\boldsymbol{r},t) \\ \boldsymbol{j}(\boldsymbol{r},t) \\ \boldsymbol{E}(\boldsymbol{r},t) \\ \boldsymbol{B}(\boldsymbol{r},t) \end{array}\right\} \quad \text{fixed sources}$$

$$\left\{\begin{array}{l} q'_j \\ \boldsymbol{F}'_j(t) \\ \boldsymbol{p}'_j(t) \\ \boldsymbol{v}'_j(t) \\ \boldsymbol{r}'_j(t) \end{array}\right\} = \left\{\begin{array}{c} -q_j \\ -\boldsymbol{F}_j(t) \\ -\boldsymbol{p}_j(t) \\ -\boldsymbol{v}_j(t) \\ -\boldsymbol{r}_j(t) \end{array}\right\} \quad \text{test charges.} \tag{7.25}$$

In a given field configuration, the motion of charges whose sign is opposite to the original ones is equivalent to a motion *antisymmetric* to the first one. For this to happen it is necessary to assume that we invert the sense of the velocities determining the initial conditions in the equations of motion. We leave the reader to consider the effect produced by a sign inversion of the sources, while keeping the sign of the test charges fixed.

7.3.4 Time Inversion

Macroscopic systems show irreversible behavior, with a monotonous tendency towards a stationary state or towards a state of equilibrium. The laws of classical physics, however, are completely symmetric under time inversion. Macroscopic irreversibility is a consequence of the presence of a great number of particles and it can only be considered when we make a statistical description of these systems.

Let us consider the effect of the transformation $t \to -t$ on the classical laws. Since time inversion induces the inversion of the velocities defining the currents, the group of equations (7.15), (7.16) and (7.17) does not vary if, at the same time, we make the following changes:

$$\left\{\begin{array}{l} \rho' \\ \boldsymbol{j}' \\ \boldsymbol{E}' \\ \boldsymbol{B}' \\ \boldsymbol{F}'_j \\ \boldsymbol{p}'_j \\ \boldsymbol{v}'_j \\ \boldsymbol{r}'_j \end{array}\right\} \to \left\{\begin{array}{c} \rho \\ -\boldsymbol{j} \\ \boldsymbol{E} \\ -\boldsymbol{B} \\ \boldsymbol{F}_j \\ -\boldsymbol{p}_j \\ -\boldsymbol{v}_j \\ -\boldsymbol{r}_j \end{array}\right\}. \tag{7.26}$$

Then, under a time inversion, the electric field remains unchanged while, due to the inversion of the currents, the magnetic field changes sign. All velocities, and in particular those of the test particles, are inverted, but due to the change in the magnetic field, the Lorentz force is not altered. The dynamical equations imply that the test particles execute a motion antisymmetric to that of particles with the opposite charge. For this it is necessary to assume, as in the previous case, that together with the time inversion we also invert the initial velocities in the laws of motion.

In next section we study the electromagnetic potentials and introduce the concept of gauge transformation, a new type of symmetry of Maxwell's equations.

7.4 Electromagnetic Potentials and Gauge Transformations

As we have shown in Chap. 4, the introduction of the scalar potential in electrostatics is justified by the irrotational properties of the electrostatic field. Similarly, the definition of a magnetic vector potential is motivated by the solenoidal character of the field $\boldsymbol{B}$, expressed by $\nabla \cdot \boldsymbol{B} = 0$.

Is it possible to extend the description in terms of potentials to time-dependent electromagnetic phenomena? According to the discussion at the end of Chap. 3, the relevance of this question resides on one hand in the fact that such a description allows the study of the interaction of charged particles within the Lagrangian formalism. On the other hand, according to our experience with static phenomena, it is to be expected that the formulation of electromagnetism in terms of potentials is simpler than through fields.

In order to answer the question, we observe first that according to (7.15) the magnetic field shows solenoidal properties also in time-dependent situations. This implies that, even in the dynamical case, we can write

$$\boldsymbol{B} = \nabla \times \boldsymbol{A}. \tag{7.27}$$

On the other hand, in general $\nabla \times \boldsymbol{E} \neq 0$, so the usual definition of the potential ϕ cannot be maintained in dynamical situations. However, replacing $\boldsymbol{B} = \nabla \times \boldsymbol{A}$ in the Faraday law, we note that

$$\nabla \times \left(\boldsymbol{E} + \frac{1}{c}\, \partial_t \boldsymbol{A} \right) = 0. \tag{7.28}$$

This implies that the vector $(\boldsymbol{E} + c^{-1}\partial_t \boldsymbol{A})$ is irrotational in the dynamical case. Therefore, the previous equation is automatically satisfied if we choose ϕ such that

$$\boldsymbol{E} + \frac{1}{c}\, \partial_t \boldsymbol{A} = -\nabla \phi. \tag{7.29}$$

This extends the definition of the electric potential. In terms of the potentials the electromagnetic fields are thus

$$\boldsymbol{E} = -\nabla\phi - \frac{1}{c}\partial_t \boldsymbol{A},$$
$$\boldsymbol{B} = \nabla \times \boldsymbol{A}. \tag{7.30}$$

It should be noticed that in relation (7.27) $\boldsymbol{A}$ is defined to within a term of zero curl, that is to say, a term which can be written as a gradient. Indeed, adding to $\boldsymbol{A}$ the gradient of an arbitrary scalar function, we obtain

$$\boldsymbol{A}' = \boldsymbol{A} + \nabla\Lambda, \tag{7.31}$$

so that $\nabla \times \boldsymbol{A}' = \nabla \times \boldsymbol{A}$. This means that the magnetic field derived from both potentials is the same. Moreover, it can be shown that this freedom in the choice of $\nabla\Lambda$ is equivalent to arbitrarily choosing the divergence of $\boldsymbol{A}$. In fact, a vector is uniquely defined if its curl and its divergence are given over all space. The definition of the vector potential determines $\nabla \times \boldsymbol{A}$ only. As we indicated in the previous chapter, a particular choice for $\nabla \cdot \boldsymbol{A}$ is called a *gauge* choice.

For the electric field to remain invariant, gauge transformations – corresponding to the change in the choice of $\nabla \cdot \boldsymbol{A}$ – should also modify the form of ϕ. If the vector potential transforms according to (7.31), the new scalar potential must be given by

$$\phi' = \phi - \frac{1}{c}\,\partial_t\Lambda. \tag{7.32}$$

Within this extension to dynamical situations, the electromagnetic potentials can be related to the sources of the fields through Gauss's law and the modified Ampère's law. It turns out that

$$-\frac{1}{c}\,\partial_t \nabla \cdot \boldsymbol{A} - \nabla^2\phi = 4\pi\rho, \tag{7.33}$$

$$\nabla \times \nabla \times \boldsymbol{A} + \frac{1}{c^2}\,\partial^2_{tt}\boldsymbol{A} + \frac{1}{c}\,\nabla\partial_t\phi = \frac{4\pi}{c}\,j. \tag{7.34}$$

These relations can be simplified by adequately choosing the gauge, that is by taking advantage of the freedom in the choice of $\nabla \cdot \boldsymbol{A}$. The invariance of electromagnetic theory under gauge transformations is a new type of symmetry, called local symmetry, since the transformation of the system implicit in (7.31) and (7.32) is a local change in the potentials leaving the fields – and, therefore the observable quantities – unaffected.

7.4.1 Coulomb Gauge

The Coulomb gauge corresponds to

$$\nabla \cdot \boldsymbol{A} = 0. \tag{7.35}$$

As a consequence, the potentials are given by the equation

$$\begin{aligned} &\nabla^2\phi = -4\pi\rho, \\ &\nabla^2\boldsymbol{A} - \tfrac{1}{c^2}\,\partial^2_{tt}\boldsymbol{A} = -\tfrac{4\pi}{c}\,j + \tfrac{1}{c}\,\nabla\partial_t\phi. \end{aligned} \tag{7.36}$$

The scalar potential ϕ thus satisfies Poisson's equation, similarly to the electrostatic case. The difference is that now ρ depends in general on position and time. The solution in free space is

$$\phi(\boldsymbol{r},t) = \int \frac{\rho(\boldsymbol{r}',t)}{|\boldsymbol{r}-\boldsymbol{r}'|}\,\mathrm{d}^3r'. \tag{7.37}$$

In this solution, the scalar potential at a certain time is given by the charge density at the same time. Variations in ρ are instantly transformed into variations of ϕ. In the Coulomb gauge, therefore, the fact that electromagnetic disturbances propagate with finite speed is not explicit.

The equation for the vector potential is a wave equation with an inhomogeneous term. This is given by a term proportional to $\boldsymbol{j}$ and another term of zero curl. This suggests dividing the current into two contributions $\boldsymbol{j} = \boldsymbol{j}_{\mathrm{T}} + \boldsymbol{j}_{\mathrm{L}}$. Here the transverse current $\boldsymbol{j}_{\mathrm{T}}$ satisfies $\nabla\cdot\boldsymbol{j}_{\mathrm{T}} = 0$ and the longitudinal current $\boldsymbol{j}_{\mathrm{L}}$ satisfies $\nabla\times\boldsymbol{j}_{\mathrm{L}} = 0$. These two components of the current are given, respectively, by

$$\boldsymbol{j}_{\mathrm{T}} = \frac{1}{4\pi}\nabla\times\int\frac{\boldsymbol{j}}{|\boldsymbol{r}-\boldsymbol{r}'|}\,\mathrm{d}^3r', \qquad \boldsymbol{j}_{\mathrm{L}} = -\frac{1}{4\pi}\nabla\cdot\int\frac{\nabla'\cdot\boldsymbol{j}}{|\boldsymbol{r}-\boldsymbol{r}'|}\,\mathrm{d}^3r'. \tag{7.38}$$

Using the equation of continuity and the solution for the scalar potential in free space we obtain

$$\partial_t\phi = \int\frac{\partial_t\rho(\boldsymbol{r}',t)}{|\boldsymbol{r}-\boldsymbol{r}'|}\,\mathrm{d}^3r' = -\int\frac{\nabla'\cdot\boldsymbol{j}}{|\boldsymbol{r}-\boldsymbol{r}'|}\,\mathrm{d}^3r'. \tag{7.39}$$

Comparing this with the definition of $\boldsymbol{j}_{\mathrm{L}}$ we get

$$\nabla\partial_t\phi = 4\pi\boldsymbol{j}_{\mathrm{L}}. \tag{7.40}$$

The equation for the vector potential reduces to

$$\nabla^2\boldsymbol{A} - \frac{1}{c^2}\,\partial^2_{tt}\boldsymbol{A} = -\frac{4\pi}{c}\,\boldsymbol{j}_{\mathrm{T}}. \tag{7.41}$$

Thus in the Coulomb gauge the sources of $\boldsymbol{A}$ are divergenceless currents.

The Coulomb gauge is particularly useful when describing electromagnetic fields in source-free regions with boundary conditions at infinity. In this situation $\phi = 0$, and $\boldsymbol{A}$ satisfies the free-wave equation

$$\nabla^2\boldsymbol{A} - \frac{1}{c^2}\,\partial^2_{tt}\boldsymbol{A} = 0. \tag{7.42}$$

The electromagnetic fields are given by

$$\begin{aligned} \boldsymbol{E} &= -\tfrac{1}{c}\partial_t \boldsymbol{A}, \\ \boldsymbol{B} &= \nabla \times \boldsymbol{A}. \end{aligned} \tag{7.43}$$

In such a case, the gauge is usually called the *radiation or transverse gauge.*

Let us indicate finally that the gauge transformation $\boldsymbol{A}' = \boldsymbol{A} + \nabla \Lambda$ brings us to a new Coulomb gauge if Λ satisfies Laplace's equation, $\nabla^2 \Lambda = 0$.

7.4.2 Lorentz Gauge

A gauge choice allowing us to make the covariance of the equations for the potentials explicit is the so-called Lorentz gauge, corresponding to

$$\nabla \cdot \boldsymbol{A} + \frac{1}{c}\partial_t \phi = 0. \tag{7.44}$$

Note that in static situations this gauge choice is identical to the Coulomb gauge.

In the Lorentz gauge, the equations for the potentials ϕ and $\boldsymbol{A}$ are decoupled and take the form

$$\begin{aligned} \nabla^2 \phi - \frac{1}{c^2}\,\partial^2_{tt}\phi &= -4\pi\rho, \\ \nabla^2 \boldsymbol{A} - \frac{1}{c^2}\,\partial^2_{tt}\boldsymbol{A} &= -\frac{4\pi}{c}\,\boldsymbol{j}. \end{aligned} \tag{7.45}$$

Therefore the potentials satisfy totally symmetric inhomogeneous wave equations. Their solutions, to be studied in Chap. 9, give rise to perturbations propagating with the velocity of light c.

The Lorentz gauge is invariant under the transformation $\boldsymbol{A}' = \boldsymbol{A} + \nabla \Lambda$, $\phi' = \phi - c^{-1}\partial_t \Lambda$, if the function Λ satisfies the homogeneous wave equation,

$$\nabla^2 \Lambda - \frac{1}{c^2}\,\partial^2_{tt}\Lambda = 0. \tag{7.46}$$

As we indicated above, gauge transformations are a special type of symmetry of Maxwell's equations and of classical dynamics. In contrast to rotations, space inversions or time inversion, a gauge transformation is a *local transformation,* since the change produced by the function $\Lambda(\boldsymbol{r}, t)$ depends on the point $\boldsymbol{r}$. This type of transformation is closely linked to the formulation of modern theories of fundamental interactions, called *gauge theories.* In Chap. 9 we introduce some elementary ideas in this connection.

7.5 Conservation Laws

In this section we study the conservation properties of the total energy and momentum involved in Maxwell's equations for an arbitrary system of charges and currents and their fields. These conservation properties, as in the case of classical mechanics, are linked to the invariance properties of the system under translations in space and time, respectively. The conservation of angular momentum is linked to invariance under rotations.

For a system of charges and fields the conserved quantities contain two contributions. The first, of mechanical origin, corresponds to the moving charges of the system. The second refers to quantities associated with the fields. Maxwell's equations and the laws of mechanics imply their combined conservation.

We first analyze energy conservation. Since, according to the form of the magnetic component of the Lorentz force – which is proportional to $\boldsymbol{v} \times \boldsymbol{B}$ – the magnetic field does not perform any work on the moving charges, the power delivered by the fields to the charges contained in a volume V is equal to the time variation of the mechanical energy W_{m}

$$\frac{\mathrm{d}W_{\mathrm{m}}}{\mathrm{d}t} = \int_V \boldsymbol{j} \cdot \boldsymbol{E}\, \mathrm{d}^3 r. \tag{7.47}$$

Using the modified Ampère's law, we can express the current in terms of the fields and write

$$\frac{\mathrm{d}W_{\mathrm{m}}}{\mathrm{d}t} = \frac{c}{4\pi} \int_V \boldsymbol{E} \cdot (\nabla \times \boldsymbol{B})\, \mathrm{d}^3 r - \frac{1}{4\pi} \int_V \boldsymbol{E} \cdot \partial_t \boldsymbol{E}\, \mathrm{d}^3 r. \tag{7.48}$$

In order to obtain a more symmetric expression in the fields we add to the previous equation the integral of the product $\boldsymbol{B} \cdot (\nabla \times \boldsymbol{E} + c^{-1} \partial_t \boldsymbol{B})$ which, according to Faraday's law, is zero. We thus obtain

$$\begin{aligned} \frac{\mathrm{d}W_{\mathrm{m}}}{\mathrm{d}t} &= -\frac{1}{4\pi} \int_V (\boldsymbol{E} \cdot \partial_t \boldsymbol{E} + \boldsymbol{B} \cdot \partial_t \boldsymbol{B})\, \mathrm{d}^3 r \\ &\quad - \frac{c}{4\pi} \int_V \nabla \cdot (\boldsymbol{E} \times \boldsymbol{B}) \mathrm{d}^3 r, \end{aligned} \tag{7.49}$$

where we have used the vector relation $\nabla \cdot (\boldsymbol{E} \times \boldsymbol{B}) = \boldsymbol{B} \cdot (\nabla \times \boldsymbol{E}) - \boldsymbol{E} \cdot (\nabla \times \boldsymbol{B})$. Transforming now the last term into a surface integral and reordering the time derivatives, we get

$$\frac{\mathrm{d}}{\mathrm{d}t} \left[W_{\mathrm{m}} + \int_V \frac{1}{8\pi} \left(E^2 + B^2\right) \mathrm{d}^3 r \right] = - \int_S \boldsymbol{S} \cdot \mathrm{d}\boldsymbol{s}. \tag{7.50}$$

The left-hand side of this equation is the time derivative of the mechanical energy W_{m} plus an integral of the energy densities obtained in previous chapters for the static fields. If we assume that such energy densities correspond also to the dynamical situation, the second term on the left-hand side of (7.50) is nothing but the energy W_{em} associated with the electromagnetic fields in the volume V. The left-hand side of the equation thus corresponds to the time variation of the total energy. The right-hand side of (7.50) can be interpreted as the energy flow incoming into V. This term involves a vector representing the density of energy flow

$$\boldsymbol{S} = \frac{c}{4\pi}\, \boldsymbol{E} \times \boldsymbol{B}, \tag{7.51}$$

called the *Poynting vector.* In such a way, Eq. (7.50) – also called Poynting's theorem – establishes the total energy balance, mechanical plus electromagnetic, in the volume V. If the limits of this volume are extended to infinity, Poynting's theorem is equivalent to the conservation of energy. Note that in the above derivation we assumed that there is no flow of particles through the surface S. Such a flow would add an additional term to the total energy variation. In differential form, Poynting's theorem can be written as

$$\frac{1}{8\pi}\partial_t \left(E^2 + B^2\right) + \boldsymbol{j} \cdot \boldsymbol{E} + \nabla \cdot \boldsymbol{S} = 0. \tag{7.52}$$

To study the conservation of linear momentum in electromagnetism we must evaluate the force on charges and currents contained in a volume V:

$$\frac{\mathrm{d}\boldsymbol{p}_\mathrm{m}}{\mathrm{d}t} = \boldsymbol{F} = \int_V \left(\rho \boldsymbol{E} + \frac{1}{c}\boldsymbol{j} \times \boldsymbol{B}\right)\, \mathrm{d}^3 r. \tag{7.53}$$

We have used the expression for the Lorentz force acting on densities of charge ρ and of current $\boldsymbol{j}$. Here $\boldsymbol{p}_\mathrm{m}$ represents the mechanical part of the momentum.

Similarly to the case of the energy conservation law, we want to express the balance of linear momentum in terms of a quantity representing the momentum associated with the field. Making use of Maxwell's equations to express the sources in terms of $\boldsymbol{E}$ and $\boldsymbol{B}$, we obtain

$$\frac{\mathrm{d}\boldsymbol{p}_\mathrm{m}}{\mathrm{d}t} = \frac{1}{4\pi}\int_V \left[\boldsymbol{E}\, \nabla \cdot \boldsymbol{E} + \left(\nabla \times \boldsymbol{B} - \frac{1}{c}\partial_t \boldsymbol{E}\right) \times \boldsymbol{B}\right] \mathrm{d}^3 r. \tag{7.54}$$

Symmetrizing this expression by adding a null contribution, we can write

$$\begin{aligned} \frac{\mathrm{d}\boldsymbol{p}_\mathrm{m}}{\mathrm{d}t} = {} & \frac{1}{4\pi}\int_V \left[\boldsymbol{E}\nabla \cdot \boldsymbol{E} + \boldsymbol{B}\,\nabla \cdot \boldsymbol{B} + (\nabla \times \boldsymbol{E}) \times \boldsymbol{E} + (\nabla \times \boldsymbol{B}) \times \boldsymbol{B}\right] \mathrm{d}^3 r \\ & - \frac{1}{c^2}\frac{\mathrm{d}}{\mathrm{d}t}\int_V \boldsymbol{S}\, \mathrm{d}^3 r. \end{aligned} \tag{7.55}$$

The first term on the right-hand side is the integral of the divergence of the *Maxwell stress tensor* whose components are

$$\mathcal{T}_{ij} = \frac{1}{8\pi}\, \left[2(E_i E_j + B_i B_j) - \delta_{ij}(E^2 + B^2)\right], \tag{7.56}$$

with $i, j \equiv x, y, z$. The integral can be transformed into a surface integral. Indeed, carrying out the corresponding calculations, the equation for the momentum variation can be rewritten as

$$\frac{\mathrm{d}}{\mathrm{d}t}\left(\boldsymbol{p}_\mathrm{m} + \frac{1}{c^2}\int_V \boldsymbol{S}\, \mathrm{d}^3 r\right) = \int_S \mathcal{T} \cdot\, \mathrm{d}\boldsymbol{s}. \tag{7.57}$$

Here S is the surface enclosing the volume V, and the term in the surface integral is the matrix product of the stress tensor times the vector element of area.

Equation (7.57), like (7.50), can also be interpreted as a balance law. Indeed, on the left-hand side we have the time variation of the mechanical momentum plus the integral of the Poynting vector over V. This vector, multiplied by c^{-2}, can be identified with the density of the electromagnetic momentum. Its integral corresponds to the total linear momentum of the electromagnetic fields in V. The right-hand side represents the *flux of momentum* entering the volume V. It should be understood as the total force acting on the volume, due to the fields outside V, which is applied through the surface S.

The situation is analogous to that found in the theory of continuous media, in which a stress tensor is defined. This allows the calculation of the force on a given portion of the material through an expression similar to (7.57). This aspect of the theory contributed historically to speculations about the existence of a continuous material medium – the ether – whose state of tension determined the propagation of electromagnetic perturbations, as discussed in Chap. 2.

In the light of Maxwell's theory, which formalizes the concepts of electromagnetic fields, the possibility of defining the stress tensor $\mathcal{T}$ reaffirms the fact that the forces between charges or currents are not due to "action at a distance," but to the existence of physical entities – the fields – filling all space around the sources.

In order to show how the stress tensor links the fields to the forces exerted on charges and currents, it is useful to consider some particular examples. We analyze here an electrostatic situation, leaving to the reader the case where currents are involved (Problem 7.8). Consider two charges, $-q$ and q, located at the points $(d, 0, 0)$ and $(-d, 0, 0)$, respectively (Fig. 7.5).

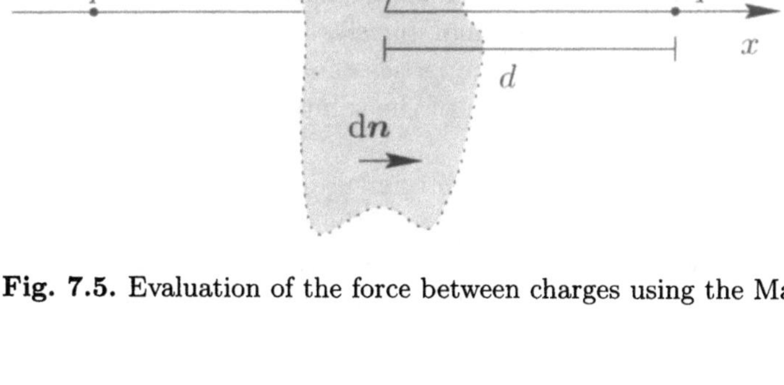

Fig. 7.5. Evaluation of the force between charges using the Maxwell stress tensor

The electrostatic field produced by these charges on the plane yz has only the component in the x direction:

$$\boldsymbol{E}(\boldsymbol{r}) = \frac{2qd}{(d^2+r^2)^{3/2}}\,\hat{\boldsymbol{x}}. \tag{7.58}$$

According to (7.56) the Maxwell stress tensor on that plane is a diagonal matrix given by

$$\mathcal{T} = \frac{1}{8\pi}\begin{pmatrix} E^2 & 0 & 0 \\ 0 & -E^2 & 0 \\ 0 & 0 & -E^2 \end{pmatrix}. \tag{7.59}$$

According to (7.57) the product of this tensor times the surface element associated with the plane yz, $\mathrm{d}\boldsymbol{s} = (\mathrm{d}s, 0, 0)$ integrated over the entire plane should give the total force applied on those charges in the half-space $x < 0$, i.e. on the charge q. Indeed, $\mathcal{T}\cdot\mathrm{d}\boldsymbol{s} = E^2\mathrm{d}s/8\pi$, and

$$\int \mathcal{T}\cdot\mathrm{d}\boldsymbol{s} = \frac{1}{8\pi}\int_0^{2\pi}\mathrm{d}\varphi\int_0^{\infty} r\,\mathrm{d}r\frac{4q^2d^2}{(d^2+r^2)^3}\hat{\boldsymbol{x}} = \frac{q^2}{4d^2}\hat{\boldsymbol{x}}. \tag{7.60}$$

This is precisely the Coulomb force due to the other charge acting on the positive charge. The force on the negative charge of the same magnitude and opposite sign can be obtained by applying the stress tensor to the element of outgoing area of the surface enclosing $-q$, that is to say, $(-\mathrm{d}s, 0, 0)$.

7.6 Symmetries in Physics

The concept of the symmetry of physical laws has played an important role in the development of modern physics. This is mainly due to the fact that quantum mechanics allows for the symmetries of the solutions to a given problem to be classified according to the rules of group theory. However, even in classical mechanics the classification of the vibration modes of a system of particles can be systematized by using the symmetry properties of the physical system. This allows us to represent the set of symmetry operations by means of matrices of different dimensions whose algebra is a representation of the algebra of the operations. Molecular physics and solid state physics make intensive use of the symmetry properties of the dynamical systems under study. This permits a classification of the solutions and can guide the search for new solutions.

In high-energy physics the symmetry properties of elementary particles are intensively used. Apart from relativity, one of the first symmetry concepts that appeared in high-energy physics corresponds to the interpretation of the proton and the neutron as being two manifestations of a single entity under rotations in an abstract *isospin* space. This type of property was exploited later on with more complex symmetries. In this way, in the 1960s, it was postulated that particles subject to strong interactions should be put in correspondence with the representations of the so called SU(3) symmetry group. An important success of this theory was the prediction of the existence of particles until then unknown, such as the resonance Ω^-.

The smallest representation of the group SU(3) corresponds to three particles of spin 1/2 and electric charge $(2/3, -1/3, -1/3)$. Even though their existence was doubtful, these particles were given a name and were called *quarks*. The Standard Model of elementary particles has embodied this classification and has completed it with new global and gauge (or local) symmetries. In this connection we recommend reading *The Concept of Symmetry in Physics* by C. Hill [7.1], and the paper by D.J. Gross, *Symmetry in Physics: Wigner's Legacy* [7.2].

Problems

7.1 In a charge-free region there is a homogeneous magnetic field which varies linearly with time. Find the induced electric field.

7.2 A potential difference $V(t) = V_0 \sin \omega t$ is applied between the plates of a plane circular capacitor. Find iteratively the electric and magnetic fields and the Poynting vector near the axis of the capacitor in the small-frequency limit. Discuss the direction of the Poynting vector. Where does the associated energy come from?

Hint: See the solution to Problem 7.9. At a distance r from the capacitor axis, the solution should be an expansion in powers of $r\omega/c$.

7.3 In the previous problem consider the effect produced in the field distribution by a metallic sheet closing the capacitor.

7.4 Consider six positive identical electric charges q placed at the vertices of a regular hexagon, and a seventh charge of value q' at the center. Use symmetry arguments to prove that the net force acting on the central charge is zero.

Hint: Under which symmetry transformations does the system remain invariant?

7.5 In the previous problem, assume that one of the charges at the vertices of the hexagon is replaced by another charge q''. Using symmetry arguments **only**, calculate the value of the net force on the central charge.

7.6 Consider a very long straight conducting wire, of resistance ρ per unit length, through which a current circulates when a potential difference is applied to its ends. Indicate the directions of the vectors $\boldsymbol{E}$ and $\boldsymbol{B}$ and of the Poynting vector. What can be concluded about the energy flow? Discuss the result in connection with Problem 7.2.

Hint: What are the energy sources in the system?

7.7 An electric field forms an angle α with a surface of differential area. (a) Use the Maxwell stress tensor to find the element of force on the surface in the absence of magnetic fields. (b) Using the previous result in the limits $\alpha = 0$ and $\alpha = \pi/2$, calculate the force between two point charges of the same absolute value.

7.8 Using the stress tensor, calculate the force between two parallel current lines carrying both the same current intensity.

Hint: Consider both directions for the currents.

7.9 Calculate the fields $\boldsymbol{E}$ and $\boldsymbol{B}$ in the process of charging the cylindrical capacitor represented in Fig. (7.6).

Solution: The solution to this problem is simplified considerably by neglecting boundary effects and assuming that near the symmetry axis the electric field is perpendicular to the plates, i.e. $\boldsymbol{E} = E_z \hat{\boldsymbol{z}}$. Another simplifying assumption is to take the lines of magnetic field circulating around those of $\boldsymbol{E}$, i.e. $\boldsymbol{B} = B_\varphi \hat{\boldsymbol{\varphi}}$.

Given the symmetry of the problem, we assume that E_z and B_φ depend only on the distance r to the axis of the capacitor. With such assumptions, the Faraday law and the modified Ampère law reduce to

$$\partial_r E_z = \frac{1}{c}\partial_t B_\varphi,$$

and

$$\partial_r (r B_\varphi) = \frac{r}{c}\partial_t E_z,$$

respectively. A particular solution to these equations is $E_z = E_0$, $B_\varphi = 0$ where E_0 is a constant. This solution corresponds to a static situation which we can associate with a charged capacitor. To describe the charging process, we take for the electric field a solution of the form

$$E_z = E_0 - E_1 \exp(-t/\tau).$$

For $t \to \infty$, this form of E_z tends to the constant value E_0. Our assumption for E_z implies, from Ampère's law,

$$B_\varphi = \frac{r}{2c\tau}\, E_1 \exp(-t/\tau).$$

According to the Faraday law this result requires E_z to depend on r. Integrating this equation, we can improve the expression for E_z:

$$E_z = E_0 - E_1 \exp(-t/\tau) - \frac{r^2}{4c^2\tau^2}\, E_1 \exp(-t/\tau).$$

This new form for E_z implies in turn that

$$B_\varphi = \frac{r}{2c\tau}\, E_1 \exp(-t/\tau) + \frac{r^3}{16c^3\tau^3} E_1 \exp(-t/\tau).$$

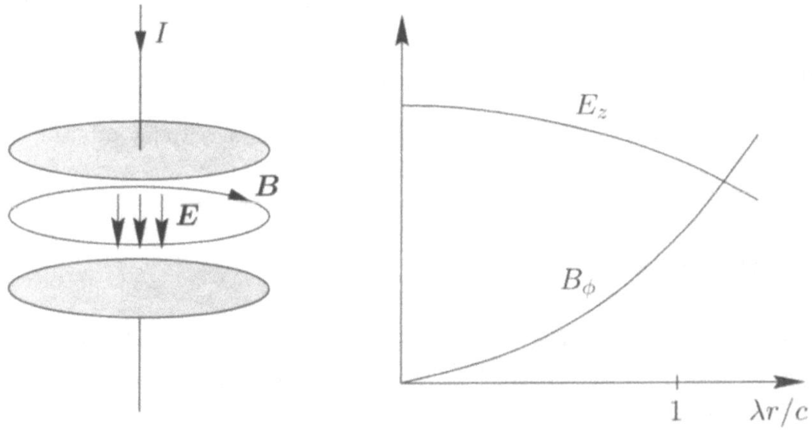

Fig. 7.6. Electromagnetic fields during the charging of a capacitor

This process can be iterated to obtain better approximations to the solutions to Faraday's and Ampère's laws. These are series expansions in powers of $r/c\tau$. Truncation at a finite order is a good description of the fields for $r \ll c\tau$.

Successive approximations obtained for E_z and B_φ suggest that the complete solution to the problem has the form

$$E_z = E_0 - \alpha(r)\exp(-t/\tau),$$

$$B_\varphi = \frac{1}{\tau}\beta(r)\exp(-t/\tau).$$

Here $\alpha(r)$ and $\beta(r)$ are functions which according to Faraday's and Ampère's laws satisfy

$$\alpha' = \beta/c\tau^2,$$

$$(r\beta)' = r\alpha/c,$$

where the primes indicate derivation with respect to r. The solutions to these equations imply

$$E_z = E_0 - E_1\ I_0(r/c\tau)\exp(-t/\tau),$$

$$B_\varphi = E_1\ I_1(r/c\tau)\exp(-t/\tau),$$

where E_1 is an arbitrary constant and $I_n(x)$ are the modified Bessel functions. These solutions are schematically plotted in Fig. 7.6. It should be kept in mind that these solutions do not take account of boundary conditions at the edges, and so they cannot be extended to arbitrarily large values of r.

References

7.1 C. Hill: *The Concept of Symmetry in Physics* (Fermi National Accelerator Laboratory, 1986).
7.2 D.J. Gross: *Symmetry in Physics: Wigner's Legacy*, Physics Today **48** (12), 46 (1995).

Further Reading

R.P. Feynman, R.B. Leighton and M. Sands: *The Feynman Lectures on Physics, Vol. II* (Addison–Wesley, New York, 1964) Cap. 18.
H. Goldstein: *Classical Mechanics* (Addison–Wesley, New York, 1980) Chap. 9.
J.D. Jackson: *Classical Electrodynamics* (Wiley, New York, 1975).
L.D. Landau and E.M. Lifshitz: *The Classical Theory of Fields* (Pergamon, Oxford and Addison–Wesley, Reading, 1971).
W.K.H. Panofsky and M. Phillips: *Classical Electricity and Magnetism* (Addison–Wesley, Massachusetts, 1955) Chaps. 9 and 10.

8. Dynamic Fields and Radiation

In order to complete the discussion of electromagnetic phenomena in vacuum, in this chapter we carry out a detailed analysis of the fields produced by moving charges. In particular we study radiation fields which are specially important in many of the applications. These include physical optics, which is formulated in terms of such fields, and telecommunications technology, which is based on the transmission, reception and processing of electromagnetic radiation. Other important applications, such as lasers and synchrotron radiation are also based on concepts to be developed in this chapter. Finally, at a more basic level, the classical theory of radiation forms the basis for the quantum description of atomic and nuclear emission and absorption processes.

We have shown in the previous chapter that in the Lorentz gauge the two electromagnetic potentials satisfy similar inhomogeneous wave equations. The inhomogeneity is directly related to the field sources:

$$\begin{aligned} \nabla^2 \phi - \frac{1}{c^2}\, \partial^2_{tt} \phi &= -4\pi \rho(\boldsymbol{r}, t), \\ \nabla^2 \boldsymbol{A} - \frac{1}{c^2}\, \partial^2_{tt} \boldsymbol{A} &= -\frac{4\pi}{c}\, \boldsymbol{j}(\boldsymbol{r}, t). \end{aligned} \tag{8.1}$$

These equations govern the dynamics of the potentials and, therefore, of electromagnetic fields in general situations. As a consequence, it is of interest to solve them and to analyze their solutions in relevant cases.

The hyperbolic character of the wave equations implies that the potentials generated by the sources describe perturbations propagating with a constant velocity c. Indeed, we recall that the one dimensional homogeneous wave equation

$$\partial^2_{xx} \psi - \frac{1}{v^2}\, \partial^2_{tt} \psi = 0 \tag{8.2}$$

has solutions

$$\psi(x,t) = f(x - vt), \quad \psi(x,t) = g(x + vt), \tag{8.3}$$

where $f(x)$ and $g(x)$ are arbitrary functions. Due to the linearity of the equation, the general solution is a combination of these two functions, where the coefficients and the particular form of f and g are determined by the initial conditions. The function $f(x - vt)$ represents a *retarded* perturbation moving with velocity of modulus $|v|$ in the positive x direction. In fact, the value of the function at a point x_0 at $t = 0$ is the same as that at time $t = t_1$ at the point $x_0 + vt_1$. Similarly, $g(x + vt)$ is an *advanced* perturbation moving with velocity $-|v|$, towards $x = -\infty$.

For the potentials satisfying the wave equations (8.1) we can expect a similar behavior in source-free regions, where the inhomogeneities vanish. We shall show that the specific form of the corresponding perturbation is determined by the space dependence of the charge and current densities. In the next sections we study the solutions of the wave equations in the absence of sources. Then, we find the general solution of (8.1) using the Green's function formalism, already introduced in Chap. 5 when studying the solutions of the Poisson equation. Finally, we analyze the dynamical electromagnetic fields in several situations of interest.

8.1 Wave Propagation in Free Space

In source-free regions, where $\rho(\boldsymbol{r}, t) = 0$ and $\boldsymbol{j}(\boldsymbol{r}, t) = 0$, equations (8.1) become homogeneous wave equations for the potentials ϕ and $\boldsymbol{A}$. From their solutions we establish some of the basic properties of electromagnetic fields in vacuum, to be found again in specific situations.

The homogeneous wave equations for the potentials admit *harmonic* solutions of the form

$$\phi(\boldsymbol{r}, t) = \phi_{\boldsymbol{k}} \exp[\mathrm{i}(\boldsymbol{k} \cdot \boldsymbol{r} - \omega t)], \qquad \boldsymbol{A}(\boldsymbol{r}, t) = \boldsymbol{A}_{\boldsymbol{k}} \exp[\mathrm{i}(\boldsymbol{k} \cdot \boldsymbol{r} - \omega t)]. \quad (8.4)$$

For mathematical convenience, we have here introduced complex solutions. It should be kept however in mind that the quantities of physical interest are the real parts of the complex quantities. In the solutions (8.4) the *wave vector* $\boldsymbol{k}$, the *frequency* ω and the complex amplitudes $\phi_{\boldsymbol{k}}$ and $\boldsymbol{A}_{\boldsymbol{k}}$ of the wave are constant, independent of space and time. As we verify at once, the wave equations impose relationships between these parameters. In particular, the frequency turns out to be a well defined function $\omega(\boldsymbol{k})$ of the wave vector. In this section, we limit ourselves to consider the case where both $\boldsymbol{k}$ and ω are real numbers, so that the potentials, and therefore the electromagnetic fields, are oscillatory quantities. In later chapters we shall show that solutions with a complex wave vector implying exponential decays are of interest in the presence of material media.

Due to the linearity of the homogeneous wave equations, a superposition principle holds allowing us to build new solutions by means of suitable linear combinations of functions of the form (8.4). In addition, since these solutions form a complete set of functions as we vary the wave vector $\boldsymbol{k}$, any solution to the homogeneous equation can be written as

$$\phi(\boldsymbol{r},t) = \int \phi_{\boldsymbol{k}} \exp[\mathrm{i}(\boldsymbol{k}\cdot\boldsymbol{r} - \omega t)]\, \mathrm{d}^3 k,$$
$$\boldsymbol{A}(\boldsymbol{r},t) = \int \boldsymbol{A}_{\boldsymbol{k}} \exp[\mathrm{i}(\boldsymbol{k}\cdot\boldsymbol{r} - \omega t)]\, \mathrm{d}^3 k. \tag{8.5}$$

The amplitudes $\phi_{\boldsymbol{k}}$ and $\boldsymbol{A}_{\boldsymbol{k}}$ are determined by the initial conditions and by the boundary conditions.

Taking advantage of the superposition principle we center our subsequent discussion on the solutions given in (8.4), which contain a single wave vector. Such solutions are called *plane waves*, since the wavefronts – where the *phase* $\boldsymbol{k}\cdot\boldsymbol{r} - \omega t$ is constant – are planes perpendicular to the vector $\boldsymbol{k}$, i.e. to the propagation direction of the wave. In this case, the wave fronts are also called *phase planes.* Since the frequency of these waves is well-defined, they are also called *monochromatic.*

The wave equation determines relations between $\boldsymbol{k}$, ω, $\phi_{\boldsymbol{k}}$ and $\boldsymbol{A}_{\boldsymbol{k}}$. In fact (8.1) imposes

$$\left(k^2 - \frac{\omega^2}{c^2}\right)\phi_{\boldsymbol{k}} = 0, \tag{8.6}$$

$$\left(k^2 - \frac{\omega^2}{c^2}\right)\boldsymbol{A}_{\boldsymbol{k}} = 0, \tag{8.7}$$

where $k = |\boldsymbol{k}|$ is the *the wave number.* In addition, the electromagnetic potentials must also satisfy the Lorentz condition, $\nabla\cdot\boldsymbol{A} + c^{-1}\partial_t\phi = 0$, which provides the additional relation

$$\boldsymbol{k}\cdot\boldsymbol{A}_{\boldsymbol{k}} - \frac{\omega}{c}\phi_{\boldsymbol{k}} = 0. \tag{8.8}$$

Equations (8.6) and (8.7) have nontrivial solutions for $\phi_{\boldsymbol{k}}$ and $\boldsymbol{A}_{\boldsymbol{k}}$ only if the condition

$$\omega^2 = c^2 k^2 \tag{8.9}$$

is satisfied. This is called the *dispersion relation* and links the frequency of the wave to the wave number k. When this relation is fixed, (8.6) to (8.8) do not have a unique solution. In fact, the absence of boundary conditions for plane waves in free space implies the existence of infinitely many solutions for the potentials. In view of this freedom it is customary to fix the additional condition $\phi_{\boldsymbol{k}} = 0$, defining the radiation gauge. As we have already mentioned in Chap. 7, this gauge is also a special case of the Coulomb gauge. According to (8.8), this choice implies $\boldsymbol{k}\cdot\boldsymbol{A}_{\boldsymbol{k}} = 0$. In the radiation gauge the scalar potential of a plane wave is thus zero, whereas the vector potential is perpendicular to the direction of propagation of the wave. The modulus of the vector $\boldsymbol{A}_{\boldsymbol{k}}$ is arbitrary and defines the amplitude of the fields associated with the wave.

Starting from the vector potential of the plane wave, we obtain the corresponding electromagnetic fields. In the radiation gauge $\boldsymbol{E} = -c^{-1}\partial_t\boldsymbol{A}$ and $\boldsymbol{B} = \nabla\times\boldsymbol{A}$. Explicitly, we have

$$\boldsymbol{E}(\boldsymbol{r},t) = \mathrm{i}k\boldsymbol{A}_{\boldsymbol{k}} \exp[\mathrm{i}(\boldsymbol{k}\cdot\boldsymbol{r} - \omega t)],$$
$$\boldsymbol{B}(\boldsymbol{r},t) = \mathrm{i}\boldsymbol{k} \times \boldsymbol{A}_{\boldsymbol{k}} \exp[\mathrm{i}(\boldsymbol{k}\cdot\boldsymbol{r} - \omega t)]. \tag{8.10}$$

Like the potentials, these quantities are complex, so that the physical fields are the corresponding real parts.

We notice from (8.10) that the electric and magnetic fields are perpendicular to the direction of wave propagation. This general property of the electromagnetic waves in vacuum is a consequence of the relation $\nabla \cdot \boldsymbol{E} = 0$ and $\nabla \cdot \boldsymbol{B} = 0$. It characterizes the waves as *transverse waves*. Moreover, the electric field is perpendicular to the magnetic vector:

$$\boldsymbol{B} = \hat{\boldsymbol{k}} \times \boldsymbol{E}. \tag{8.11}$$

The two fields and the wave vector form thus an orthogonal system. We also notice that both fields have the same modulus, $|\boldsymbol{E}| = |\boldsymbol{B}|$. It is worth emphasizing that these properties of the fields associated with plane waves are independent of the gauge choice used to fix the potentials, since they are a direct consequence of Maxwell equations. We encourage the reader to verify these relations explicitly starting from Maxwell's equations. Later on we shall rederive these relations in more general situations.

At this point it is appropriate to calculate the real physical fields starting from the complex quantities given in (8.10). To do this we first recall that the constant vector $\boldsymbol{A}_{\boldsymbol{k}}$ is in general a complex quantity that we write, in Cartesian coordinates, $\boldsymbol{A}_{\boldsymbol{k}} = (a_x \exp(\mathrm{i}\psi_x), a_y \exp(\mathrm{i}\psi_y), 0)$. Here we have conventionally taken the z axis in the direction of propagation of the wave, $\boldsymbol{k} = k\hat{\boldsymbol{z}}$, and we have expressed each complex Cartesian component of $\boldsymbol{A}_{\boldsymbol{k}}$ in polar form. Starting from this form for the amplitude of the complex vector potential, we can write the Cartesian components of the real vector potential in the plane perpendicular to the direction of propagation of the wave as

$$A_x = a_x \cos(kz - \omega t + \psi_x), \qquad A_y = a_y \cos(kz - \omega t + \psi_y). \tag{8.12}$$

For the real electric field we have

$$E_x = ka_x \cos(kz - \omega t + \psi_x + \pi/2),$$
$$E_y = ka_y \cos(kz - \omega t + \psi_y + \pi/2). \tag{8.13}$$

Similarly, for the magnetic field, we get

$$B_x = -ka_y \cos(kz - \omega t + \psi_y + \pi/2),$$
$$B_y = ka_x \cos(kz - \omega t + \psi_x + \pi/2). \tag{8.14}$$

The electric and magnetic fields thus oscillate in phase and keep their mutual orthogonality. In turn, both are out of phase by $\pi/2$ with respect to the vector potential.

It is interesting to analyze the relative phase in the oscillations of each of the field components determining the polarization state of the wave. For both the electric and magnetic fields the phase between the oscillations of their components is given by the difference $\delta = \psi_x - \psi_y$. When $\delta = 0$ or $\delta = \pi$, the vectors $\boldsymbol{E}$ and $\boldsymbol{B}$ oscillate in the plane (x, y) along straight lines whose slopes are determined by the ratio a_y/a_x. As the wave propagates in space, both fields oscillate therefore in two well-defined planes perpendicular to each other. In this situation, we say that the wave is *linearly* polarized. The components of each field oscillate *in phase* if $\delta = 0$ and in *counterphase* if $\delta = \pi$.

For other values of δ both $\boldsymbol{E}$ and $\boldsymbol{B}$ describe ellipses on the (x, y) plane, in which case the waves are called *elliptically polarized.* In this situation, as the wave propagates the fields rotate around the direction of propagation, while their moduli vary harmonically. In the particular case when $\delta = \pi/2$ or $\delta = 3\pi/2$ and $a_x = a_y$ the waves are *circularly* polarized. When $\delta = \pi/2$ we have right circular polarization, and for $\delta = -\pi/2$ we have left circular polarization. The possible polarization states of the electromagnetic fields obtained as the dephasing δ varies have been schematized in Fig. 8.1.

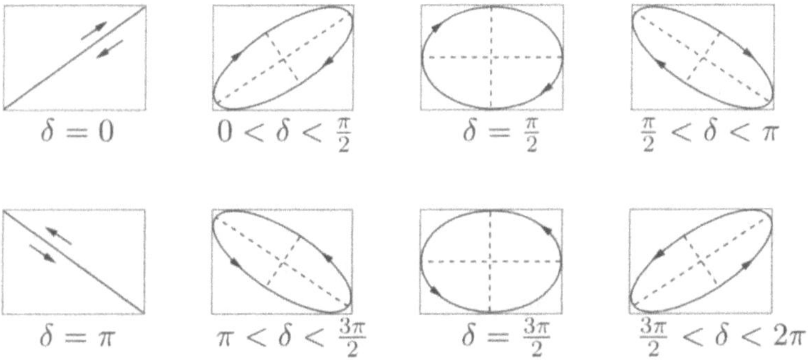

Fig. 8.1. Polarization states of a plane wave

It is important to notice that an arbitrary polarization state can always be obtained as a linear superposition of two waves with independent polarizations and conveniently selected amplitudes. For example, it is possible to choose as a basis for describing an arbitrary polarization state two waves linearly polarized in perpendicular directions. A pair of right and left circularly polarized waves can also be used.

To conclude the discussion of the properties of plane waves it is worthwhile to evaluate the associated Poynting vector, namely, the energy flow transported by these waves. Recalling that $\boldsymbol{S} = c\boldsymbol{E} \times \boldsymbol{B}/4\pi$ we obtain

$$\boldsymbol{S} = \frac{ck^2}{4\pi} \Big[a_x^2 \cos^2(\boldsymbol{k}\cdot\boldsymbol{r} - \omega t + \psi_x + \pi/2) + a_y^2 \cos^2(\boldsymbol{k}\cdot\boldsymbol{r} - \omega t + \psi_y + \pi/2)\Big]\hat{\boldsymbol{k}}. \quad (8.15)$$

The Poynting vector must be evaluated using the real physical fields since, being a nonlinear function of the fields, it cannot be obtained directly as the real part of the product of the complex fields. We notice in (8.15) that, as was to be intuitively expected, the energy flow associated with a plane wave has the same direction of propagation as the wave. Also, it oscillates with twice the frequency of the fields. Since the period of these oscillations is in general much smaller than the characteristic observation times, it is useful to take the time average of $\boldsymbol{S}$ over one period. The result is

$$\langle \boldsymbol{S} \rangle = \frac{ck^2}{8\pi}(a_x^2 + a_y^2)\,\hat{\boldsymbol{k}}. \tag{8.16}$$

Similarly it is possible to calculate the energy density, $w = (E^2 + B^2)/8\pi$, whose time average is

$$\langle w \rangle = \frac{k^2}{8\pi}(a_x^2 + a_y^2). \tag{8.17}$$

The *velocity of energy flow* is thus

$$\boldsymbol{v} = \frac{\langle \boldsymbol{S} \rangle}{\langle w \rangle} = c\hat{\boldsymbol{k}}. \tag{8.18}$$

For a monochromatic plane wave the velocity of energy propagation coincides with the phase velocity c.

Plane waves are not the only complete set of solutions to the wave equations, although their detailed study is particularly justified in view of their applications. Indeed, sufficiently far from the sources radiation fields can always be approximated by superpositions of plane waves. Another complete group of solutions to the wave equations of importance in some applications are the spherical waves, given by

$$\begin{aligned} \phi(\boldsymbol{r}, t) &= \phi_k\, r^{-1} \exp[\mathrm{i}(kr - \omega t)], \\ \boldsymbol{A}(\boldsymbol{r}, t) &= \boldsymbol{A}_k\, r^{-1} \exp[\mathrm{i}(kr - \omega t)]. \end{aligned} \tag{8.19}$$

Solutions with spherical symmetry are characterized by a position-dependent wave vector, $\boldsymbol{k} = k\hat{\boldsymbol{r}}$. The wavefronts are in this case radially propagating spherical surfaces. Far from the origin, spherical waves resemble plane waves when observed within a small solid angle. In these conditions plane waves are sometimes used as an approximation to describe spherical waves. We leave it as an exercise for the reader to study the properties of the fields associated with spherical waves, similarly to what we did above for plane waves.

In the laboratory, many experiments involve the use of collimated beams such as those obtained by passing light through a diaphragm or as those produced by a laser. In contrast to plane or spherical waves, which describe electromagnetic perturbations whose energy is distributed over all space, in a collimated beam electromagnetic fields are appreciably different from zero in a narrow region along the axis of the beam only. The energy density is also concentrated in that region. Next we show how, due to interference effects, the superposition of plane waves can be used to describe collimated beams.

Consider the combination of two plane waves with wave vectors $\boldsymbol{k}^+ = \boldsymbol{k} + \Delta\boldsymbol{k}$ and $\boldsymbol{k}^- = \boldsymbol{k} - \Delta\boldsymbol{k}$, with $\Delta\boldsymbol{k}\cdot\boldsymbol{k} = 0$ and such that $\Delta k \ll k$. The frequencies of these waves are $\omega^+ = \omega^- = ck\sqrt{1+(\Delta k/k)^2} \approx ck[1+(\Delta k/k)^2/2]$. They differ from the value $\omega = ck$ by a quantity of second order in Δk. We can write the complex electric fields associated with the two waves as

$$\boldsymbol{E}^+(\boldsymbol{r},t) = \boldsymbol{E}_0 \exp[\mathrm{i}(\boldsymbol{k}\cdot\boldsymbol{r} - \omega t)] \exp(\mathrm{i}\Delta\boldsymbol{k}\cdot\boldsymbol{r}) \tag{8.20}$$

$$\boldsymbol{E}^-(\boldsymbol{r},t) = \boldsymbol{E}_0 \exp[\mathrm{i}(\boldsymbol{k}\cdot\boldsymbol{r} - \omega t)] \exp(-\mathrm{i}\Delta\boldsymbol{k}\cdot\boldsymbol{r}), \tag{8.21}$$

where $\boldsymbol{E}_0$ is the amplitude. The sum of both fields with equal weights is

$$\boldsymbol{E}^+(\boldsymbol{r},t) + \boldsymbol{E}^-(\boldsymbol{r},t) = 2\boldsymbol{E}_0 \exp[\mathrm{i}(\boldsymbol{k}\cdot\boldsymbol{r} - \omega t)] \cos(\Delta\boldsymbol{k}\cdot\boldsymbol{r}). \tag{8.22}$$

This represents a plane wave whose amplitude, equal to twice that of each of the original waves, is modulated on the phase plane by the function $\cos(\Delta\boldsymbol{k}\cdot\boldsymbol{r})$. This modulation implies a concentration of the wave energy in regions of width Δk^{-1}. To achieve higher collimation it is necessary in general to superimpose an infinite number of plane waves with appropriate coefficients.

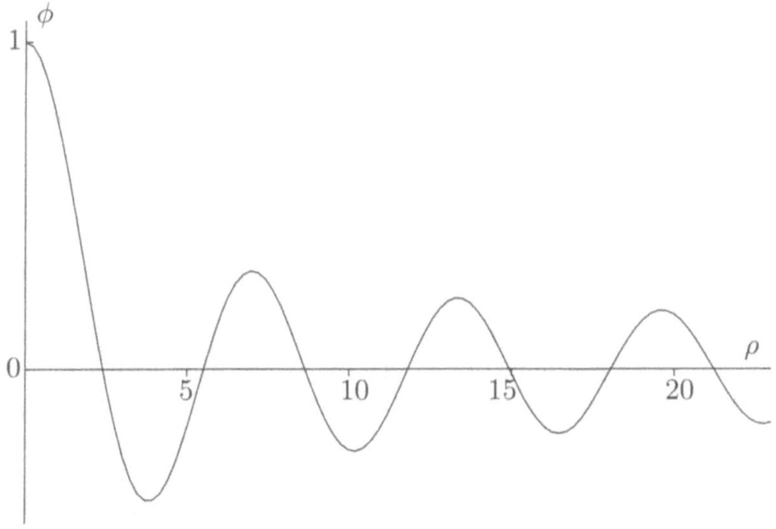

Fig. 8.2. Transverse profile of the electric potential in the collimated light beam described in the text

Due to *diffraction* effects collimated beams generally spread out as they propagate, so that collimation is more imperfect as the wave moves away from the diaphragm. One possibility to overcome this difficulty is to build in the laboratory a wave describing a collimated beam which corresponds to a particular solution of the wave equation with cylindrical symmetry around the propagation direction. It can be shown that one such solution is

$$\phi(\boldsymbol{r},t) \propto \exp[\mathrm{i}(\beta z - \omega t)]\, J_0(\alpha\rho), \tag{8.23}$$

where J_0 is a Bessel function, $\alpha^2 + \beta^2 = \omega^2/c^2$ and ρ is the polar coordinate transverse to the beam direction of propagation (Fig. 8.2). The resulting electromagnetic perturbation represents a wave propagating in the z direction such that, if the condition $0 < \alpha < \omega^2/c^2$ is met, it propagates without diffraction. In the paper by J. Durnin, J.J. Miceli and J.H. Eberly, *Diffraction-Free Beams* [8.1], the experimental realization of a simple optical device generating this type of waves is presented. We leave to the reader to show how the wave (8.23) can be expressed in terms of plane waves.

8.2 Green's Function for the Wave Equation

In this section we find the general solution to the wave equations (8.1) for the potentials by means of the Green function formalism. By definition, the Green's function $G(\boldsymbol{r},t;\boldsymbol{r}',t')$ for the wave equation satisfies

$$\left(\nabla^2 - \frac{1}{c^2}\,\partial^2_{tt}\right) G(\boldsymbol{r},t;\boldsymbol{r}',t') = -4\pi\delta(t-t')\;\delta(\boldsymbol{r}-\boldsymbol{r}'), \tag{8.24}$$

where the inhomogeneities of the original equation have been replaced by delta functions in space and time. Exploiting the linearity of the equation, the solution for the scalar potential can be obtained using the Green's function as

$$\phi(\boldsymbol{r},t) = \int_{-\infty}^{+\infty} \mathrm{d}t' \int \mathrm{d}^3r'\; G(\boldsymbol{r},t;\boldsymbol{r}',t')\;\rho(\boldsymbol{r}',t'). \tag{8.25}$$

A similar expression is obtained for the vector potential by replacing ρ by $\boldsymbol{j}/c$.

The Green's function can be interpreted as the electric potential produced at point $\boldsymbol{r}$ at time t by a unit point charge placed at $\boldsymbol{r}'$ existing at time t' only. This source does not correspond to any physical situation, since it violates charge conservation and it is used here only as a mathematical tool. This interpretation of the Green function allows us to ensure that, if no other boundary conditions besides those at infinity are fixed, $G(\boldsymbol{r},t;\boldsymbol{r}',t')$ depends on the space variables through the modulus of the vector $\boldsymbol{R} = \boldsymbol{r} - \boldsymbol{r}'$, i.e. on the distance between the point source and the observation point. Moreover, homogeneity of time implies that the Green's function depends on t and t' through the difference $t - t'$ only. Therefore $G(\boldsymbol{r},t;\boldsymbol{r}',t') \equiv G(R,t-t')$.

To find the Green's function for the wave equation, we first study (8.24) for $\boldsymbol{r} \neq \boldsymbol{r}'$. In this case the right hand side vanishes. Away from the point $\boldsymbol{R} = 0$ the Laplace operator in the wave equation can be written as $\nabla^2 G \equiv R^{-2}\partial_R\left(R^2\partial_R G\right)$. Assuming for $G(R,t-t')$ a solution of the form

$$G(R, t-t') = \frac{g(R, t-t')}{R}, \tag{8.26}$$

we obtain for $g(R, t-t')$ the radial one-dimensional wave equation. As we have indicated in the introduction to this chapter this equation has solutions $g(t \pm R/c)$. Since the perturbation generated by the source originates at $\boldsymbol{r} = 0$ and should spread outwards, we choose the outward or retarded solution $g(t - R/c)$ only. The Green's function $G(R, t-t')$ turns out to be

$$G(R, t-t') = \frac{g(t-t'-R/c)}{R}. \tag{8.27}$$

This solution is to be compared with (8.19).

Equation (8.27) is a solution for the Green's function outside the point $\boldsymbol{r} = \boldsymbol{r}'$, whatever the form of the function g. In order to determine $g(t-t'-R/c)$ it is necessary to study the solution for $\boldsymbol{r} \to \boldsymbol{r}'$. The calculation should be carried out carefully, since both the spherical coordinates and the solution (8.27) are singular at the origin. The solution (8.27) can be rewritten as

$$G(R, t-t') = \frac{g(t-t'-R/c) - g(t-t')}{R} + \frac{g(t-t')}{R}. \tag{8.28}$$

Here, the first term is regular for $R \to 0$ and the singularity has been absorbed by the second term. Applying the wave operator to this expression, the second term on the right-hand side produces a delta-like contribution, coming from the Laplacian of R^{-1}, and an additional term canceling the contribution of the non singular part of the Green's function. As a consequence, we have

$$\nabla^2 G - \frac{1}{c^2}\,\partial^2_{tt} G = -4\pi g(t-t')\delta(\boldsymbol{r} - \boldsymbol{r}'). \tag{8.29}$$

Comparing with (8.24) we can identify

$$g(t-t') = \delta(t-t'). \tag{8.30}$$

Finally the Green's function can be written as

$$G(\boldsymbol{r}, t; \boldsymbol{r}', t') = \frac{\delta\left(t - |\boldsymbol{r} - \boldsymbol{r}'|/c - t'\right)}{|\boldsymbol{r} - \boldsymbol{r}'|}. \tag{8.31}$$

From (8.25) this Green's function allows us to express the electric potential as

$$\phi(\boldsymbol{r}, t) = \int \frac{\rho\left(\boldsymbol{r}', t - |\boldsymbol{r} - \boldsymbol{r}'|/c\right)}{|\boldsymbol{r} - \boldsymbol{r}'|}\, \mathrm{d}^3 r', \tag{8.32}$$

where we have used the Dirac delta in $G(\boldsymbol{r}, t; \boldsymbol{r}', t')$ to perform the integration over the time variable. This solution indicates that the potential is given by contributions generated by each charge element at a previous time. This time is such that the perturbation, moving at a velocity c, arrives at $\boldsymbol{r}$ at time t. The solution for the vector potential is totally similar:

$$\boldsymbol{A}(\boldsymbol{r}, t) = \frac{1}{c} \int \frac{\boldsymbol{j}\left(\boldsymbol{r}', t - |\boldsymbol{r} - \boldsymbol{r}'|/c\right)}{|\boldsymbol{r} - \boldsymbol{r}'|}\, \mathrm{d}^3 r'. \tag{8.33}$$

8.3 Fields of a Charge in Arbitrary Motion

As a first application of the solution of the wave equations derived in the previous section, we study the potentials and fields produced by a point charge q in arbitrary motion. Assuming the position of the charge to be given by the vector $\boldsymbol{r}_0(t)$, the associated charge and current densities are, respectively,

$$\begin{aligned} \rho(\boldsymbol{r},t) &= q\,\delta(\boldsymbol{r}-\boldsymbol{r}_0(t)), \\ \boldsymbol{j}(\boldsymbol{r},t) &= q\boldsymbol{v}(t)\delta(\boldsymbol{r}-\boldsymbol{r}_0(t)), \end{aligned} \tag{8.34}$$

where $\boldsymbol{v}=\dot{\boldsymbol{r}}_0$ is the velocity of the charge (Fig. 8.3). According to the definition of the Green's function given in the previous section, the scalar potential generated by the charge is

$$\phi(\boldsymbol{r},t) = q\int_{-\infty}^{+\infty} \frac{\delta\left(t-t'-|\boldsymbol{r}-\boldsymbol{r}_0(t')|/c\right)}{|\boldsymbol{r}-\boldsymbol{r}_0(t')|}\,\mathrm{d}t'. \tag{8.35}$$

The argument of the delta function in the integral vanishes for

$$t = t' + \frac{|\boldsymbol{r}-\boldsymbol{r}_0(t')|}{c}. \tag{8.36}$$

This equation allows us to find t', the *retarded time* at which the source emits the perturbation affecting point $\boldsymbol{r}$ at time t.

Keeping in mind the integration properties of the delta function (Appendix A), the scalar potential can be written as

$$\phi(\boldsymbol{r},t) = q\left[\frac{1}{KR}\right]_{\mathrm{ret}}, \tag{8.37}$$

where $\boldsymbol{R}(t') = \boldsymbol{r}-\boldsymbol{r}_0(t')$ and

$$K = 1 - \frac{\boldsymbol{R}(t')\cdot\boldsymbol{v}(t')}{cR(t')}. \tag{8.38}$$

The subscript "ret" indicates that t' should be evaluated at the retarded time, given in terms of $\boldsymbol{r}$ and t by the solution to (8.36).

With a similar procedure, we can find the vector potential generated by a moving charge:

$$\boldsymbol{A}(\boldsymbol{r},t) = \frac{q}{c}\left[\frac{\boldsymbol{v}}{KR}\right]_{\mathrm{ret}}. \tag{8.39}$$

Equations (8.37) and (8.39) define the so-called *Liénard–Wiechert potentials.*

To evaluate the electromagnetic fields derived from these potentials it is convenient to operate directly on the integral expressions for ϕ and $\boldsymbol{A}$. In these we can differentiate independently with respect to the variables $\boldsymbol{r}$ and t. Using again the properties of the delta function and its derivatives we obtain for the electric field

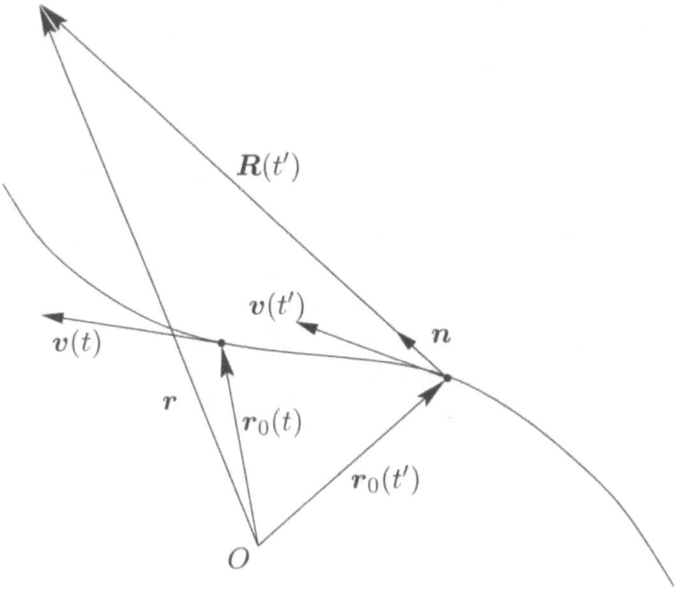

Fig. 8.3. A moving charge

$$\boldsymbol{E}(\boldsymbol{r},t) = \left[\frac{q}{K^3}\left\{\frac{\boldsymbol{n}-\boldsymbol{v}/c}{R^2}\left(1-\frac{v^2}{c^2}\right)+\frac{\boldsymbol{n}\times[(\boldsymbol{n}-\boldsymbol{v}/c)\times\boldsymbol{a}]}{c^2R}\right\}\right]_{\text{ret}} \tag{8.40}$$

and for the magnetic field

$$\boldsymbol{B}(\boldsymbol{r},t) = \left[\frac{q}{K^3}\boldsymbol{n}\times\left\{-\frac{\boldsymbol{v}}{cR^2}\left(1-\frac{v^2}{c^2}\right)+\frac{\boldsymbol{n}\times[(\boldsymbol{n}-\boldsymbol{v}/c)\times\boldsymbol{a}]}{c^2R}\right\}\right]_{\text{ret}} \tag{8.41}$$

Here, $\boldsymbol{a} = \dot{\boldsymbol{v}}$ is the acceleration of the charge, and $\boldsymbol{n}$ is the unit vector in the direction of $\boldsymbol{R}$ (Fig. 8.3).

Note that the fields of a moving charge given above satisfy the relation

$$\boldsymbol{B} = \boldsymbol{n}\times\boldsymbol{E}. \tag{8.42}$$

As in the case of plane waves in free space, the magnetic field is always perpendicular to the electric field and to the direction of *retarded observation.* Now, however, as can be seen from (8.41) the modulus of the magnetic field is always smaller than or equal to that of the electric field, $B \leq E$.

Both the electric and the magnetic field can be separated into two contributions. The first one, corresponding to the first terms in (8.40) and (8.41), decays as R^{-2}, i.e. as the inverse of the square of the distance between the observation point and the charge. The second contribution is proportional to R^{-1} and therefore dominates at sufficiently large distances. As we show in the next section, this second term of the electromagnetic fields is associated with a net energy flow moving away from the charge with velocity c. For this reason, the terms falling off as R^{-1} are called *radiation fields*. In contrast, the contributions proportional to R^{-2} are called *bound* or *convection fields*.

The bound fields are of the same magnitude as the electrostatic field. To compare these fields with the radiation fields we notice that the latter prevail at distances $R > c^2/a$, where a is the modulus of the charge acceleration. For example, for an acceleration similar to that of gravity, $a \sim g \sim 10\,\mathrm{m/s}^2$, $c^2/a \sim 10^{16}\,\mathrm{m}$ so that for all practical purposes the relevant fields are the convection fields. Let us consider, on the other hand, an electron in an accelerator like the LINAC of Centro Atómico Bariloche. In this case after almost reaching the velocity of light, electrons are stopped by collisions in a target until coming practically to rest in a distance $d \sim 5\,\mathrm{cm}$. The (negative) acceleration of this particle during the stopping process is approximately given by $a \sim c^2/d$, so that $c^2/a \sim d$. Consequently, radiation fields dominate for distances larger than $R = 5\,\mathrm{cm}$.

In the following section, we examine the radiation fields in some interesting limits. These fields vanish when the charge moves with constant velocity. In this situation the charge produces convection fields only, given by

$$\boldsymbol{E}_{\mathrm{conv}} = \left[\frac{q}{K^3}\frac{(\boldsymbol{n}-\boldsymbol{v}/c)(1-v^2/c^2)}{R^2}\right]_{\mathrm{ret}}, \tag{8.43}$$

and

$$\boldsymbol{B}_{\mathrm{conv}} = \left[\frac{q}{cK^3}\frac{(\boldsymbol{v}\times\boldsymbol{n})(1-v^2/c^2)}{R^2}\right]_{\mathrm{ret}}. \tag{8.44}$$

Notice that the principle of relativity gives a plausibility argument to explain the non-existence of radiation terms in the fields generated by a uniformly moving charge. If such terms were present they would describe a perturbation moving at the speed of light which could be eliminated by performing a Lorentz transformation on the particle rest system, where the field is Coulombian. The invariance of the speed of light under such transformations implies that the fields of a uniformly moving charge cannot have a radiative part.

8.4 Radiation Fields of a Moving Charge

As we have shown in the previous section, fields produced by an accelerated charge are dominated at sufficiently large distances by the radiation terms,

$$\boldsymbol{E}_{\mathrm{rad}} = \left[\frac{q}{K^3}\frac{\boldsymbol{n}\times[(\boldsymbol{n}-\boldsymbol{v}/c)\times\boldsymbol{a}]}{c^2R}\right]_{\mathrm{ret}} \tag{8.45}$$

and $\boldsymbol{B}_{\mathrm{rad}} = \boldsymbol{n}\times\boldsymbol{E}_{\mathrm{rad}}$. Note that the electric radiation field is perpendicular to the direction $\boldsymbol{n}$ to the observation point. As in the case of plane waves, the vectors $\boldsymbol{n}$, $\boldsymbol{E}_{\mathrm{rad}}$ and $\boldsymbol{B}_{\mathrm{rad}}$ form an orthogonal system.

Associated with the radiation fields there is a net flow of energy moving away from the sources. In fact, this condition defines such fields. The energy flow density associated with radiation is given by the Poynting vector

$$\boldsymbol{S}_{\mathrm{rad}} = \frac{c}{4\pi}\, \boldsymbol{E}_{\mathrm{rad}} \times \boldsymbol{B}_{\mathrm{rad}} = \frac{c}{8\pi}\, \left(E_{\mathrm{rad}}^2 + B_{\mathrm{rad}}^2\right)\, \boldsymbol{n} = \frac{c}{4\pi}\, E_{\mathrm{rad}}^2\, \boldsymbol{n}. \tag{8.46}$$

The modulus of this vector is the product of the speed of light times the density of electromagnetic energy ($E_{\mathrm{rad}}^2 = B_{\mathrm{rad}}^2$). Its direction coincides with that of observation. The dependence on E_{rad}^2 implies that it decays proportional to R^{-2}. As a consequence, the net energy flow through a sufficiently large surface does not depend on the radius of such a surface. Associated with the radiation fields there is therefore a net amount of *radiated energy* that definitely leaves the surroundings of the charge producing them. On the other hand, energy flow corresponding to the convection fields is restricted to regions close to the source. This fact justifies the name *bound* given to the fields that dominate at small distances.

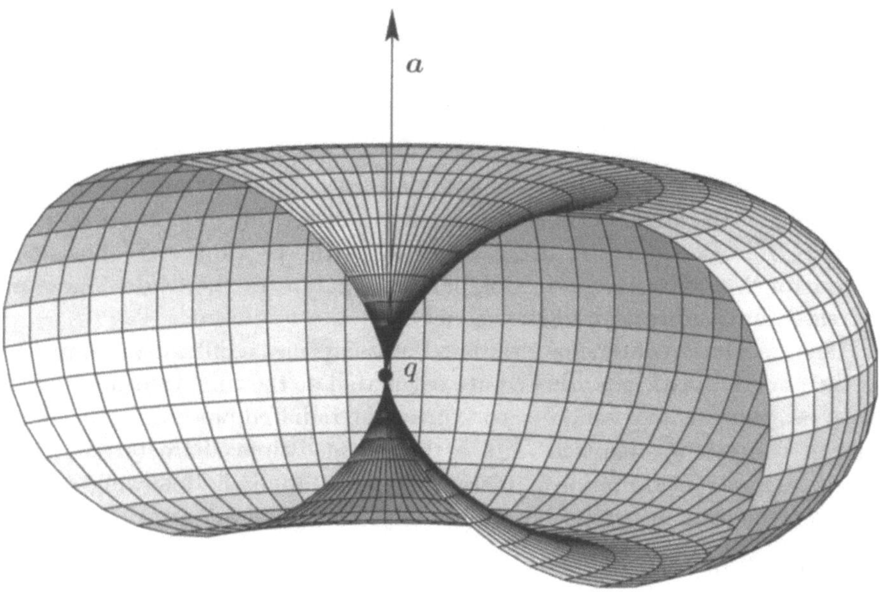

Fig. 8.4. Radiation of an accelerated nonrelativistic charge

The detailed form of the radiation fields depends on the motion performed by the source charge. Here we study some representative cases. In the first place we consider a charge in nonrelativistic motion, so that we can take $v/c \approx 0$ and $K \approx 1$. In this limit, the electric field is

$$\boldsymbol{E}_{\mathrm{rad}} = \left[\frac{q}{c^2 R}\, \boldsymbol{n} \times (\boldsymbol{n} \times \boldsymbol{a})\right]_{\mathrm{ret}}, \tag{8.47}$$

It does not depend on the velocity of the charge. The power radiated per unit solid angle is given by

$$\frac{\mathrm{d}W_{\mathrm{rad}}}{\mathrm{d}t\ \mathrm{d}\Omega} = R^2 S_{\mathrm{rad}} = \frac{q^2 a^2}{4\pi c^3} \sin^2\theta \tag{8.48}$$

where θ is the angle between the acceleration $\boldsymbol{a}$ and the observation direction $\boldsymbol{n}$. Integrating this quantity over a spherical surface centered at the charge, we obtain the total energy irradiated per unit time

$$\frac{\mathrm{d}W_{\mathrm{rad}}}{\mathrm{d}t} = R^2 \int_S \boldsymbol{S}_{\mathrm{rad}} \cdot \mathrm{d}\boldsymbol{s} = \frac{2}{3c^3}\, q^2 a^2, \tag{8.49}$$

This expression is known as the *Larmor formula.* The radiation pattern associated with an accelerated nonrelativistic charge is shown in Fig. 8.4. The fields and energy radiated by the charge are symmetric around the direction of acceleration. Radiation is then concentrated in the direction perpendicular to $\boldsymbol{a}$; the energy radiated in the direction of acceleration vanishes.

As a second example we consider the case in which the velocity and the acceleration of the charge are parallel, as is the case in rectilinear motion, without restrictions on the magnitude of the velocity. In this situation the electric field is given by

$$\boldsymbol{E}_{\mathrm{rad}} = \left[\frac{q}{c^2 K^3 R}\ \boldsymbol{n} \times (\boldsymbol{n} \times \boldsymbol{a})\right]_{\mathrm{ret}}, \tag{8.50}$$

which differs from the field of the nonrelativistic charge by a factor K^{-3}.

Calling θ the angle between the velocity and the direction of observation, we have $K = 1 - v\cos\theta/c$. Comparison with the previous case shows that the factor K^{-3} introduces an angular deformation of the fields. These are increased in the direction of the velocity.

The calculation of the power radiated presents now a difficulty, not present in the nonrelativistic case. Fields are calculated at the time measured by the observer, whereas we would like to express the radiated power in terms of the time at which it was emitted. This is the retarded time, characteristic of the charge. This transformation is made keeping in mind the following relation between the total energy radiated in a fixed time interval

$$S_{\mathrm{rad}}(t)\ \mathrm{d}t = S_{\mathrm{rad}}(t')\ \mathrm{d}t'. \tag{8.51}$$

This follows from the fact that the total radiated energy cannot depend on the way of measuring time along the interval. The Poynting vector referred to the retarded time is thus connected to $S_{\mathrm{rad}}(t)$ through the derivative $\mathrm{d}t/\mathrm{d}t'$, which follows from (8.36). The result is in fact $\mathrm{d}t/\mathrm{d}t' = K$. Therefore, the energy radiated per unit time and per unit solid angle, referred to the time t' is

$$\frac{\mathrm{d}W_{\mathrm{rad}}}{\mathrm{d}t'\ \mathrm{d}\Omega} = K\ R^2 S_{\mathrm{rad}} = \frac{q^2 a^2}{4\pi c^3} \frac{\sin^2\theta}{(1 - v\cos\theta/c)^5}. \tag{8.52}$$

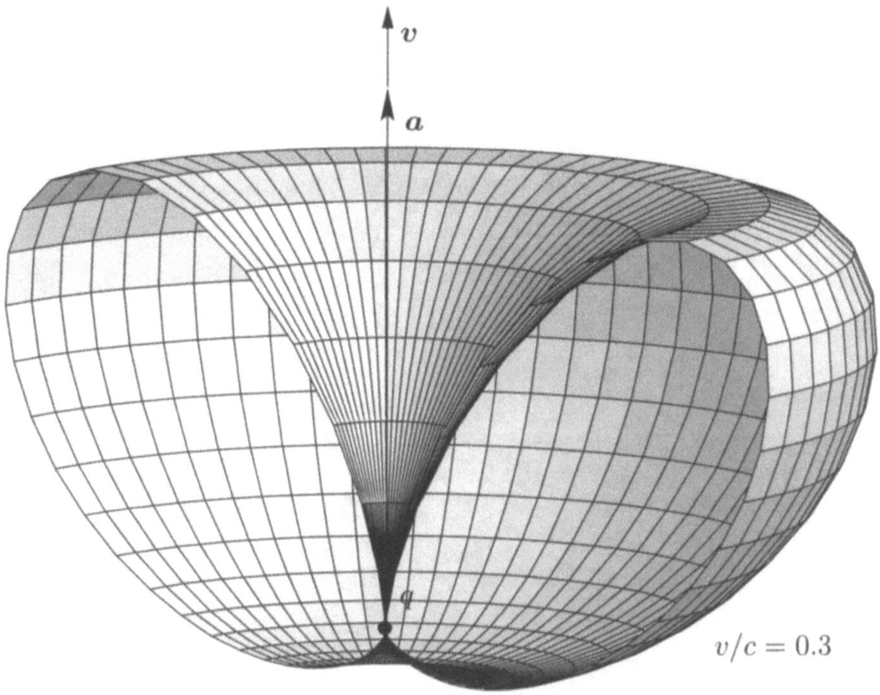

Fig. 8.5. Radiation of a charge with acceleration parallel to velocity

The angular dependence in the denominator implies that the surfaces of equal intensity "lean forward" (Fig. 8.5). In fact, the radiated energy reaches its maximum at an angle

$$\theta_{\max} = \arccos\left[\frac{c}{3v}\left(\sqrt{1 + 15v^2/c^2} - 1\right)\right]. \tag{8.53}$$

It tends to zero as $v/c \to 1$. Therefore, in the ultrarelativistic limit most of the energy is radiated along the direction of motion of the charge. In the LINAC of Centro Atómico Bariloche $\theta_{\max}$ is approximately 10^{-2}, i.e. about half a degree. In this case, the width of the angular distribution of energy around the maximum is 1.5°.

Finally, we study the fields produced by a charge accelerated perpendicularly to the velocity, as in uniform circular motion. In order to simplify the calculation we assume that $\boldsymbol{v}$ is parallel to the z axis and $\boldsymbol{a}$ parallel to the x axis. The electric field turns out to be

$$\boldsymbol{E}_{\text{rad}} = \left[\frac{qa}{c^2K^3R}\{(-n_y^2 - n_z^2 + vn_z/c)\hat{\boldsymbol{x}} + n_xn_y\hat{\boldsymbol{y}} + (n_xn_z - vn_x/c)\hat{\boldsymbol{z}}\}\right]_{\text{ret}} \tag{8.54}$$

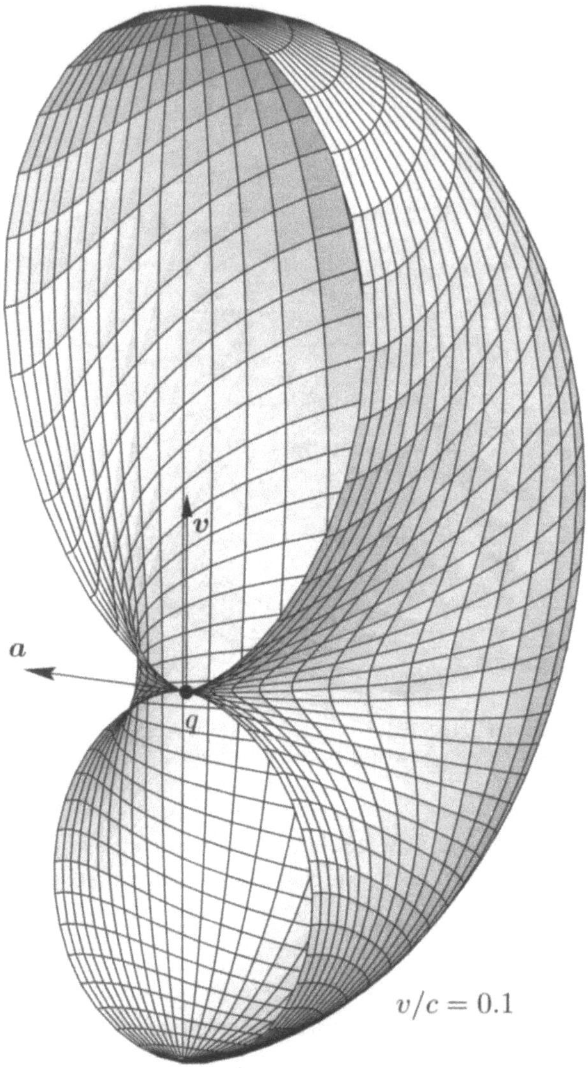

Fig. 8.6. Radiation of a charge accelerated perperdicularly to the velocity

where $\boldsymbol{n} = (n_x, n_y, n_z)$. This result can be analyzed for two representative situations. In the first one, the observation point is placed in the plane formed by $\boldsymbol{v}$ and $\boldsymbol{a}$ so that $n_y = 0$. If θ is the angle between $\boldsymbol{n}$ and the velocity, $n_z = \cos\theta$, the power radiated per unit solid angle is

$$\frac{\mathrm{d}W_{\mathrm{rad}}}{\mathrm{d}t'\,\mathrm{d}\Omega} = \frac{q^2 a^2}{4\pi c^3} \frac{(v/c - \cos\theta)^2}{(1 - v\cos\theta/c)^5} . \tag{8.55}$$

According to this expression, the intensity of radiation is a maximum for $\theta = 0$ and vanishes for $\cos\theta = v/c$. The radiation is distributed in two asymmetric lobes with the largest one in the direction of the motion. This lobe increases in size as $v \to c$ so that in the ultrarelativistic limit radiation is emitted essentially forward.

If now the observation point is placed in a plane perpendicular to the acceleration, $n_x = 0$, we have

$$\frac{\mathrm{d}W_{\mathrm{rad}}}{\mathrm{d}t'\,\mathrm{d}\Omega} = \frac{q^2a^2}{4\pi c^3}\left(1 - \frac{v}{c}\cos\theta\right)^{-3}. \tag{8.56}$$

The radiation is also directed "forwards" and varies monotonously as θ grows. Again, the focusing effect of the radiated energy increases as the velocity increases. The radiation pattern for $\boldsymbol{a}$ perpendicular to $\boldsymbol{v}$ is shown in Fig. 8.6.

The detailed analysis of the ultrarelativistic limit for the radiation emitted by a particle with acceleration parallel or perpendicular to the velocity indicates that, in the latter case, the radiated intensity is larger than in the first one by a factor $(1 - v^2/c^2)^{-1}$, which becomes very large as $v \to c$. Therefore, for a charge in arbitrary motion and velocity close to that of light, the main contribution to radiation is given by the component of the acceleration perpendicular to $\boldsymbol{v}$. This property is used to generate radiation in a synchrotron, where by means of magnetic fields a beam of particles is forced to rotate in a circular orbit at ultrarelativistic velocities.

8.5 Dipole Radiation

Up to this point, we have studied the electromagnetic fields generated by a single moving charge. Due to the superposition principle fields of more complex sources can be built up by adding the fields corresponding to each of the individual charges in the system. The structure of the solutions (8.32) and (8.33) for the electromagnetic potentials in the Lorentz gauge suggest that when the sources are bounded in space it is possible to make an expansion similar to the multipolar expansions studied for static situations. Such an expansion gives the form of the potentials at points outside of the sources.

Let us consider in the first place the vector potential, given by (8.33). To the lowest order in powers of r'/r we have $|\boldsymbol{r} - \boldsymbol{r}'| \approx r$, so that $\boldsymbol{A}$ reduces to

$$\boldsymbol{A}(\boldsymbol{r}, t) \approx \frac{1}{rc}\int \boldsymbol{j}(\boldsymbol{r}', t - r/c)\,\mathrm{d}^3r'. \tag{8.57}$$

In contrast to what happens in magnetostatics, in this case the integral of the current density over all space does not vanish. Integrating by parts and using the continuity equation (4.6), we find

$$\boldsymbol{A}(r, t) = \dot{\boldsymbol{p}}(t - r/c)/rc, \tag{8.58}$$

where $\boldsymbol{p}$ is the dipole moment of the source distribution, defined exactly as in (4.38). In general this quantity is time-dependent. Note that $\dot{\boldsymbol{p}}$ is evaluated at the retarded time $t' = t - r/c$, which is a solution to (8.36) in the approximation we are considering here.

It can be proved that to the same order of approximation (8.32) implies

$$\phi(\boldsymbol{r}, t) = \frac{Q}{r}, \tag{8.59}$$

Q being the total charge of the distribution. This scalar potential describes only the electrostatic component of the fields. This order of approximation for ϕ is therefore not satisfactory. In fact, for the approximation (8.58) which we found for the vector potential $\boldsymbol{A}$, the result $\phi = Q/r$ does not satisfy the condition $\nabla \cdot \boldsymbol{A} + c^{-1}\partial_t\phi = 0$ of the Lorentz gauge. The lowest order for the scalar potential corresponds rather to $\boldsymbol{A} = 0$. Consequently, it is necessary to go one step further in the expansion of $\phi(\boldsymbol{r}, t)$ in order to find an approximation consistent with (8.58). However, instead of starting from the expansion (8.32) in powers of r'/r, we take advantage of the Lorentz condition and calculate ϕ starting from (8.58). With that expression for $\boldsymbol{A}(\boldsymbol{r}, t)$ we write

$$\partial_t\phi = -c\ \nabla \cdot \boldsymbol{A} = \dot{\boldsymbol{p}}(t - r/c) \cdot \frac{\boldsymbol{r}}{r^3} + \frac{1}{c}\ddot{\boldsymbol{p}}(t - r/c) \cdot \frac{\boldsymbol{r}}{r^2}, \tag{8.60}$$

from which we obtain

$$\phi(\boldsymbol{r}, t) = \frac{Q}{r} + \left[\boldsymbol{p}(t - r/c) + \frac{r}{c}\dot{\boldsymbol{p}}(t - r/c)\right] \cdot \frac{\boldsymbol{r}}{r^3}. \tag{8.61}$$

In this expression we have used the electrostatic contribution to ϕ as an "integration constant." The first consistent nontrivial approximation for the electromagnetic potentials generated by a bounded source distribution can thus be written as

$$\begin{aligned} \phi(\boldsymbol{r}, t) &= \frac{Q}{r} + \frac{\boldsymbol{r}}{r^3} \cdot \left[\boldsymbol{p} + \frac{r}{c}\ \dot{\boldsymbol{p}}\right]_{\text{ret}}, \\ \boldsymbol{A}(\boldsymbol{r}, t) &= \left[\frac{1}{rc}\ \dot{\boldsymbol{p}}\right]_{\text{ret}}. \end{aligned} \tag{8.62}$$

Since in this approximation the potentials depend exclusively on the dipole moment of the source distribution, we call it the *dipole approximation.*

The electric field is given by

$$\boldsymbol{E}(\boldsymbol{r}, t) = Q\ \frac{\boldsymbol{r}}{r^3} + \frac{1}{r^3}\left[\frac{3(\boldsymbol{p}^* \cdot \boldsymbol{r})\boldsymbol{r}}{r^2} - \boldsymbol{p}^* + \frac{1}{c^2}\ (\ddot{\boldsymbol{p}} \times \boldsymbol{r}) \times \boldsymbol{r}\right]_{\text{ret}}, \tag{8.63}$$

with $\boldsymbol{p}^* = \boldsymbol{p} + r\dot{\boldsymbol{p}}/c$. The first term in this expression corresponds clearly to the electrostatic field associated with the total charge at the sources. The vector $\boldsymbol{p}^*$ has been introduced to stress the fact that the first two terms inside the bracket, which dominate the field at small distances from the source, have exactly the form of a dipole field with a moment $\boldsymbol{p}^*$. Finally we note that, by comparison with the field of an accelerated charge studied in the previous section, the last term represents a radiation contribution.

The corresponding magnetic field can be written as

$$\boldsymbol{B}(\boldsymbol{r},t) = \frac{1}{cr^3}\left[\dot{\boldsymbol{p}}^* \times \boldsymbol{r}\right]_{\text{ret}} . \tag{8.64}$$

We notice that, in contrast to the fields generated by a single charge, dipole fields **do not** satisfy $\boldsymbol{B} = \hat{\boldsymbol{r}} \times \boldsymbol{E}$. Thus $\boldsymbol{B}$ is not necessarily perpendicular either to the electric field or to the observation direction. We also notice that the separation between convective and radiative contributions of the magnetic field is clearer if we write out the explicit form of $\boldsymbol{p}^*$:

$$\boldsymbol{B}(\boldsymbol{r},t) = \frac{1}{cr^3}\left[\dot{\boldsymbol{p}} \times \boldsymbol{r} + \frac{r}{c}\,\ddot{\boldsymbol{p}} \times \boldsymbol{r}\right]_{\text{ret}} . \tag{8.65}$$

The radiative contributions to the electric and magnetic fields, called *dipole radiation fields*, and given by the last terms in Eqs. (8.63) and (8.65), satisfy $\boldsymbol{B}_{\text{rad}} = \hat{\boldsymbol{r}} \times \boldsymbol{E}_{\text{rad}}$. This generalizes the geometric characterization of the radiation fields introduced in the previous section. The Poynting vector associated with the dipole radiation fields has the same direction and form as for an accelerated nonrelativistic charge. To compare both cases, the product qa must be replaced by $\ddot{p}$, so that the power radiated per unit solid angle is

$$\frac{\mathrm{d}W_{\text{rad}}}{\mathrm{d}t\,\mathrm{d}\Omega} = \frac{\ddot{p}^2}{4\pi c^3}\sin^2\theta. \tag{8.66}$$

Here θ is the angle between the observation direction and $\ddot{\boldsymbol{p}}$. The intensity of dipole radiation is thus a maximum in the direction perpendicular to $\ddot{\boldsymbol{p}}$, as shown in Fig. 8.4.

The relative importance of the nonradiative terms depends on the details of the time evolution of $\boldsymbol{p}$. Indeed, $\boldsymbol{p}^* = \boldsymbol{p} + r\dot{\boldsymbol{p}}/c$ involves a space dependence weighted with the time derivative of the dipole moment. Here we analyze the relative magnitude of each term in the dipole fields in a case of particular importance, i.e. when the dipole moment of the sources varies harmonically. The result applies for instance to the fields produced by an atom where the positive and negative charges perform an oscillatory relative motion. It is also a good first approximation for the fields of an antenna.

We set $\boldsymbol{p}(t) = \boldsymbol{p}\exp(\mathrm{i}\omega t)$, where $\boldsymbol{p}$ is now a constant vector. The complex form of $\boldsymbol{p}(t)$ has been introduced for simplicity in the calculation: the corresponding physical quantity is its real part. With this choice, we have $\dot{\boldsymbol{p}}(t) = \mathrm{i}\omega\boldsymbol{p}(t)$, $\ddot{\boldsymbol{p}}(t) = -\omega^2\boldsymbol{p}(t)$ and $\boldsymbol{p}^*(t) = (1+\mathrm{i}kr)\boldsymbol{p}(t)$, where $k = \omega/c$ is the wave number associated with the propagation of the radiation of the oscillating dipole. In this situation, the complex electric field is given by

$$\boldsymbol{E}(\boldsymbol{r},t) = Q\frac{\boldsymbol{r}}{r^3} + \frac{\exp(\mathrm{i}\omega t - \mathrm{i}kr)}{r^3} \times \left\{\left[\frac{3\boldsymbol{r}(\boldsymbol{p}\cdot\boldsymbol{r})}{r^2} - \boldsymbol{p}\right](1+\mathrm{i}kr) - k^2\,(\boldsymbol{p}\times\boldsymbol{r})\times\boldsymbol{r}\right\} \tag{8.67}$$

and the corresponding magnetic field is

$$\boldsymbol{B}(\boldsymbol{r},t) = \frac{\mathrm{i}\omega \exp(\mathrm{i}\omega t - \mathrm{i}kr)}{cr^3} (\boldsymbol{p} \times \boldsymbol{r})(1 + \mathrm{i}kr). \tag{8.68}$$

These expressions show that a new length scale is defined by the inverse of the wave number k. This length scale is associated with the motion of the sources and to the time dependence of its dipole moment. We define it as the *wavelength* $\lambda = 2\pi/k$. The relative importance of each term in the expression for the fields depends on how r and λ compare.

Here we limit ourselves to the case in which the wavelength is much larger than the dimensions of the sources, as happens when these perform nonrelativistic motion. In this situation, in the convection region, where $r \ll \lambda$, the dipole fields are

$$\begin{aligned} \boldsymbol{E}_{\text{conv}} &= Q\boldsymbol{r}/r^3 + [3\boldsymbol{r}(\boldsymbol{p}\cdot\boldsymbol{r})/r^2 - \boldsymbol{p}] \exp(\mathrm{i}\omega t)/r^3, \\ \boldsymbol{B}_{\text{conv}} &= \mathrm{i}k\ (\boldsymbol{p}\times\boldsymbol{r}) \exp(\mathrm{i}\omega t)/r^3. \end{aligned} \tag{8.69}$$

Apart from the electrostatic term, the electric field has exactly the form of a dipole field, modulated by the time dependence of the dipole moment. The magnetic field also corresponds to a time-dependent electric dipole.

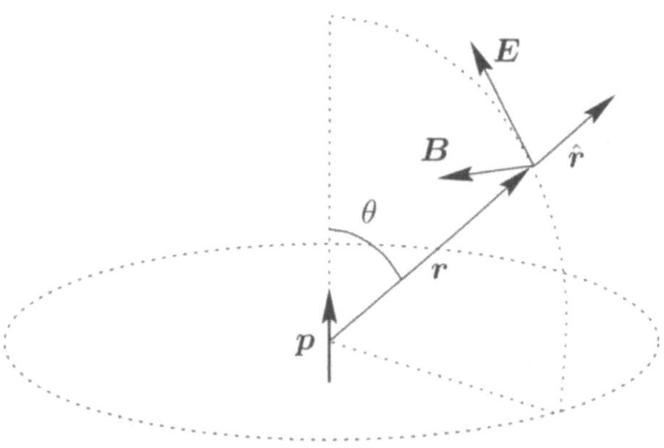

Fig. 8.7. Electric and magnetic dipole fields

Beyond the intermediate region where $r \sim \lambda$, the dominant fields are the radiation fields. For $r \gg \lambda$, in fact, $\boldsymbol{E}$ and $\boldsymbol{B}$ reduce to

$$\begin{aligned} \boldsymbol{E}_{\text{rad}} &= -r^{-1} \exp(\mathrm{i}\omega t - \mathrm{i}\boldsymbol{k}\cdot\boldsymbol{r})\ (\boldsymbol{p}\times\boldsymbol{k})\times\boldsymbol{k}, \\ \boldsymbol{B}_{\text{rad}} &= -kr^{-1} \exp(\mathrm{i}\omega t - \mathrm{i}\boldsymbol{k}\cdot\boldsymbol{r})\ (\boldsymbol{p}\times\boldsymbol{k}), \end{aligned} \tag{8.70}$$

where we have introduced the wave vector $\boldsymbol{k} = k\hat{\boldsymbol{r}}$. According to this we can write $\boldsymbol{B}_{\text{rad}} = \hat{\boldsymbol{k}} \times \boldsymbol{E}_{\text{rad}}$. Note that the vector $\boldsymbol{E}$ is always within the plane formed by $\boldsymbol{p}$ and $\boldsymbol{k}$. The electric field is polarized in that plane and, seen from the observation point $\boldsymbol{r}$, points in the same direction as the dipole moment of the sources. On the other hand, the magnetic field is perpendicular to the electric field (Fig. 8.7).

The power radiated per unit solid angle is calculated taking the real part of the radiation fields, leading to

$$\frac{\mathrm{d}W_{\mathrm{rad}}}{\mathrm{d}t\,\mathrm{d}\Omega} = \frac{p^2\omega^4}{4\pi c^3}\sin^2\theta\,\cos^2(\omega t - kr). \tag{8.71}$$

The total power radiated in all directions and averaged over one period of oscillation of the dipole is

$$\langle\frac{\mathrm{d}W_{\mathrm{rad}}}{\mathrm{d}t}\rangle = \frac{p^2\omega^4}{3c^3}, \tag{8.72}$$

an expression similar to the Larmor formula (8.49). According to this result, two essential characteristics of the dipole radiation are its dependence on the square of the oscillation amplitude p and on the fourth power of frequency ω.

This last property provides the basis for an explanation of the blue color of the sky, proposed by Lord Rayleigh at the end of the nineteenth century. Essentially white sun light illuminates the atmosphere exciting the charges contained in air molecules. The latter, when oscillating, emit radiation. According to (8.72) this process is more effective for high frequencies, i.e. frequencies in the blue region of the spectrum. The radiation so dispersed is received on earth from all directions, giving rise to the blue color of the sky.

The mechanism proposed by Lord Rayleigh is verified by observing the polarization of sky light. Sunlight is not polarized, so charges excited by it oscillate in random directions, always staying in a plane perpendicular to the direction to the Sun. In this way, when the light coming from the sky in directions near to the sun or opposite to it is analyzed, no polarization is detected. On the other hand, when the scattered light comes from directions perpendicular to that to the Sun, the observer is in the plane of oscillation of the excited charges and the light he or she receives is polarized in that plane. This fact can be proved by observing the clear sky through a polarizer. This phenomenon is used by some insects for orientation when the Sun is not directly visible, as for example, in a forest.

As we mentioned at the beginning of this section, we can go beyond the dipole approximation by considering terms of higher order in powers of r'/r, as in the expansion of static multipoles. Such an extension, which we will not analyze here, gives rise to contributions to the radiation fields characterized by more complicated angular distributions of radiated energy. In some applications these fields are combined to produce radiation sources which are especially efficient in energy emission in certain directions. This is the case of some antennas.

8.6 Inadequacy of the Planetary Model of the Atom

From the point of view of the development of modern physics, one of the most relevant consequences of the fact that accelerated charges radiate is the inadequacy of the classical model for the atom. This model assumes that the electrons perform orbits around the atomic nucleus in a form similar to a planetary system. Indeed, due to this accelerated motion electrons should radiate and therefore continually lose energy. As a consequence, they would end up collapsing onto the nucleus. The experimental evidence that this does not happen was one of the reasons that led to a complete reformulation of the mechanical laws at the microscopic level, giving rise to quantum mechanics.

Using the results obtained in this chapter we can evaluate, within reasonable approximations, the time it would take for an electron to collapse onto the nucleus due to energy loss by radiation. To simplify, we consider the classical model of a hydrogen atom consisting of an electron orbiting around a proton. In the nonrelativistic limit, either starting from the results for charges in circular motion or in the framework of the dipole approximation, we obtain for the total energy radiated by the electron per unit time

$$\frac{\mathrm{d}W_{\mathrm{rad}}}{\mathrm{d}t} = \frac{2e^6}{3c^3m^2} r^{-4}, \tag{8.73}$$

where e is the electronic charge and r is the radius of the orbit.

The radiated power should be equal to the energy loss per unit time of the electron. Considering the electrostatic interaction only, we have

$$\frac{\mathrm{d}}{\mathrm{d}t}\left(\frac{1}{2}mv^2 - \frac{e^2}{r}\right) = -\frac{2e^6}{3c^3m^2} r^{-4}, \tag{8.74}$$

where m is the mass of the electron and v its orbital velocity.

Assuming the radiated energy per period to be much smaller than the initial energy of the electron – or, equivalently, assuming that the electron is able to perform many revolutions of its orbit before collapsing onto the proton – we can take the orbit as essentially circular, so that

$$\frac{mv^2}{r} = \frac{e^2}{r^2}. \tag{8.75}$$

Substituting this relation into the energy balance equation (8.74), we obtain a differential equation for the radius of the orbit $r(t)$ as a function of time. The solution is

$$r(t) = \left[r(0)^3 - 4\frac{e^4}{c^3m^2}\, t\right]^{1/3}. \tag{8.76}$$

As a consequence, r decreases and becomes zero at a time

$$t_{\mathrm{c}} = \frac{r(0)^3 m^2 c^3}{4e^4}. \tag{8.77}$$

We associate this time with the time for the atom to collapse. Taking $r(0)$ as the Bohr radius, $r(0) = 0.5 \times 10^{-8}$ cm, the time of collapse $t_c \approx 10^{-10}$ s.

To verify the hypothesis on which we have based this calculation, we can compare t_c with the classical orbital period of the electron around the proton. This turns out to be of the order of 10^{-16} s, much smaller indeed than t_c.

If the classical model for the atom were correct, matter would collapse in extremely short times, a fact in open contradiction with experience. It is interesting to note, however, that the problem of the collapse of the electron onto the nucleus can be solved, in the framework of classical theories, by representing the electron as an evenly distributed charge around a rotating ring. Indeed, a loop carrying stationary current does not radiate.

Problems

8.1 Using the superposition principle, show that the electromagnetic fields generated by a system of charges in arbitrary motion **do not** satisfy $\boldsymbol{B} \perp \boldsymbol{E}$. Show that if the system of charges is bounded, sufficiently far from it the fields **are** perpendicular.

Hint: Consider the fields as sum of fields proportional to the velocity and fields proportional to the acceleration.

8.2 Show that the derivative of $t = t' + |\boldsymbol{r} - \boldsymbol{r}_0(t')|/c$ with respect to t' is $K = 1 - \boldsymbol{n} \cdot \boldsymbol{v}/c$.

8.3 Consider the equation for the retarded time t'

$$t = t' + \frac{|\boldsymbol{r} - \boldsymbol{r}_0(t')|}{c}.$$

(a) Show that it has at most **one** solution. (b) Discuss in detail the cases in which it has **no** solution (Hint: analyze the case of one dimensional motion where $r_0(t) = \sqrt{c^2t^2 + a^2}$).

Hint: Begin by trying a graphical solution.

8.4 Show by direct calculation that the curl of

$$\boldsymbol{A}(\boldsymbol{r}, t) = \frac{q}{c} \int_{-\infty}^{\infty} \mathrm{d}t' \frac{\boldsymbol{v}(t')}{|\boldsymbol{r} - \boldsymbol{r}_0(t')|} \delta\left(t' + \frac{|\boldsymbol{r} - \boldsymbol{r}_0(t')|}{c} - t\right),$$

is the magnetic field of a charge q with trajectory $\boldsymbol{r}_0(t)$.

8.5 Draw the lines of equal radiation intensity in the plane normal to $\boldsymbol{v}$, for a charge with acceleration perpendicular to the velocity.

8.6 A plane wave with electric field $\boldsymbol{E} = \boldsymbol{E}_0 \exp(-i\omega t)$ is incident on a free electron initially at rest. (a) Determine the motion of the particle in the nonrelativistic limit. (b) Study the radiation fields of the particle. (c) Calculate the cross section for radiation dispersion, defined as the energy radiated per unit time and unit solid angle, divided by the total flow of energy incident per unit time. (d) Integrating over the angular variable, calculate the total cross section. Compare with the classical electron radius.

Hint: Use the dipole approximation to determine the radiation fields.

8.7 In the dipole approximation, study the radiation fields of a charge performing uniform circular motion. Compare with the exact radiation fields.

8.8 A system of two identical charges of equal mass is impacted upon by a linearly polarized plane wave propagating in the direction of the line joining both charges. Calculate the radiated energy in the dipole approximation, at distances much larger than the separation a between charges, assuming that the mutual interaction of the charges can be neglected. Discuss the result in terms of the relation between a and the wavelength of the incident field. Why does the result not depend on the sign of the charges?

Hint: Consider separately the cases of polarization along the line joining the charges and polarization perpendicular to this line.

8.9 Two charges of the same absolute value describe a circular orbit with the same velocity, but dephased by an angle φ. Find the time average of radiated power per unit solid angle in the dipole approximation. Schematize the surfaces of constant radiation intensity. Study and interpret the cases in which $\varphi = 0$, $\varphi = \pi$ and consider also whether the signs of the charges are equal or opposite.

8.10 In the planetary model of the hydrogen atom, determine the position of the electron as a function of time. Plot the trajectory.

8.11 Determine the fields of a charged particle in uniform motion. Compare with the result given in next chapter, obtained by means of a Lorentz transformation applied to the Coulomb field of a particle at rest.

Solution: For a charge in uniform motion we can take $\boldsymbol{r}_0(t) = \boldsymbol{v}t$. This implies $\boldsymbol{r}(t') = \boldsymbol{r} - \boldsymbol{v}t'$. Equation (8.36), defining the retarded time, implies

$$t = t' + \frac{1}{c}|\boldsymbol{r} - \boldsymbol{v}t'|.$$

Setting $t = 0$ and $\boldsymbol{v} = v\hat{\boldsymbol{x}}$, the solution to this equation is given by

$$t' = -\frac{vx}{c^2 - v^2} - \frac{1}{\sqrt{c^2 - v^2}}\sqrt{\frac{x^2}{1 - v^2/c^2} + y^2 + z^2}.$$

This retarded time is negative, since the fields at $t = 0$ were generated at a previous time by the charge. The electric field can thus be written

$$\boldsymbol{E}(t = 0) = q\frac{1 - v^2/c^2}{K^3 R^3}[\boldsymbol{r} - R\boldsymbol{v}/c]_{\text{ret}},$$

where $R = |\boldsymbol{r} - \boldsymbol{v}t'| = -ct'$, while the magnetic field is simply $\boldsymbol{B} = (\boldsymbol{r} \times \boldsymbol{E})/R$. We notice that at the retarded time

$$\boldsymbol{R} - R\boldsymbol{v}/c = \boldsymbol{r} - \boldsymbol{v}t' + ct' \ \boldsymbol{v}/c = \boldsymbol{r}.$$

This implies that the electric field at $t = 0$ is parallel to $\boldsymbol{r}$, $\boldsymbol{E} \propto \boldsymbol{r}$, that is to say, it points in the direction of the observation point as seen from the present position of the charge. We see that, although the electric field at a given point and at a given time is generated at a previous time – when the charge is at the retarded position – its orientation is radial, pointing towards the position of the charge at the time at which the field is measured. This is a special case of a general property of the convection electric field (8.43): it can be shown that $\boldsymbol{E}_{\mathrm{conv}}$ is, at all times, a radial field, with origin at the position the charge would have if, starting from the retarded position, it had moved with constant velocity. Using the explicit form of the retarded time and generalizing the result to any time, we find that the electric field generated by a charge in uniform motion in the direction x is

$$\boldsymbol{E}(\boldsymbol{r}, t) = \frac{q}{\sqrt{1 - v^2/c^2}} \frac{(x - vt, y, z)}{\left[(x - vt)^2/(1 - v^2/c^2) + y^2 + z^2\right]^{3/2}}.$$

It is interesting to notice that, except for a factor, this expression is identical to that for the electrostatic field of a charge located at $x = vt$, to which a scale change has been applied – namely, a contraction by factor a $\sqrt{1 - v^2/c^2}$ – in the x direction. In Fig. 8.8 the corresponding field lines have been plotted.

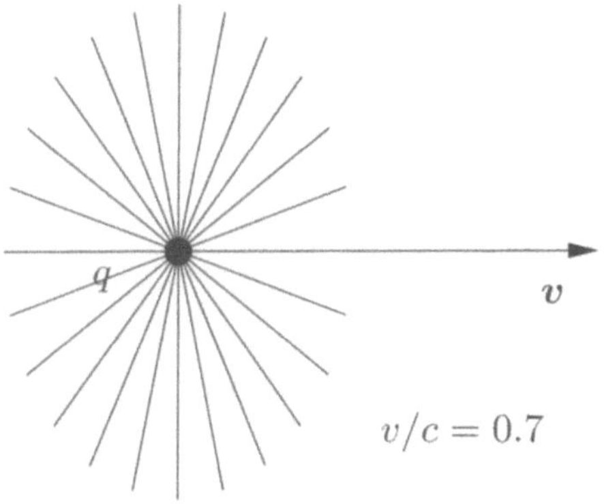

Fig. 8.8. Electric field of a charge in uniform motion

The electric field is more intense in the direction perpendicular to the motion. On the x axis, the field has intensity $E(x, t) = q(1 - v^2/c^2)/(x - vt)^2$, which, in fact, decreases and tends to zero as v comes closer to the speed of light. This result corresponds to the transformation of the electrostatic field to a reference frame moving relative to the charge. We shall study this in the next chapter.

References

8.1 J. Durnin, J.J. Miceli and J.H. Eberly: *Diffraction-Free Beams*, Phys. Rev. Lett. **58**, 1499 (1987).

Further Reading

R. Becker and F. Sauter: *Electromagnetic Fields and Interactions* (Blaisdell, New York, 1964) Part D.

R.P. Feynman, R.B. Leighton and M. Sands: *The Feynman Lectures on Physics, Vol. II* (Addison–Wesley, New York, 1964) Chap. 21.

J.M. Stone: *Radiation and Optics* (McGraw–Hill, New York, 1963) Chap. 5.

9. Covariant Formulation of Electromagnetism

Having studied in previous chapters the classical formulation of electromagnetism, whose fundamental laws are summarized in Maxwell's equations, it is now appropriate to set the theory within the framework of special relativity. As we have already mentioned, the principles of relativity underlie electromagnetic theory, so it is to be expected that the laws governing the dynamics of potentials and fields can be expressed in covariant form relative to Lorentz transformations.

In several previous chapters we made reference to the covariant structure of the theory under study. For example, in Chap. 3 we noticed that electromagnetic interactions can be described through a Lagrangian depending on the potential four-vector $A^\mu = (\phi, \boldsymbol{A})$ whose components are nothing else but the scalar and vector potentials. In Chap. 4 it was indicated that the group of quantities associated with the field sources $J^\mu = (c\rho, \boldsymbol{j})$, is a strong candidate for forming a four-vector of zero four-divergence, according to the principle of charge conservation. Moreover, we have stressed the fact that the apparent symmetry between the laws of electrostatics and those of magnetostatics should be associated with the covariance of electromagnetic theory.

In this chapter we develop the covariant formulation of Maxwell equations starting from the assumption that $J^\mu = (c\rho, \boldsymbol{j})$ and $A^\mu = (\phi, \boldsymbol{A})$ are relativistic four-vectors. This requires introducing a covariant entity describing the electromagnetic fields. As we will show, this is a second rank tensor. We also study some consequences of this formulation, such as the description of the fields associated with a moving charge and a moving dipole and those of those associated with free radiation.

9.1 Covariant Formulation of Potentials and Fields

The principle of charge conservation, a basic assumption in the theory of special relativity and endorsed by incontrovertible experimental evidence, is expressed in differential form through the continuity equation (4.6). Introducing the current density four-vector $J^\mu = (c\rho, \boldsymbol{j})$, the continuity equation can be written as

$$\partial_\mu J^\mu = 0, \tag{9.1}$$

where $\partial_\mu = \partial/\partial x^\mu$. Since x^μ is a contravariant vector, the vector ∂_μ is covariant and plays the role of a four-dimensional gradient. The previous equation indicates that the contraction of the four-dimensional gradient with the current density four-vector, i.e. its four-divergence, vanishes. We recall that, in agreement with the Einstein notation, repeated indices should be summed (see Appendix C).

In the Lorentz gauge, the potential four-vector $A^\mu = (\phi, \boldsymbol{A})$ is also a four-vector with zero four-divergence. Indeed, the condition defining the Lorentz gauge (7.44) can be written in covariant form as

$$\partial_\mu A^\mu = 0. \tag{9.2}$$

In fact, we had already pointed out that this choice of gauge explicitly shows the covariant character of the theory.

The wave equations (7.45) linking the potentials to their sources in the Lorentz gauge can be written explicitly in covariant form as

$$\Box A^\mu = \frac{4\pi}{c} J^\mu, \tag{9.3}$$

where the D'Alembert operator or *Dalembertian*, $\Box = c^{-2}\partial^2_{tt} - \nabla^2$, is the differential operator of the wave equation. This operator can be written, in explicitly covariant form, as the contraction of the four-dimensional gradient with itself

$$\Box = g^{\mu\nu}\partial_\mu\partial_\nu = \partial^\mu\partial_\mu, \tag{9.4}$$

where $g^{\mu\nu}$ is the metric tensor, defined in (3.20).

In order to obtain the covariant form of the fields, we notice that the magnetic field is given by the curl of the space part of the four-vector potential:

$$\begin{aligned} B_x &= \partial_y A_z - \partial_z A_y = -\partial^2 A^3 + \partial^3 A^2, \\ B_y &= \partial_z A_x - \partial_x A_z = -\partial^3 A^1 + \partial^1 A^3, \\ B_z &= \partial_x A_y - \partial_y A_x = -\partial^1 A^2 + \partial^2 A^1. \end{aligned} \tag{9.5}$$

According to these identities, to each component of the magnetic field a component F^{jk} of an antisymmetric tensor can be associated. The magnetic field $\boldsymbol{B}$ is thus an antisymmetric tensor of second rank in three dimensions. With its three independent components one can build a pseudovector $\boldsymbol{B}$. Recall that, in contrast to vectors, pseudovectors are not invariant under reflections of the coordinate system.

The form of the magnetic field components suggests extending the definition of the tensor F^{jk} to four dimensions by introducing the *electromagnetic field tensor* $F^{\mu\nu}$ through

$$F^{\mu\nu} = \partial^\mu A^\nu - \partial^\nu A^\mu, \tag{9.6}$$

This is an antisymmetric four-dimensional tensor of the second rank whose space part provides the magnetic field. As for the space-time part let us calculate one element, say F^{01}:

$$F^{01} = \partial^0 A^1 - \partial^1 A^0 = \frac{1}{c}\partial_t A_x + \partial_x \phi = -E_x. \tag{9.7}$$

This component thus corresponds to the x component of the electric field. In general, $F^{0i} = -E_i$. The complete form of the field tensor is

$$F^{\mu\nu} = \begin{pmatrix} 0 & -E_x & -E_y & -E_z \\ E_x & 0 & -B_z & B_y \\ E_y & B_z & 0 & -B_x \\ E_z & -B_y & B_x & 0 \end{pmatrix}. \tag{9.8}$$

The covariant components are

$$F_{\mu\nu} = g_{\mu\alpha} g_{\nu\beta} F^{\alpha\beta} = \begin{pmatrix} 0 & E_x & E_y & E_z \\ -E_x & 0 & -B_z & B_y \\ -E_y & B_z & 0 & -B_x \\ -E_z & -B_y & B_x & 0 \end{pmatrix}. \tag{9.9}$$

The field tensor $F^{\mu\nu}$ is a covariant object with respect to the Lorentz transformations. As a consequence, its image through such transformations,

$$F'^{\mu\nu} = a^{\mu}_{\alpha} a^{\nu}_{\beta} F^{\alpha\beta}, \tag{9.10}$$

has as components the electromagnetic fields in a new reference system in uniform motion with respect to the first one. The explicit form of the fields in a system S′ moving with velocity $\boldsymbol{v}$ parallel to the x axis of the original system S becomes

$$\begin{cases} E'_x = E_x, & B'_x = B_x, \\ E'_y = \gamma(E_y - \beta B_z), & B'_y = \gamma(B_y + \beta E_z), \\ E'_z = \gamma(E_z + \beta B_y), & B'_z = \gamma(B_z - \beta E_y), \end{cases} \tag{9.11}$$

where $\beta = v/c$ and $\gamma = (1-\beta^2)^{-1/2}$.

When applying the Lorentz transformations to the electromagnetic fields it should be kept in mind that these, in general, depend on the space coordinates and time and that these variables should also be transformed. For example, let us consider the transformation of the electrostatic field of a point charge q placed at the origin to a reference system in uniform motion with velocity $\boldsymbol{v}$ along the x axis. As there is no magnetic field in the original system, the component of the electric field E'_x in the direction of motion is

$$E'_x = E_x = q\,\frac{x}{(x^2+y^2+z^2)^{3/2}} = q\,\frac{(x'+vt')/\sqrt{1-\beta^2}}{\left(\frac{(x'+vt')^2}{1-\beta^2} + y'^2 + z'^2\right)^{3/2}}. \tag{9.12}$$

Proceeding similarly with the remaining components of the field tensor it is possible to obtain the full electromagnetic fields of a charge moving with constant velocity.

Once the field tensor has been defined, the covariant form of Maxwell's equations must involve $F^{\mu\nu}$. For example, let us evaluate the divergence of the field tensor:

$$\partial_\mu F^{\mu\nu} = \partial_\mu(\partial^\mu A^\nu - \partial^\nu A^\mu) = \Box A^\nu - \partial^\nu(\partial_\mu A^\mu). \tag{9.13}$$

Due to the Lorentz condition, the second term vanishes, and in view of the wave equation for the four-dimensional potentials we have

$$\partial_\mu F^{\mu\nu} = \frac{4\pi}{c} J^\nu. \tag{9.14}$$

This equation corresponds, in the classical formulation, to Gauss's equation and to the modified Ampère's equation, which relate the fields to their sources.

The other two Maxwell's equations are obtained from the following property

$$\partial^\alpha F^{\beta\gamma} + \partial^\beta F^{\gamma\alpha} + \partial^\gamma F^{\alpha\beta} = 0. \tag{9.15}$$

This identity is the four-dimensional equivalent of the divergence of a curl, as can be proved in direct form using the definition of the field tensor. If at least two of the indices (α, β, γ) are equal, the previous equation becomes a trivial identity, due to the antisymmetry of $F^{\mu\nu}$. If the three indices are different, on the other hand, the equation is equivalent to Faraday's law and to $\nabla \cdot \boldsymbol{B} = 0$.

In summary, the fundamental equations of electromagnetism are given, in covariant form, by

$$\begin{cases} \partial_\mu J^\mu = 0, & \text{Charge conservation} \\ \partial_\mu A^\mu = 0, & \text{Lorentz condition} \\ \Box A^\mu = \frac{4\pi}{c} J^\mu, & \text{Wave equation} \\ F^{\mu\nu} = \partial^\mu A^\nu - \partial^\nu A^\mu, & \text{Field tensor} \\ \partial_\mu F^{\mu\nu} = \frac{4\pi}{c} J^\mu, & \text{Gauss-Ampère} \\ \partial^\lambda F^{\mu\nu} + \partial^\mu F^{\nu\lambda} + \partial^\nu F^{\lambda\mu} = 0. & \text{Faraday}-\nabla \cdot \boldsymbol{B} = 0 \end{cases} \tag{9.16}$$

The last equation in (9.16) can be rewritten in a simpler form defining the dual tensor of $F^{\mu\nu}$. For this, we introduce the completely antisymmetric fourth-rank tensor $\varepsilon^{\alpha\beta\gamma\delta}$, given by

$$\varepsilon^{\alpha\beta\gamma\delta} = \begin{cases} 1, & \text{if } (\alpha,\beta,\gamma,\delta) \text{ is an even permutation of } (0,1,2,3;); \\ -1, & \text{if } (\alpha,\beta,\gamma,\delta) \text{ is an odd permutation of } (0,1,2,3); \\ 0, & \text{if there are two repeated indices.} \end{cases} \tag{9.17}$$

The dual tensor of $F^{\mu\nu}$ is

$$G^{\mu\nu} = \frac{1}{2}\varepsilon^{\mu\nu\gamma\delta}F_{\gamma\delta} = \begin{pmatrix} 0 & -B_x & -B_y & -B_z \\ B_x & 0 & E_z & -E_y \\ B_y & -E_z & 0 & E_x \\ B_z & E_y & -E_x & 0 \end{pmatrix}. \tag{9.18}$$

In terms of this tensor the last equation in (9.16) can be written as

$$\partial_\mu G^{\mu\nu} = 0. \tag{9.19}$$

As a first application of the covariant formulation of electromagnetism, let us calculate two relativistic invariants associated with the fields. Contracting the field tensor with itself and with the dual tensor, we obtain

$$\begin{aligned} F^{\mu\nu}F_{\mu\nu} &= 2\,(B^2 - E^2), \\ G^{\mu\nu}F_{\mu\nu} &= -4\,\boldsymbol{B}\cdot\boldsymbol{E}. \end{aligned} \tag{9.20}$$

This means that the quantities $B^2 - E^2$ and $\boldsymbol{B}\cdot\boldsymbol{E}$ must have the same value in all inertial reference frames. Being the scalar product of a pseudovector and a vector, $\boldsymbol{B}\cdot\boldsymbol{E}$ is in fact a pseudoscalar whose sign depends on the relative orientation of the coordinate axis.

These results have several interesting implications. For example, they imply that a charge in uniform motion cannot give rise to radiation fields. Indeed in a system at rest relative to the charge $\boldsymbol{B} = 0$ and thus $B^2 - E^2 < 0$. But, as we have shown in Chap. 8, in a radiation field $B = E$. Consequently the electrostatic field cannot give rise to radiative components under a Lorentz transformation.

Another implication of these invariant quantities, also referred to radiation fields, is that whatever the inertial reference system from which such fields are observed, they must always satisfy $B = E$ and $\boldsymbol{B} \perp \boldsymbol{E}$. Later on, we will develop in more detail the covariant characteristics of the radiation fields.

9.2 Covariant Form of the Field of Charges and Dipoles

The covariant formulation of electromagnetism manifests all its formal beauty if it is accepted as a general rule that the relevant quantities of the theory should be the simplest compatible with the symmetry implied by the covariance. In this way, for example, the potential four-vector A^ν associated with a moving point charge should be determined only by the covariant elements associated with the motion. For a charge in uniform motion the only four-vectors at our disposal are the positions of the charge q, x_q^ν, and of the observation point P, x_P^ν, and the velocity four-vector of the charge, V^ν.

Given the isotropy of space, the field can only depend on the relative position of P and q, that is to say, on $X^\nu = x_P^\nu - x_q^\nu$. The potential four-vector must thus be a function of these four-vectors and, possibly, of the scalars that can be derived from them: $X^\nu X_\nu$, $X^\nu V_\nu$, and $V^\nu V_\nu$.

It can be shown that the four-vector

$$A^\nu(x) = \frac{q}{c} \frac{V^\nu}{\left[(V^\lambda X_\lambda)^2/c^2 - X^\lambda X_\lambda\right]^{1/2}} \tag{9.21}$$

reproduces the field of a charge at rest and corresponds to the vector potential of a charge in uniform motion.

The electromagnetic field $F^{\mu\nu}$ is derived from (9.21) by application of the definition (9.6). The exercise solved in the Problem section shows that the field tensor $F^{\mu\nu}$ can also be obtained using an argument similar to the one used here to determine A^ν.

In Chap. 6 we have indicated that the symmetry between electrostatics and magnetostatics is apparent in the form of the respective dipole fields. We show now how the potential of a magnetic dipole can be written in covariant form, making evident the origin of the symmetry mentioned. Let us assume that at the origin of the reference system S there is a particle having an electric dipole of magnitude $\boldsymbol{p}$ and a magnetic dipole of magnitude $\boldsymbol{m}$. This situation could represent, for instance, the case of a neutron. This is a particle without net charge possessing, however, a weak electric dipole moment and an intrinsic magnetic dipole moment.

We can postulate the existence of a moment tensor $m^{\mu\nu}$ having as components $\boldsymbol{p}$ and $\boldsymbol{m}$, in the same way as $F^{\mu\nu}$ has as components the electromagnetic fields $\boldsymbol{E}$ and $\boldsymbol{B}$. Given the vector character of each of the components, this tensor should have the form

$$m^{\mu\nu} = \begin{pmatrix} 0 & -p_x & -p_y & -p_z \\ p_x & 0 & m_z & -m_y \\ p_y & -m_z & 0 & m_x \\ p_z & m_y & -m_x & 0 \end{pmatrix}. \tag{9.22}$$

The simplest form of the potential four-vector obtained from the requirement of covariance used in the previous example, is

$$A^\mu = \frac{m^{\mu\nu} X_\nu}{\left[(V^\lambda X_\lambda)^2/c^2 - X^\lambda X_\lambda\right]^{3/2}}. \tag{9.23}$$

This form satisfies the requirement of being proportional to the moment tensor and reproduces the potential for a dipole moment at rest. In the reference system S it takes the form:

$$\begin{aligned} A^0 &= (p_x x + p_y y + p_z z)/r^3, \\ A^1 &= \quad (m_y z - m_z y)/r^3, \\ A^2 &= \quad (m_z x - m_x z)/r^3, \\ A^3 &= \quad (m_x y - m_y x)/r^3. \end{aligned} \tag{9.24}$$

Using the transformation laws for the potential A^μ, we have

$$A'^\nu = a^\nu{}_\mu A^\mu. \tag{9.25}$$

From this the potentials or the fields for charges or dipoles in uniform motion can be obtained.

9.3 Lorentz Force and Energy–Momentum Tensor

As we have mentioned several times above, the force acting on a charge q moving with velocity $\boldsymbol{v}$ in the presence of electromagnetic fields is given by the *Lorentz force*:

$$\boldsymbol{F} = q\left(\boldsymbol{E} + \frac{\boldsymbol{v} \times \boldsymbol{B}}{c}\right). \tag{9.26}$$

If this force acts on a system of charges described by a charge density $\rho(r,t)$, it is possible to define a density of force $\boldsymbol{f}$, such that the total force on the charge in a volume element d^3r is

$$\mathrm{d}\boldsymbol{F} = \boldsymbol{f}\ \mathrm{d}^3r. \tag{9.27}$$

The reader can check that the density of force is

$$\boldsymbol{f} = \rho\boldsymbol{E} + \frac{\boldsymbol{j} \times \boldsymbol{B}}{c}. \tag{9.28}$$

In terms of the components of the field tensor and the current density four-vector, the density of force is given by

$$\boldsymbol{f} = \frac{1}{c}\ \left(F^{1\nu}J_\nu, F^{2\nu}J_\nu, F^{3\nu}J_\nu\right). \tag{9.29}$$

Generalizing this result to four-dimensional space we define the force density four-vector f^μ as

$$f^\mu = \frac{1}{c}\ F^{\mu\nu}J_\nu, \tag{9.30}$$

whose three space components coincide with $\boldsymbol{f}$. The time component is

$$f^0 = \frac{1}{c}\boldsymbol{j} \cdot \boldsymbol{E}. \tag{9.31}$$

This represents the power transferred by the field to the charge distribution per unit volume.

In order to define the concept of electromagnetic energy density in the covariant formulation, we try to write the force density four-vector as the divergence of a tensor, that is to say, as

$$f^\mu = -\partial_\nu T^{\mu\nu}. \tag{9.32}$$

Making use of the covariant Maxwell equations and after some manipulations, we find as solution for the tensor $T^{\mu\nu}$:

$$T^{\mu\nu} = -\frac{1}{4\pi}g^{\mu\alpha}\left(F^{\beta\nu}F_{\beta\alpha} - \frac{1}{4}\ \delta^\nu_\alpha F^{\sigma\tau}F_{\sigma\tau}\right), \tag{9.33}$$

as a function of the electromagnetic fields. Here δ^ν_μ is the unit tensor in four-dimensional space.

Let us analyze the components of $T^{\mu\nu}$. In the first place, for $\mu = \nu = 0$, we have

$$T^{00} = \frac{1}{8\pi}(E^2 + B^2), \tag{9.34}$$

which coincides with the electromagnetic energy density w. As for the space–time components, we have

$$T^{01} = T^{10} = \frac{1}{4\pi}(\boldsymbol{E} \times \boldsymbol{B})_x = \frac{1}{c}\, S_x, \tag{9.35}$$

and similarly $T^{02} = S_y/c$, $T^{03} = S_z/c$, where $\boldsymbol{S} = (S_x, S_y, S_z)$ is the Poynting vector. The space–time part of the tensor $T^{\mu\nu}$ is thus proportional to the density of linear momentum of the electromagnetic field. Finally, the space components of $T^{\mu\nu}$ coincide, except for a sign, with the corresponding components of the Maxwell stress tensor $\mathcal{T}$, introduced in Chap. 7. The tensor $T^{\mu\nu}$ turns out to be symmetric. In matrix form it reads

$$T^{\mu\nu} = \begin{pmatrix} w & S_x/c & S_y/c & S_z/c \\ S_x/c & -\mathcal{T}_{xx} & -\mathcal{T}_{xy} & -\mathcal{T}_{xz} \\ S_y/c & -\mathcal{T}_{yx} & -\mathcal{T}_{yy} & -\mathcal{T}_{yz} \\ S_z/c & -\mathcal{T}_{zx} & -\mathcal{T}_{zy} & -\mathcal{T}_{zz} \end{pmatrix}. \tag{9.36}$$

This tensor thus contains information related to the energy and momentum of the electromagnetic field. In view of this, it is called *electromagnetic energy–momentum tensor.*

9.4 Covariant Properties of the Free Radiation Field

In this section we study, from the point of view of the covariant formulation, some of the solutions of the wave equation for the electromagnetic potentials in free space

$$\Box\, A^\mu = 0, \tag{9.37}$$

whose classical formulation has already been considered in Sect. 8.1. A particular solution of this equation can be obtained from two four-vectors independent of the coordinates, a^μ and k^μ, in the form

$$A^\mu = a^\mu f(k^\nu x_\nu). \tag{9.38}$$

Independently of the function f, this is a solution to (9.37) if the relation

$$k^\mu\, k_\mu = 0 \tag{9.39}$$

is satisfied. That is to say, the four-vector k^μ, called the *wavenumber four-vector*, must have vanishing modulus. Therefore, this four-vector can be written in its ordinary components as $k^\mu = (k, \boldsymbol{k})$, where $k = |\boldsymbol{k}|$. In addition, it can be shown that the Lorentz condition requires

$$a^\mu k_\mu = 0 \tag{9.40}$$

to be satisfied by the four-vector a^μ.

As a function of the coordinates and time, each component of the potential four-vector turns out to be $A^\mu = a^\mu f(kct - \boldsymbol{k}\cdot\boldsymbol{r})$. This expression describes a perturbation propagating in the direction of the vector $\boldsymbol{k}$ with the speed of light c. The amplitude of the perturbation is determined by the four-vector a^μ, while its phase is given by the product $k^\nu x_\nu = kct - \boldsymbol{k}\cdot\boldsymbol{r}$. Since at a fixed time this phase is constant on the plane orthogonal to the wave vector $\boldsymbol{k}$, the perturbation is a plane wave. It should be noted that the phase is given by the scalar product of two four-vectors; therefore, it is a relativistic invariant, independent of the reference system in which it is measured.

In order to simplify the discussion, from here on we consider harmonic perturbations, taking

$$A^\mu = a^\mu \exp(\mathrm{i}k^\nu x_\nu), \tag{9.41}$$

in analogy with the solutions studied in the previous chapter. In these conditions the wave has a well-defined frequency, given by $\omega = ck$. We consider that the wave vector $\boldsymbol{k}$ has components in the x direction only, so that $k^\mu = (k, k, 0, 0) = (\omega/c, k, 0, 0)$. In this situation, the Lorentz condition (9.40) imposes $a^0 = a^1$. Calculating the field tensor $F^{\mu\nu}$ from (9.6), we obtain

$$F^{\mu\nu} = \mathrm{i}k\ \exp(\mathrm{i}k^\nu x_\nu)\begin{pmatrix} 0 & 0 & a^2 & a^3 \\ 0 & 0 & a^2 & a^3 \\ -a^2 & -a^2 & 0 & 0 \\ -a^3 & -a^3 & 0 & 0 \end{pmatrix}. \tag{9.42}$$

From this we can write the electromagnetic fields explicitly as

$$\begin{aligned} \boldsymbol{E}(\boldsymbol{r},t) &= -\mathrm{i}k\,(0, a^2, a^3)\ \exp[\mathrm{i}(\omega t - \boldsymbol{k}\cdot\boldsymbol{r})], \\ \boldsymbol{B}(\boldsymbol{r},t) &= -\mathrm{i}k\,(0, a^3, -a^2)\ \exp[\mathrm{i}(\omega t - \boldsymbol{k}\cdot\boldsymbol{r})]. \end{aligned} \tag{9.43}$$

Here (a^1, a^2, a^3) indicate vectors with three space components. The fields associated with the free plane wave are orthogonal to the wave vector $\boldsymbol{k}$ and to each other. Also, their amplitudes, given by the constants a^2 and a^3, are the same, implying $E = B$. We have shown above that these characteristics are relativistic invariants and, as pointed out in Chap. 8, correspond to general properties of the radiation fields.

For a radiation field occupying a source-free bounded volume V, it can be shown that the energy–momentum tensor has zero divergence

$$\partial_\mu T^{\mu\nu} = 0. \tag{9.44}$$

This equation is equivalent to the energy and momentum conservation laws. Applying the theorem studied in Chap. 4 when discussing charge conservation, which can be extended to higher rank tensors, it can be shown that the four quantities

$$p^\mu = \frac{1}{c}\int_V T^{0\mu}\ \mathrm{d}^3r, \tag{9.45}$$

form a four-vector. Explicitly, their components are

$$p^0 = \frac{1}{c}\int_V w \, \mathrm{d}^3r, \qquad \boldsymbol{p} = \frac{1}{c^2}\int_V \boldsymbol{S} \, \mathrm{d}^3r, \tag{9.46}$$

that is, they correspond to the energy and the total momentum of the field. The fact that these quantities form a four-vector suggests associating a particle character to such a field.

If $\boldsymbol{n}$ is the direction of propagation of the wave, its energy and momentum are given respectively by $W = V(E^2 + B^2)/8\pi$ and $\boldsymbol{p} = V(E^2 + B^2)\boldsymbol{n}/8\pi c$. The momentum four-vector corresponding to the field is

$$p^\mu = \frac{VE^2}{4\pi c}\,(1, \boldsymbol{n}), \tag{9.47}$$

since $E = B$. The modulus of this four-vector is $p^\mu p_\mu = 0$. Recalling that the invariant associated with the modulus of the momentum is proportional to the rest mass of the particle, we can say that the particle associated with the radiation field must have zero mass. In fact, according to the theory of the relativity, this is the only way in which such a particle could propagate with velocity c.

It can be proved that, according to the Lorentz transformations, the energies W and W' of the radiation field observed from two different systems S and S′ moving with relative velocity $v = \beta c$ are related according to

$$W' = W\sqrt{\frac{1-\beta}{1+\beta}}, \tag{9.48}$$

when the relative motion of the reference systems has the same direction as that of the radiation propagation. Moreover, if the radiation field has a well defined frequency ω in S, the Lorentz transformations allow us to relate ω to the frequency ω' observed in S′, since ω/c is the time component of the four-vector wave number:

$$\omega' = \omega\sqrt{\frac{1-\beta}{1+\beta}}. \tag{9.49}$$

From this we conclude that

$$\frac{W}{\omega} = \frac{W'}{\omega'}. \tag{9.50}$$

We invite the reader to prove that this result is also valid for arbitrary orientation of the wave propagation and the relative motion between the two reference systems.

The proportionality between energy and frequency of the radiation field is of particular importance from the point of view of the quantum theory of radiation. According to this theory, a field of finite extension with well defined frequency – a wave train – consists of a certain number N of particles of zero rest mass, the photons. The proportionality between energy and frequency can be written as

$$W = N\hbar\omega. \tag{9.51}$$

Here $\hbar\omega$ represents the energy of each photon, where $\hbar = h/2\pi$, and h is the Planck constant. The invariance of W/ω implies that of $N\hbar$. As a consequence, since the number of particles associated with the wave train should be the same in any reference system, the Planck constant is invariant under Lorentz transformations.

9.5 Electromagnetic Theory of the Electron

As we have already indicated in Chap. 4, the mass and charge of the electron have been obtained experimentally. From such data it is known that this particle behaves in total agreement with the predictions of the theory of relativity. In particular, according to the Bucherer experiment, its mass depends on velocity in the form predicted by the theory. Moreover, we have mentioned in Chap. 4 that, when at rest, the electron is surrounded by its electrostatic field. The energy of this field can be associated with an inertial mass, called the *electromagnetic mass*. Explicitly evaluating the energy of the field of a charged sphere at rest, and equating this result to m_0c^2, where m_0 is the rest mass of the electron, the concept of electromagnetic mass allows us to associate a "classical" radius to this particle.

Once the covariant formulation of the electromagnetic theory has been developed, the assumption that all the mass of the electron can be associated with the inertia of its electrostatic field naturally suggests the covariance hypothesis. Such a hypothesis constitutes the basis of the Abraham–Lorentz model for the electron, developed in the first decades of the twentieth century. According to this model, the energy and momentum of the electron are given by space integrals of the corresponding densities associated with its electromagnetic field:

$$\begin{aligned} W &= \int T^{00}\ \mathrm{d}^3r = \int w\ \mathrm{d}^3r, \\ \boldsymbol{p} &= c^{-1}\int (T^{01}, T^{02}, T^{03})\ \mathrm{d}^3r = c^{-2}\int \boldsymbol{S}\ \mathrm{d}^3r. \end{aligned} \tag{9.52}$$

Through the transformation laws of the energy–momentum tensor $T^{\mu\nu}$, the energy W and the momentum $\boldsymbol{p}$ of the electron measured in a system S can be linked to the quantities W' and $\boldsymbol{p}'$ in a system S′.

If in the original system the electron is at rest and its energy is W_0, and the system S′ moves relative to S with velocity v in the x direction, the Lorentz transformations imply

$$W' = W_0(1+\beta^2/3)/\sqrt{1-\beta^2},$$
$$p'_x = \tfrac{4}{3}m_0 v/\sqrt{1-\beta^2}, \tag{9.53}$$

where $\beta = v/c$ and $m_0 = W_0/c^2$ is the electromagnetic mass of the electron. According to the second of these relations, in the Abraham–Lorentz model the momentum is not simply the product of the electromagnetic mass times the velocity, but rather it includes a spurious factor 4/3. Moreover, the transformation of energy does not correspond to that of the fourth component of a four-vector.

In the context of the Abraham–Lorentz model it was at this point assumed that this inconsistency in the definition of the electromagnetic energy–momentum four-vector of the electron was related to the stability of the particle. In fact, in order for the charge distribution making up the electron to be stable, it was necessary to assume the existence of a force different from the electromagnetic interaction. This would have an associated energy whose inertia would give rise to an additional, nonelectromagnetic contribution to the mass. In the best case, such an ingredient should solve both the stability problem and the question of covariance of the energy–momentum of the electron.

The problem has been discussed in more modern publications, especially by Rohrlich in his paper *Self-Energy and Stability of the Classical Electron* [9.1]. This author considered that starting from the tensor $T^{\mu\nu}$ it is possible to define a four-vector

$$p^\mu = \frac{1}{c}\int_\Sigma T^{\mu\nu}\mathrm{d}\sigma_\nu. \tag{9.54}$$

According to (9.32) this four-vector is independent of the integration hypersurface Σ, a three-dimensional manifold, if and only if

$$\partial_\mu T^{\mu\nu} = 0. \tag{9.55}$$

This relation, which corresponds to the conservation of electromagnetic energy and momentum, is satisfied at all points where there is no charge. In the Abraham–Lorentz model p^μ and p'^μ are calculated with two different hypersurfaces $\Sigma = (x, y, z)$ and $\Sigma' = (x', y', z')$, respectively, each containig points where there are charges. Therefore, the four-divergence of the tensor $T^{\mu\nu}$ is non-zero, and p^μ and p'^μ are in fact two different four-vectors. It is not surprising then that they are not related by a Lorentz transformation.

In order to define a unique four-vector p^μ under conditions in which (9.55) is not satisfied we must choose a single integration hypersurface Σ. For an electron in uniform motion the four-dimensional velocity V^μ is a contravariant four-vector associated with the motion. This suggests that a hypersurface Σ normal to V^μ is the only hypersurface singled out by the motion. This hypersurface coincides with the three-dimensional space in the reference system in which the electron is at rest. Explicitly calculating the components of p^μ gives for the energy and momentum:

$$W = \int w \, d^3r - c^{-2} \int \boldsymbol{S} \cdot \boldsymbol{v} \, d^3r,$$
$$\boldsymbol{p} = c^{-2} \left(\int \boldsymbol{S} \, d^3r - \int \mathcal{T} \cdot \boldsymbol{v} \, d^3r \right). \tag{9.56}$$

In both cases, the expressions present additional terms to those of the Abraham–Lorentz model.

It is possible to prove rigorously that the additional terms solve the inconsistency of the model in relation to its covariance. Here we illustrate this fact in the special case where the electron moves at low speed. Under these conditions we can write

$$\boldsymbol{E} = e\frac{\boldsymbol{r}}{r^3}, \qquad \boldsymbol{B} = \frac{1}{c}\boldsymbol{v} \times \boldsymbol{E}. \tag{9.57}$$

The term modifying the component of the linear momentum in the direction of motion along x is, neglecting terms of order β^2,

$$-\frac{1}{4\pi c^2} \int \sum_{j=1}^{3} \mathcal{T}_{1j} v_j d^3r = -\frac{v}{24\pi c^2} \int E^2 \, d^3r = \frac{W_0 v}{3c^2} = -\frac{1}{3} m_0 v. \tag{9.58}$$

Thus, this term corrects exactly the spurious factor 4/3 affecting the momentum in the Abraham–Lorentz model.

Although the choice of a single integration hypersurface Σ solves the problem of covariance of the momentum four-vector in the electromagnetic model of the electron, the stability of the associated charges is not solved. In the present formulation, this second problem appears totally separate to the first one. The stability problem can be solved by means of renormalization techniques. In the classical model the electron is unstable due to the mutual repulsion between its different parts. The renormalization process consists of separating the field, in a covariant way, into a contribution acting on the electron itself and another acting on the external charges. Only the second part contributes to the field of the "renormalized electron." As we have already mentioned in Chap. 4, this idea lies in the spirit of defining the electron as an elementary particle. Within the framework of quantum mechanics, it is meaningless to consider nonobservable quantities, and the field of an elementary particle acting on itself is one such meaningless quantity.

9.6 A "Derivation" of Maxwell's Theory

The electromagnetic field is responsible for the interaction between charges. In this sense it is an *intermediary field*, since it plays the role of mediating the force between the elementary particles having electric charge. Its properties are known as a result of long years of experimentation and of theoretical analysis and are summarized in Maxwell's equations, whose covariant form we show in this chapter. In its quantum mechanical version, the electromagnetic field is represented by *light quanta* or *photons*, one of the elementary particles.

Modern theories of the forces between elementary particles predict the existence of other *intermediary fields* for the transmission of weak and strong forces. In contrast to electromagnetic forces, subatomic forces provide far fewer elements for setting up a theoretical framework to explain known phenomena and predict others. In particular, with such forces it is not possible to perform macroscopic experiments of the same type as those of Ampère or Faraday, which led to the formulation of electromagnetic theory by Maxwell.

To solve these difficulties, modern theoretical physics makes use of general principles allowing the formulation of the theory of intermediary fields. Within the limits set by the scope of this book, we shall explain the type of reasoning leading to the so-called *gauge field theories*, based on an analogy with Maxwell's theory. These theories introduce the so-called *gauge bosons*, which are quanta associated with the intermediary fields of the other forces between elementary particles.

In classical mechanics the electromagnetic interaction acts through the Lorentz force, which depends on the fields and not on the potentials. For this reason, in the classical formulation it is not so important to stress the role of the gauge invariance of electromagnetic theory. This situation contrasts with what happens in quantum theory, where the interaction with the electromagnetic field is necessarily expressed through the vector potential A^ν.

The gauge invariance of classical physics, discussed in Chap. 7, can be expressed in covariant form noting that (9.6) is invariant under gauge or *gradient* transformations:

$$A'^\nu = A^\nu + \partial^\nu \Lambda, \tag{9.59}$$

where Λ is a function of the coordinates and $\partial^\nu \Lambda$ represents the four-dimensional gradient of Λ. The choice of the Lorentz gauge $\partial_\nu A^\nu = 0$ implies that Λ satisfies $\Box \Lambda = 0$.

To illustrate the point of view of the gauge theories, we present a "derivation" of Maxwell's theory starting from similar principles to those used in modern theoretical physics. To do this we assume that we know classical mechanics but we do not know anything about electromagnetism. Is it possible under these conditions to derive Maxwell's theory starting from some general principles?

Classical mechanics is contained in the principle of minimum action introduced in Chap. 3. This can also be written in terms of a Lagrangian density in the form

$$\mathcal{S} = \int \mathcal{L}_0 \, \mathrm{d}\tau. \tag{9.60}$$

Let us assume that we know that the particles involved possess an intrinsic quality, the charge, which has associated with it a current, which we write in covariant form as

$$J^\mu(x) = qV^\mu \delta(x - x_0(\tau)), \tag{9.61}$$

where $x_0(\tau)$ is the position of the particle. The equations of motion are not altered if we add to the Lagrangian $\mathcal{L}_0$ a four-divergence in the form

$$\mathcal{L}_0' = \mathcal{L}_0 + \partial_\nu (J^\nu \Lambda), \tag{9.62}$$

where Λ is an arbitrary function of the coordinates. Indeed, the added four-divergence gives rise, in (3.32), to a surface contribution that can be eliminated. The Lagrangian (9.62) takes the value

$$\mathcal{L}_0' = \mathcal{L}_0 + \Lambda \partial_\nu J^\nu + J^\nu \partial_\nu \Lambda, \tag{9.63}$$

so that the form of $\mathcal{L}_0'$ is different from that of $\mathcal{L}_0$.

Gauge theories make use of the so called *principle of form invariance* requiring that, when a four-divergence like that in (9.62) is added, the Lagrangian should preserve its form. This is achieved by introducing a *compensating field* represented by a four-potential A^ν that couples to the particle current J^ν so that the complete Lagrangian of the system is

$$\mathcal{L}_{0,I} = \mathcal{L}_0 - J^\nu A_\nu . \tag{9.64}$$

If now the four-divergence $\partial_\nu (J^\nu \Lambda)$ is added to this Lagrangian, we obtain

$$\mathcal{L}_{0,I}' = \mathcal{L}_0' - J^\nu A_\nu' . \tag{9.65}$$

This has the same form as the original Lagrangian if the current J^ν is conserved and if the new four-potential is

$$A_\nu' = A_\nu + \partial_\nu \Lambda . \tag{9.66}$$

We recognize here the gauge transformation of Maxwell's theory. From the four-potential A_ν it is possible to define an antisymmetric tensor

$$F^{\mu\nu} = \partial^\mu A^\nu - \partial^\nu A^\mu , \tag{9.67}$$

which automatically satisfies the relationship

$$\partial^\alpha F^{\beta\gamma} + \partial^\beta F^{\gamma\alpha} + \partial^\gamma F^{\alpha\beta} = 0. \tag{9.68}$$

As we have shown, this relationship corresponds to Faraday's law and to the non-existence of monopoles. In order to obtain the other Maxwell's equations in this way it is necessary to consider the complete Lagrangian of the system of particles and field, and to apply the principle of minimum action. This procedure shows that starting from the principle of form invariance classical mechanics allows us to derive the existence of a compensating field, a role performed by the electromagnetic field. We recommend reading the paper by Donald H. Kobe: *Derivation of Maxwell's Equations from the Gauge Invariance of Classical Mechanics* [9.2].

Problems

9.1 Using the covariant form of Maxwell's equations, show that the equation for the Lorentz force

$$f^\mu = \frac{1}{c} F^{\mu\nu} J_\nu$$

can be expressed as

$$f^\mu = -\partial_\nu T^{\mu\nu},$$

where

$$T^{\mu\nu} = -\frac{1}{4\pi} g^{\mu\alpha} (F^{\beta\nu} F_{\beta\alpha} - \frac{1}{4} \delta^\nu_\alpha F^{\sigma\tau} F_{\sigma\tau}).$$

Hint: Follow the steps indicated in the text to arrive at Eq. (7.57).

9.2 Show that the condition of zero divergence for $T^{\mu\nu}$ implies the laws of energy and momentum conservation.

9.3 Consider a finite region of space, free of charges and currents, occupied by electromagnetic radiation. (a) Show that the electromagnetic momentum and energy associated with the field transform as a four-vector. (b) Show that in the case of a plane wave this four-vector has zero length but that this is not so for other possible configurations of the field (for example, for an outgoing spherical wave).

9.4 Consider the tensor

$$M^{\alpha\beta\gamma} = T^{\alpha\beta} x^\gamma - T^{\alpha\gamma} x^\beta.$$

(a) Show that it has zero divergence, $\partial_\alpha M^{\alpha\beta\gamma} = 0$. (b) Identify the components of the tensor $M^{\alpha\beta\gamma}$ associated with the angular momentum density of the electromagnetic field:

$$\boldsymbol{l} = \frac{1}{c^2} \boldsymbol{r} \times \boldsymbol{S}.$$

(c) Relate the result of (a) to the conservation of angular momentum. What other conservation laws can be derived from (a)?

Hint: Write down the tensor M explicitly in non-covariant notation.

9.5 Calculate the fields $\boldsymbol{E}$ and $\boldsymbol{B}$ between the plates of a plane capacitor with charge densities $\pm\sigma$ (Fig. 9.1) as seen from a moving reference frame, when (a) it moves parallel to the plates; (b) it moves perpendicular to the plates. Discuss and compare the results.

9.6 Verify that the potential four-vector introduced in the text for uniformly moving charges and dipoles satisfies the Lorentz gauge condition.

9.7 Write in covariant form the equations of motion for a particle in the presence of an electromagnetic field. Study the laws of energy and momentum conservation in the laboratory system and in the particle system.

Hint: The equations simplify considerably when written in terms of the proper time of the particle.

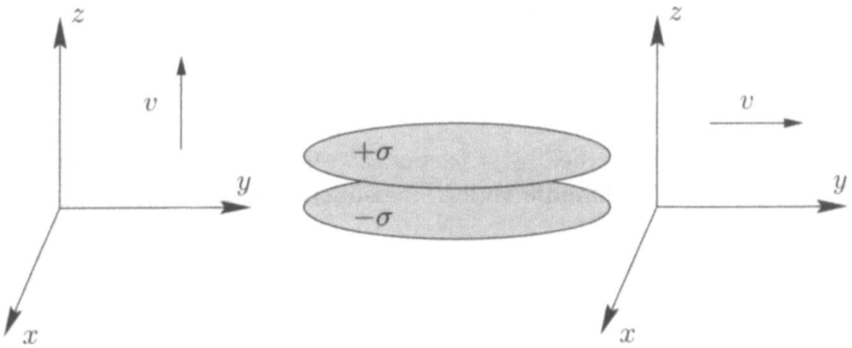

Fig. 9.1. A plane capacitor seen from moving reference frames

9.8 We have shown in the text that $k^\mu = (\omega/c, \boldsymbol{k})$ is a four-vector, where ω is the angular frequency of the wave and $k = |\boldsymbol{k}|$ is the wave number. Use this result to study the transformation properties of the frequency and of the wave vector $\boldsymbol{k}$, when an electromagnetic wave is observed from two inertial reference systems. In particular, pay attention to (a) the transformation law for the frequency when the wave moves parallel to the relative velocity of the inertial systems (Doppler effect); (b) the transformation law for the frequency when the wave moves perpendicularly to the relative motion of the two systems (transverse Doppler effect); (c) the change in the direction of propagation of a wave (aberration). Compare with the result obtained starting from the velocity addition law.

Hint: Consider first the cases of propagation parallel and perpendicular to the direction of relative motion of the two systems.

9.9 Determine the electromagnetic field tensor associated with a point charge in uniform motion using covariance arguments only.

Solution: The only antisymmetric tensor that can be built with the covariant elements at our disposal, that is X^μ and V^ν, is $X^\mu V^\nu - X^\nu V^\mu$. The electromagnetic field must be proportional to this tensor and the proportionality factor must be given by the electric charge of the particle and by a combination of the invariants. By analogy with (9.21) we consider the expression

$$F^{\mu\nu} = \frac{q}{c} \frac{(X^\mu V^\nu - X^\nu V^\mu)}{\left[(V^\lambda X_\lambda)^2/c^2 - X^\lambda X_\lambda\right]^{3/2}},$$

where we have chosen the power $3/2$ in the denominator in analogy with the classical (non-covariant) expression for the electric field. In order to verify that this expression is the correct one, we particularize it to the case of a charge at rest at the origin. In such a case we have $X^\nu = x_p^\nu - x_q^\nu = x_p^\nu$ and, therefore, from the above expression we obtain

$$F^{0i} = \frac{q}{c} \frac{(X^0V^i - X^iV^0)}{\left[(V^\lambda X_\lambda)^2/c^2 - X^\lambda X_\lambda\right]^{3/2}} = q\frac{-x_p^i}{r^{3/2}} = -E^i,$$

$$F^{ij} = 0.$$

This corresponds to the field of a particle at rest, (4.13). It turns out that the general expression found here for $F^{\mu\nu}$ is identical to (9.12).

References

9.1 F. Rohrlich: *Self-Energy and Stability of the Classical Electron*, Am. J. Phys. **28**, 639 (1960).
9.2 Donald H. Kobe: *Derivation of Maxwell's Equations from the Gauge Invariance of Classical Mechanics*, Am. J. Phys. **48**, 348 (1980).

Further Reading

L.D. Landau and E.M. Lifshitz: *The Classical Theory of Fields* (Pergamon, Oxford and Addison–Wesley, Reading, 1971).
H.A. Lorentz: *Theory of Electrons* (Dover, New York, 1952).
F. Rohrlich: in *Lectures in Theoretical Physics, Vol. II* (Interscience, New York, 1960).

10. Fields in Material Media

In previous chapters we have studied the fundamental laws of electromagnetism and some of their most important applications in free space. With this, we have developed the appropriate tools to deal with the analysis of electromagnetic phenomena in material media. From a historical point of view, it is important to recall that both electric and magnetic phenomena were originally studied mainly in material media. Thus, for instance, basic experiments in electrostatics and magnetism, like those based on the interaction of material bodies charged by rubbing and of magnets, involve such media. Our presentation departs from historical development and will allow us to analyze the electromagnetic properties of matter with all the elements provided by Maxwell's theory in vacuum.

It is not expected that the electrodynamics of material media contributes new basic laws to the theory. Rather, the theory developed so far must be adapted to a situation of particular interest. In each problem, fundamental laws are supplemented with additional relations describing the characteristics of the material under study, called *constitutive equations*.

To be able to elaborate the electromagnetic theory of material media we should begin by understanding that electric and magnetic phenomena really take place in the empty space surrounding the atoms or molecules of the material. This becomes apparent if we notice that atomic radii are of the order of the Bohr radius, $a_0 \sim 10^{-8}$ cm, while we can estimate the nuclear radii to the order of 10^{-13} cm. This means that the volume inside an atom or a molecule is practically empty. As a matter of fact, the atomic nucleus occupies only about 10^{-15} of the total atomic volume. In addition, even in the most dense bodies found in an average experiment, interatomic distances are of the order of 10^{-8} cm. This means that also in these cases large fractions of space are empty. In these empty volumes between atomic and nuclear charges and between atoms, *microscopic fields* $\mathbf{e}(\boldsymbol{r}, t)$ and $\boldsymbol{b}(\boldsymbol{r}, t)$ exist. If measured, these fields would present abrupt space variations, since they are very large near the charges and decay fast in the intermediate regions. The dynamics of electrons, moreover, makes these microscopic fields vary very rapidly in time.

The electromagnetic theory of material media can be formulated starting from the fundamental laws of electromagnetism in vacuum by introducing fields averaged over a large volume compared to the molecular volume. These averages are called *macroscopic fields.* The averaging volume should still be small compared to distances accessible to experimentation. Naturally, the interest in the definition of such fields resides in the fact that these are the fields measured in an experiment with an instrument of macroscopic dimensions. In fact, microscopic fields are very difficult to determine experimentally. They are also of relatively low interest, since they would be relevant in regions of space where the classical formulation of electromagnetism must be modified by quantum theory.

10.1 Macroscopic Fields

We are interested in studying the quantities directly associated with observable macroscopic electromagnetic fields $\boldsymbol{E}(\boldsymbol{r},t)$ and $\boldsymbol{B}(\boldsymbol{r},t)$, defined as averages of the microscopic fields $\boldsymbol{e}$ and $\boldsymbol{b}$:

$$\begin{aligned} \boldsymbol{E}(\boldsymbol{r},t) &= \langle \boldsymbol{e}(\boldsymbol{r},t)\rangle, \\ \boldsymbol{B}(\boldsymbol{r},t) &= \langle \boldsymbol{b}(\boldsymbol{r},t)\rangle. \end{aligned} \tag{10.1}$$

The average $\langle\cdot\rangle$ can be carried out in several ways. Any given form of averaging is acceptable if the resulting macroscopic theory is insensitive to the details of the averaging process.

The possible definitions of the average are discussed by G. Russakoff, in his paper *A Derivation of the Macroscopic Maxwell Equations* [10.1]. It is appropriate to introduce the macroscopic electric field as

$$\boldsymbol{E}(\boldsymbol{r},t) = \int w(\boldsymbol{r}')\ \boldsymbol{e}(\boldsymbol{r}-\boldsymbol{r}',t)\ \mathrm{d}^3r'. \tag{10.2}$$

A similar expression should be used for $\boldsymbol{B}(\boldsymbol{r},t)$, where $w(\boldsymbol{r})$ is a positive weight function practically constant in a volume much larger than the atomic volume around $\boldsymbol{r}=0$. The characteristic radius of this volume can be associated with the size of a probe used to measure the macroscopic fields. Outside that volume the weight function decays more or less smoothly (Fig. 10.1). The function $w(\boldsymbol{r})$ must satisfy the normalization

$$\int w(\boldsymbol{r})\ \mathrm{d}^3r = 1. \tag{10.3}$$

A possible choice for the weight function is

$$w(\boldsymbol{r}) = \frac{N}{1+(r/r_0)^n}, \tag{10.4}$$

where $N = n\ \sin(3\pi/n)/4\pi^2$ is the normalization constant, and $n > 3$ is an integer. For n sufficiently large, this function is practically constant and equal to N in a radius approximately equal to r_0, and it decays to zero in a region of width r_0/n.

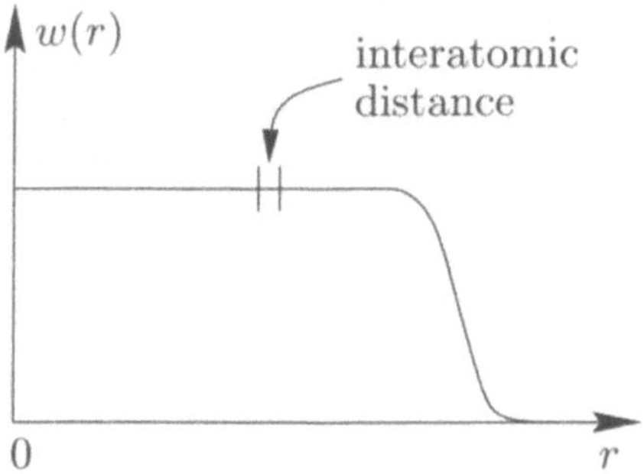

Fig. 10.1. Weight function used to average microscopic fields

The condition of smooth decay in a region of the order of several interatomic distances ensures that averages do not show abrupt spatial fluctuations as a result of variations in the microscopic charge density.

Russakoff has argued against carrying out time averages, since in that case it would be impossible to describe dynamic phenomena such as those involving electromagnetic waves.

Maxwell's equations for the microscopic fields not involving the sources are

$$\begin{aligned} &\nabla \cdot \boldsymbol{b} = 0, \\ &\nabla \times \boldsymbol{e} + \tfrac{1}{c}\partial_t \boldsymbol{b} = 0. \end{aligned} \tag{10.5}$$

Similar equations for the macroscopic fields are immediately obtained from the above equations,

$$\begin{aligned} &\nabla \cdot \boldsymbol{B} = 0, \\ &\nabla \times \boldsymbol{E} + \tfrac{1}{c}\partial_t \boldsymbol{B} = 0, \end{aligned} \tag{10.6}$$

since the derivatives of the averages equal the averages of the derivatives. Indeed, for example,

$$\begin{aligned} \langle \nabla \cdot \boldsymbol{b} \rangle &= \int w(\boldsymbol{r}')\nabla \cdot \boldsymbol{b}(\boldsymbol{r}-\boldsymbol{r}',t)\ \mathrm{d}^3r' = \nabla \cdot \int w(\boldsymbol{r}')\boldsymbol{b}(\boldsymbol{r}-\boldsymbol{r}',t)\ \mathrm{d}^3r' \\ &= \nabla \cdot \boldsymbol{B}. \end{aligned} \tag{10.7}$$

The average of Maxwell's equations involving field sources requires a more careful procedure. In particular, it is necessary that the weight function $w(\boldsymbol{r})$ admits a series expansion of the type

$$w(\boldsymbol{r}+\boldsymbol{r}_0) = w(\boldsymbol{r}) + (\boldsymbol{r}_0 \cdot \nabla)w(\boldsymbol{r}) + \frac{1}{2}(\boldsymbol{r}_0 \cdot \nabla)^2 w(\boldsymbol{r}) + \cdots . \tag{10.8}$$

As for the charge density, we can write it in general as

$$\rho(\boldsymbol{r},t) = \sum_i q_i \delta(\boldsymbol{r} - \boldsymbol{r}_i(t)), \tag{10.9}$$

where $\boldsymbol{r}_i$ are the positions of the point charges q_i. These charges can be divided into two groups, qualitatively different from the macroscopic point of view. We have on the one hand the charges associated with atoms and molecules, and on the other hand the *free*, additional charges not associated with any particular atom or molecule. The density can be rewritten as

$$\begin{aligned}\rho(\boldsymbol{r},t) &= \sum_m \sum_{i(m)} q_i \delta(\boldsymbol{r}-\boldsymbol{r}_i(t)) + \sum_f q_f \delta(\boldsymbol{r}-\boldsymbol{r}_f(t)) \\ &= \rho_{\mathrm{M}}(\boldsymbol{r},t) + \rho_{\mathrm{F}}(\boldsymbol{r},t).\end{aligned} \tag{10.10}$$

The sum over m runs over the atoms or molecules of the medium, and the sum over i runs over the charges of each atom or molecule. In turn, the sum over f involves the free charges only.

The position of each charge in a molecule can be written as $\boldsymbol{r}_i = \boldsymbol{r}_m + \boldsymbol{d}_{im}$, where $\boldsymbol{r}_m$ is the position of the center of charge of the molecule, and $\boldsymbol{d}_{im}$ is the position of the charge relative to that center (Fig. 10.2). Clearly, $|\boldsymbol{d}_{im}|$ is smaller than the characteristic distance within which $w(\boldsymbol{r})$ varies. Therefore, we can use $\boldsymbol{d}_{im}$ as a parameter for a series expansion of the weight function.

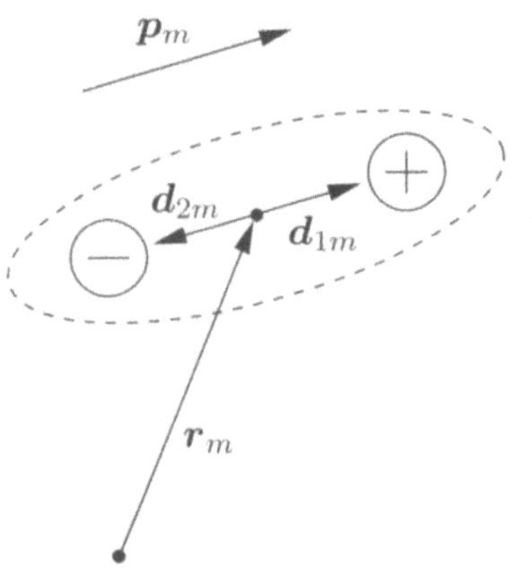

Fig. 10.2. Internal coordinates in a dipole molecule

The macroscopic average of the charges of the medium is

$$\begin{aligned}\langle \rho_{\mathrm{M}}(\boldsymbol{r},t)\rangle &= \int w(\boldsymbol{r}')\rho_{\mathrm{M}}(\boldsymbol{r}-\boldsymbol{r}',t)\ \mathrm{d}^3r' \\ &= \sum_m \sum_{i(m)} q_i \int w(\boldsymbol{r}')\delta(\boldsymbol{r}-\boldsymbol{r}_m-\boldsymbol{d}_{im}-\boldsymbol{r}')\ \mathrm{d}^3r'.\end{aligned} \tag{10.11}$$

Defining a new integration variable $\boldsymbol{r}'' = \boldsymbol{r}' + \boldsymbol{d}_{im}$, this expression becomes

$$\langle \rho_{\mathrm{M}}(\boldsymbol{r},t)\rangle = \sum_m \sum_{i(m)} q_i \int w(\boldsymbol{r}''-\boldsymbol{d}_{im})\delta(\boldsymbol{r}-\boldsymbol{r}_m-\boldsymbol{r}'')\ \mathrm{d}^3r''. \tag{10.12}$$

Now, it is possible to expand $w(\boldsymbol{r}'' - \boldsymbol{d}_{im})$ in powers of $\boldsymbol{d}_{im}$. Taking into account that the integral over $\boldsymbol{r}''$ extends to all space, we have

$$\langle \rho_{\mathrm{M}}(\boldsymbol{r},t)\rangle = \int \sum_m \sum_{i(m)} q_i \left[1 - \boldsymbol{d}_{im} \cdot \nabla'' + \frac{1}{2} \sum_{\alpha\beta} d_{im}^{\alpha} d_{im}^{\beta} \partial''_{\alpha} \partial''_{\beta} + \cdots \right]$$

$$\times w(\boldsymbol{r}'')\delta(\boldsymbol{r} - \boldsymbol{r}_m - \boldsymbol{r}'')\, \mathrm{d}^3 r'', \tag{10.13}$$

where the indices α and β run over the three components (x, y, z). In this expression we identify

$$q_m = \sum_{i(m)} q_i, \tag{10.14}$$

as the total charge of the molecule m;

$$\boldsymbol{p}_m = \sum_{i(m)} q_i \boldsymbol{d}_{im}, \tag{10.15}$$

as the dipole moment of the molecule, and finally

$$q_m^{\alpha\beta} = \frac{1}{2} \sum_{i(m)} q_i d_{im}^{\alpha} d_{im}^{\beta}, \tag{10.16}$$

as the molecular quadrupole moment tensor. With these three quantities, we define the molecular charge density

$$\rho_{\mathrm{M}}^0(\boldsymbol{r}) = \sum_m q_m \delta(\boldsymbol{r} - \boldsymbol{r}_m), \tag{10.17}$$

the dipole moment density

$$\boldsymbol{p}(\boldsymbol{r}) = \sum_m \boldsymbol{p}_m \delta(\boldsymbol{r} - \boldsymbol{r}_m), \tag{10.18}$$

and the quadrupole moment density

$$q_{\alpha\beta}(\boldsymbol{r}) = \sum_m q_m^{\alpha\beta} \delta(\boldsymbol{r} - \boldsymbol{r}_m). \tag{10.19}$$

The macroscopic molecular charge density can be written as

$$\langle \rho_{\mathrm{M}}(\boldsymbol{r},t)\rangle = \langle \rho_{\mathrm{M}}^0(\boldsymbol{r},t)\rangle - \nabla \cdot \boldsymbol{P}(\boldsymbol{r},t) + \sum_{\alpha,\beta} \partial_\alpha \partial_\beta Q_{\alpha\beta}(\boldsymbol{r},t) + \cdots, \tag{10.20}$$

where $\boldsymbol{P} = \langle \boldsymbol{p} \rangle$ is the average molecular dipole moment density or *polarization* of the medium, and $Q_{\alpha\beta} = \langle q_{\alpha\beta} \rangle$ is the average quadrupole moment density.

Starting from Gauss's law for the microscopic fields

$$\nabla \cdot \boldsymbol{e} = 4\pi\rho, \tag{10.21}$$

we obtain, carrying out the average,

$$\nabla \cdot \boldsymbol{E} = 4\pi(\rho_{\rm F} + \langle \rho_{\rm M}^0 \rangle) - 4\pi \nabla \cdot \boldsymbol{P} + 4\pi(\nabla\nabla) \cdot \mathcal{Q}, \tag{10.22}$$

where $\mathcal{Q} = \{Q_{\alpha\beta}\}$. Note that we have truncated the series at the quadrupole contribution. In order to simplify the form of this equation, we define an auxiliary vector, called the *electric displacement*, and given by

$$\boldsymbol{D} = \boldsymbol{E} + 4\pi \boldsymbol{P} - 4\pi \nabla \cdot \mathcal{Q}. \tag{10.23}$$

The electric displacement satisfies

$$\nabla \cdot \boldsymbol{D} = 4\pi(\rho_{\rm F} + \langle {\rho_{\rm M}}^0 \rangle). \tag{10.24}$$

We see that the source of the displacement vector is the charge density not associated with the polarization or with the quadrupole moment – and, in general, to higher order moments – of the charge distribution. This vector plays a similar role to that of the electric field in vacuum, since it satisfies an equation similar to Gauss's equation. However it should be kept in mind that it is an auxiliary quantity without direct physical meaning.

In general, the influence of the quadrupole term is small. It appears only when studying phenomena on the atomic scale, for example near an atom in a crystal. On the other hand, the total charge of each molecule of the material medium is usually zero, so that $\langle \rho_{\rm M}^0 \rangle = 0$. Under these conditions the macroscopic Gauss's law is

$$\nabla \cdot \boldsymbol{D} = 4\pi \rho_{\rm F}, \tag{10.25}$$

where $\boldsymbol{D} = \boldsymbol{E} + 4\pi \boldsymbol{P}$.

The practical advantage of the use of the vector $\boldsymbol{D}$ resides in the particularly simple form taken by Gauss's law in terms of it. However, it should be clear that the form of this law, indicating that the displacement is determined by the free charge distribution, **does not imply** that $\boldsymbol{D}$ is the same in the presence of material media as in their absence. Indeed, the solution of (10.25) depends on the boundary conditions imposed by such media.

In a way similar to the expansion made for Gauss's law, but with somewhat heavier mathematical work, it is possible to find the macroscopic equivalent of the modified Ampère's law involving the currents

$$\boldsymbol{j}(\boldsymbol{r}, t) = \sum_i q_i \dot{\boldsymbol{r}}_i(t)\, \delta(\boldsymbol{r} - \boldsymbol{r}_i(t)). \tag{10.26}$$

These currents can be separated into the contributions due to free charges not associated with a given molecule in the material, such as electrons in a metal, and those due to the molecular charges:

$$\boldsymbol{j} = \boldsymbol{j}_{\rm F} + \sum_m \sum_{i(m)} q_i \boldsymbol{v}_i \delta(\boldsymbol{r} - \boldsymbol{r}_i), \tag{10.27}$$

where $\boldsymbol{v}_i = \dot{\boldsymbol{r}}_i$ is the velocity of the charge q_i. To evaluate the average of the molecular contribution, we divide the coordinate of each charge again as $\boldsymbol{r}_i = \boldsymbol{r}_m + \boldsymbol{d}_{im}$, so that the velocities also separate as $\boldsymbol{v}_i = \boldsymbol{v}_m + \dot{\boldsymbol{d}}_{im}$.

The macroscopic average of the total current is

$$\langle \boldsymbol{j} \rangle = \boldsymbol{j}_{\mathrm{F}} + \partial_t \boldsymbol{P} + c\nabla \times \boldsymbol{M}, \tag{10.28}$$

where $\boldsymbol{P}(\boldsymbol{r}) = \langle \boldsymbol{P}(\boldsymbol{r}) \rangle$, and

$$\boldsymbol{M}(\boldsymbol{r}) = \sum \boldsymbol{m}_m \delta(\boldsymbol{r} - \boldsymbol{r}_m) = \frac{1}{2c} \sum_m \sum_{i(m)} q_i \boldsymbol{r}_i \times \boldsymbol{v}_i \delta(\boldsymbol{r} - \boldsymbol{r}_m), \tag{10.29}$$

is the density of molecular magnetic dipole moments, or *magnetization*. This result is valid when the material medium is at rest in the laboratory, so that the average of the molecular velocities vanishes. If this is not the case, additional terms appear, equivalent to those obtained by formulating the relativistic theory of material media in slow motion.

Taking the average of the microscopic law,

$$\nabla \times \boldsymbol{b} - \frac{1}{c}\, \partial_t \boldsymbol{e} = \frac{4\pi}{c}\, \boldsymbol{j}, \tag{10.30}$$

and defining the auxiliary vector

$$\boldsymbol{H} = \boldsymbol{B} - 4\pi \boldsymbol{M}, \tag{10.31}$$

we get

$$\nabla \times \boldsymbol{H} - \frac{1}{c}\, \partial_t \boldsymbol{D} = \frac{4\pi}{c} \boldsymbol{j}_{\mathrm{F}}. \tag{10.32}$$

This is the macroscopic version of the modified Ampère's law.

Maxwell's equations in material media can thus be written as

$$\begin{cases} \nabla \cdot \boldsymbol{D} = 4\pi \rho_{\mathrm{F}}, \\ \nabla \times \boldsymbol{H} - \frac{1}{c} \partial_t \boldsymbol{D} = \frac{4\pi}{c} \boldsymbol{j}_{\mathrm{F}}, \\ \nabla \cdot \boldsymbol{B} = 0, \\ \nabla \times \boldsymbol{E} + \frac{1}{c} \partial_t \boldsymbol{B} = 0. \end{cases} \tag{10.33}$$

From the last two relations we see that the phenomenon of magnetic induction and the property that the lines of $\boldsymbol{B}$ are closed stay unaffected when considering material media.

On the other hand, the equations involving the sources of the fields have been considerably modified. Defining the vectors $\boldsymbol{D} = \boldsymbol{E} + 4\pi \boldsymbol{P}$ and $\boldsymbol{H} = \boldsymbol{B} - 4\pi \boldsymbol{M}$, it has been possible to obtain new laws having the form of Gauss's and Ampère's laws. The sources of the fields $\boldsymbol{D}$ and $\boldsymbol{H}$ are the free charges and currents, which are not associated with the charge distributions linked to the molecules of the material. Information about the molecular sources is of course not irrelevant and, in fact, it is implicit in the relations connecting $\boldsymbol{D}$ and $\boldsymbol{H}$ to the electromagnetic fields through the quantities $\boldsymbol{P}$ and $\boldsymbol{M}$. We discuss this point in more detail in the next section.

In order to complete the group of Maxwell's laws for material media it is still necessary to specify the polarization $\boldsymbol{P}$ and the magnetization $\boldsymbol{M}$ of the material in question, as we indicated in the introduction, in terms of the fields $\boldsymbol{E}$ and $\boldsymbol{B}$. This implies giving the so-called constitutive equations characterizing the material. Later on, we study in more detail the properties of some material media of importance for applications.

10.2 Sources of the Macroscopic Fields

The last two equations in (10.33) imply that in material media it is still possible to define potentials ϕ and $\boldsymbol{A}$, from which the fields $\boldsymbol{E}$ and $\boldsymbol{B}$ can be obtained in the usual form, $\boldsymbol{E} = -\nabla\phi - c^{-1}\partial_t \boldsymbol{A}$ and $\boldsymbol{B} = \nabla \times \boldsymbol{A}$. According to the discussion of the previous section, in the presence of material media the sources of the electromagnetic fields are not only given by the distributions of external charges and currents but also by the induced electric and magnetic dipole moments. In order to show this explicitly we analyze in the static case the potentials ϕ and $\boldsymbol{A}$ generated by the distribution of electric and magnetic dipole moments.

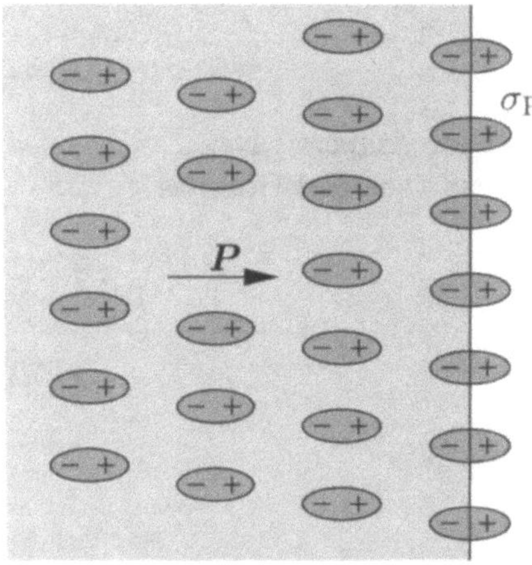

Fig. 10.3. Surface charge density induced on a polarized dielectric

In the electrostatic case, a differential volume d^3r inside the material has an associated total electric dipole moment $\boldsymbol{P}(\boldsymbol{r})\mathrm{d}^3r$. According to (4.32), the electrostatic potential generated by the electric polarization of the medium is given by

$$\phi(\boldsymbol{r}) = \int_V \frac{\boldsymbol{P}(\boldsymbol{r}') \cdot (\boldsymbol{r} - \boldsymbol{r}')}{|\boldsymbol{r} - \boldsymbol{r}'|^3}\, \mathrm{d}^3r' = \int_V \boldsymbol{P}(\boldsymbol{r}') \cdot \nabla' |\boldsymbol{r} - \boldsymbol{r}'|^{-1} \mathrm{d}^3r', \quad (10.34)$$

where V is the volume of the material. Using a vector identity, this relationship can be transformed into

$$\phi(\boldsymbol{r}) = \int_S \frac{\boldsymbol{P}(\boldsymbol{r}') \cdot \mathrm{d}\boldsymbol{s}'}{|\boldsymbol{r} - \boldsymbol{r}'|} + \int_V \frac{-\nabla' \cdot \boldsymbol{P}(\boldsymbol{r}')}{|\boldsymbol{r} - \boldsymbol{r}'|} \, \mathrm{d}^3 r', \tag{10.35}$$

where S is the surface enclosing V. This last expression shows us that the electric polarization gives rise to two clearly differentiated contributions. The first one, given by the surface integral, corresponds to a surface charge density

$$\sigma_{\mathrm{P}} = \boldsymbol{P} \cdot \boldsymbol{n}, \tag{10.36}$$

where $\boldsymbol{n}$ is the normal to S (Fig. 10.3). This contribution appears if the polarization of the medium changes abruptly at the surface, going from a finite value inside to zero outside. Making a more realistic model of the surface, we can assume the polarization to decrease smoothly from the internal to the external value. In that case the potential is given by the second term only. This is a volume contribution which can be associated with a volume charge density

$$\rho_{\mathrm{P}} = -\nabla \cdot \boldsymbol{P}. \tag{10.37}$$

When the polarization is uniform, this volume charge vanishes. In such a case the only contribution to the field generated by the polarization is given by the surface charge density σ_{P}, which measures the charge induced at the surface when the positive charges move relative to the negative charges to generate polarization. If $\boldsymbol{P}$ is inhomogeneous, the relative displacement of the charges inside the medium is not homogeneous and, therefore, in each volume element the net density of induced charge is $-\nabla \cdot \boldsymbol{P}$.

Similarly, in the magnetostatic case the density of dipole moments $\boldsymbol{M}$ gives rise to a vector potential which, according to (6.26), is

$$\boldsymbol{A} = \int_V \boldsymbol{M}(\boldsymbol{r}') \times \nabla' |\boldsymbol{r} - \boldsymbol{r}'|^{-1} \mathrm{d}^3 r'. \tag{10.38}$$

Proceeding as in the previous case, we obtain

$$\boldsymbol{A}(\boldsymbol{r}) = \frac{1}{c} \int_S \frac{c\boldsymbol{M}(\boldsymbol{r}') \times \mathrm{d}\boldsymbol{s}'}{|\boldsymbol{r} - \boldsymbol{r}'|} + \frac{1}{c} \int \frac{c\nabla' \times \boldsymbol{M}(\boldsymbol{r}')}{|\boldsymbol{r} - \boldsymbol{r}'|} \, \mathrm{d}^3 r'. \tag{10.39}$$

We see here that the magnetization contributes a surface current density

$$\boldsymbol{K}_{\mathrm{M}} = c\, \boldsymbol{M} \times \boldsymbol{n}, \tag{10.40}$$

and a volume current density

$$\boldsymbol{j}_{\mathrm{M}} = c \nabla \times \boldsymbol{M}. \tag{10.41}$$

In the case of dynamic fields, the potentials associated with the microscopic distributions of electric and magnetic dipole moment are given by expressions equivalent to (10.35) and (10.39). The sources are now given by the densities $\boldsymbol{P}$ and $\boldsymbol{M}$, which in general are time-dependent.

10.3 Interfaces and Boundary Conditions

In the calculation of electromagnetic fields in the presence of material media it is important to determine the behavior of the fields in the vicinity of the interfaces between two different media. The relation between the fields at each side of the separation surface constitutes a group of boundary conditions that must be satisfied at all points on that surface. Of course, these conditions should follow from Maxwell's equations.

Let us first consider a circuit C crossing the interface, as in Fig. 10.4. Integrating the Faraday equation on the surface $S(C)$ determined by the circuit and letting the length of the sides normal to the interface go to zero, the Stokes theorem implies that

$$0 = \int_{S(C)} \left(\nabla \times \boldsymbol{E} + \frac{1}{c} \partial_t \boldsymbol{B} \right) \cdot \mathrm{d}\boldsymbol{s} = \int_1 \boldsymbol{E} \cdot \mathrm{d}\boldsymbol{l} - \int_2 \boldsymbol{E} \cdot \mathrm{d}\boldsymbol{l}. \tag{10.42}$$

Here the line integrals are evaluated on the pieces of C parallel to the interface in each material. We have assumed the time derivative of the magnetic field to be finite. From this relation we find that the components of the electric field parallel to the interface are equal on each side of it,

$$\boldsymbol{E}_{\|1} - \boldsymbol{E}_{\|2} = 0. \tag{10.43}$$

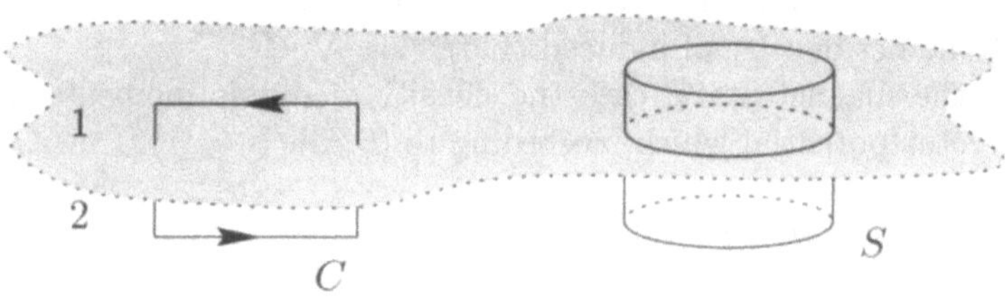

Fig. 10.4. Boundary conditions in material media

If we now consider the closed surface S shown in Fig. 10.4 and integrate the Gauss's equation in the volume $V(S)$ determined by S, the divergence theorem, in the limit in which the lateral surface of S vanishes, implies

$$\int_{V(S)} \nabla \cdot \boldsymbol{D} \, \mathrm{d}^3 r = \int_2 \boldsymbol{D} \cdot \mathrm{d}\boldsymbol{s} - \int_1 \boldsymbol{D} \cdot \mathrm{d}\boldsymbol{s} = 4\pi s \sigma_{\mathrm{F}}. \tag{10.44}$$

The integrals are evaluated on the pieces of the surface S on each side of the interface; s is the area of these pieces and σ_{F} is the free surface charge density present at the separation surface. This relationship can be written in differential form in terms of the normal components of $\boldsymbol{D}$ as

$$D_{\perp 2} - D_{\perp 1} = 4\pi \sigma_{\mathrm{F}}. \tag{10.45}$$

This indicates that the discontinuity in the normal components of the displacement is proportional to the surface density of free charge.

Proceeding similarly with the equations involving the fields $\boldsymbol{B}$ and $\boldsymbol{H}$, we obtain the conditions

$$B_{\perp 1} - B_{\perp 2} = 0, \qquad \boldsymbol{H}_{\parallel 1} - \boldsymbol{H}_{\parallel 2} = \frac{4\pi}{c}\boldsymbol{K}_{\mathrm{F}} \cdot \boldsymbol{n}, \tag{10.46}$$

where $\boldsymbol{K}_{\mathrm{F}}$ are the free surface currents and $\boldsymbol{n}$ is the unit vector perpendicular to the circuit C.

10.4 Electromagnetic Energy in Material Media

The energy balance in a material medium in the presence of electromagnetic fields can be analyzed similarly to the case of fields in vacuum, studied in Chap. 7. In this section we consider energy conservation in a general situation, without specifying the type of material or its properties.

The power delivered by the electromagnetic fields to the free charges contained in a volume V is, according to Maxwell's equations for material media (10.33),

$$\int_V \boldsymbol{j}_{\mathrm{F}} \cdot \boldsymbol{E}\,\mathrm{d}^3 r = \frac{c}{4\pi}\int_V \boldsymbol{E} \cdot (\nabla \times \boldsymbol{H})\,\mathrm{d}^3 r - \frac{1}{4\pi}\int_V \boldsymbol{E} \cdot \partial_t \boldsymbol{D}\,\mathrm{d}^3 r. \tag{10.47}$$

Just as in Chap. 7, we now subtract from the above equation the integral of $\boldsymbol{H} \cdot (\nabla \times \boldsymbol{E} + c^{-1}\partial_t \boldsymbol{B})$ which, according to Faraday's law for the average fields, vanishes. We obtain

$$\begin{aligned}\int_V \boldsymbol{j}_{\mathrm{F}} \cdot \boldsymbol{E}\,\mathrm{d}^3 r = &-\frac{1}{4\pi}\int_V (\boldsymbol{E} \cdot \partial_t \boldsymbol{D} + \boldsymbol{H} \cdot \partial_t \boldsymbol{B})\,\mathrm{d}^3 r \\ &+ \frac{c}{4\pi}\int_V \nabla \cdot (\boldsymbol{E} \times \boldsymbol{H})\,\mathrm{d}^3 r. \end{aligned} \tag{10.48}$$

Locally, this identity can be expressed as

$$\partial_t w + \boldsymbol{j}_{\mathrm{F}} \cdot \boldsymbol{E} + \nabla \cdot \boldsymbol{S} = 0, \tag{10.49}$$

where

$$\partial_t w = \frac{1}{4\pi}(\boldsymbol{E} \cdot \partial_t \boldsymbol{D} + \boldsymbol{H} \cdot \partial_t \boldsymbol{B}), \tag{10.50}$$

and $\boldsymbol{S} = c(\boldsymbol{E} \times \boldsymbol{H})/4\pi$ is the Poynting vector associated with the energy flow in the material medium.

Equation (10.50) allows us to calculate the energy change in a differential time interval δt, $\delta w = (\boldsymbol{E} \cdot \delta \boldsymbol{D} + \boldsymbol{H} \cdot \delta \boldsymbol{B})/4\pi$. The total energy change per unit volume necessary to build up a given field configuration can be calculated as

$$w = \frac{1}{4\pi}\int_{\boldsymbol{D}_0}^{\boldsymbol{D}} \boldsymbol{E} \cdot \delta \boldsymbol{D} + \frac{1}{4\pi}\int_{\boldsymbol{B}_0}^{\boldsymbol{B}} \boldsymbol{H} \cdot \delta \boldsymbol{B}. \tag{10.51}$$

It should be mentioned, as an interesting historical fact, that the form of the first integral on the right-hand side of this expression – where the electric power variation is expressed as the product $\boldsymbol{E}$ times $\delta\boldsymbol{D}$, similar to a force times a displacement – is the origin of the name *displacement vector* for $\boldsymbol{D}$. In agreement with (10.51) we see that the value of w can not only depend on the initial and final values of the fields, but also on the path followed to build up the configuration, which can give rise to the phenomenon of *hysteresis*. Linear materials, to be studied in more detail in the following chapters, do not exhibit this phenomenon.

The term $\partial_t w$ in (10.49) contains two contributions. One is purely electromagnetic and corresponds to the macroscopic fields $\boldsymbol{E}$ and $\boldsymbol{B}$. The other, involving the properties of the material medium through the polarization and the magnetization, is associated with the energy variation due to the action of the fields on the medium. In order to make apparent this separation it is convenient to rewrite (10.49), making explicit the polarization and the magnetization both in $\partial_t w$ and in the Poynting vector. From this substitution it follows that

$$\partial_t \frac{E^2 + B^2}{8\pi} + \frac{c}{4\pi} \nabla \cdot (\boldsymbol{E} \times \boldsymbol{B}) + \boldsymbol{E} \cdot (\boldsymbol{j}_{\mathrm{F}} + \partial_t \boldsymbol{P} + c\nabla \times \boldsymbol{M}) = 0. \quad (10.52)$$

The first two terms correspond to the variation of the purely electromagnetic energy which is compensated by the third term. The latter is the scalar product of the electric field times the total current, made up by the free currents and those associated with the polarization, $\partial_t \boldsymbol{P}$, and with the magnetization, $c\nabla \times \boldsymbol{M}$. It represents the work done by the electromagnetic fields on all currents.

Problems

10.1 Show that the total charge associated with the polarization of a body, with surface density $\sigma_{\mathrm{P}} = \boldsymbol{P} \cdot \boldsymbol{n}$ and volume density $\rho_{\mathrm{P}} = -\nabla \cdot \boldsymbol{P}$ vanishes.

Hint: Use vector identities to simplify the expressions for the total charge.

10.2 Calculate the magnetic field $\boldsymbol{B} = \nabla \times \boldsymbol{A}$ from (10.38) and show that it can be written as $\boldsymbol{B} = 4\pi\boldsymbol{M} + \boldsymbol{H}$, with $\boldsymbol{H} = -\nabla\phi_{\mathrm{M}}$, where the sources of the scalar magnetic potential are the surface density of "magnetic charges" and the volume density.

Hint: Convert the derivatives to a single variable where appropriate.

10.3 Consider a gas of particles having a permanent dipole moment filling a vessel. Study the macroscopic behavior of this gas as a function of temperature, when an uniform magnetic field is applied. (N.B.: The resolution of this problem requires basic knowledge of statistical mechanics.)

10.4 Consider a material cube inside which the electric polarization varies according to the law $\boldsymbol{P} = A\boldsymbol{r}$, where A is a constant and $\boldsymbol{r}$ is the position vector from the origin of coordinates, which coincides with the center of the cube. Calculate the surface and volume charge densities.

10.5 Determine the field produced by a material sphere with constant electric polarization. Show that outside the sphere the field coincides with that of an electric dipole placed at the center. Evaluate the result using PhysicSolver.

Hint: Assume a uniform electric field inside the sphere and prove the consistency of this assumption.

10.6 Determine the magnetic field produced by a material cylinder with uniform polarization $\boldsymbol{M}$, directed along the axis of the cylinder. Use also PhysicSolver.

Solution: Equation (10.40) indicates that a uniform magnetization produces the same field as a surface current $\boldsymbol{K}_{\mathrm{M}} = c\boldsymbol{M} \times \boldsymbol{n}$, where $\boldsymbol{n}$ is a unit vector normal to the surface of the cylinder. This equivalent current circulates in laminar form around the surface and therefore the evenly magnetized cylinder is equivalent to a solenoid transporting such a current.

Now, according to (6.18) the field produced on the axis by a current loop is $B_z = (2\pi I/a)\sin^3\alpha$, where α is the angle subtended by the loop from the point on the axis where the field is evaluated. Using this result, the magnetic field of a solenoid with N turns per unit length carrying a current I can be determined by summing the contributions of each loop. We notice that a piece of solenoid of length $\mathrm{d}z$ corresponds to an angle $\mathrm{d}\alpha$ such that $\mathrm{d}z = -a\,\mathrm{d}\alpha/\sin^2\alpha$. As a consequence, the contribution of that piece is $\mathrm{d}B_z(\alpha) = -(2\pi NI/a)\sin^3\alpha\,\mathrm{d}z = -2\pi NI\sin\alpha\mathrm{d}\alpha$.

The contribution of the whole solenoid is given by

$$B_z = \int_{\alpha_1}^{\alpha_2} \mathrm{d}B_z(\alpha) = 2\pi NI\,(\cos\alpha_1 - \cos\alpha_2).$$

The field created on the z axis by the material with uniform magnetization is then

$$B_z = 2\pi M_z(\cos\alpha_1 - \cos\alpha_2).$$

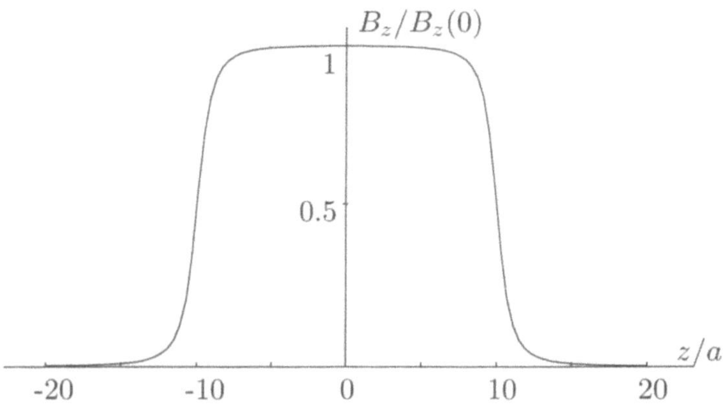

Fig. 10.5. Magnetic field of a cylindrical magnet along its axis

In terms of the coordinate z measured along the axis of the magnet, or of the solenoid, this field can be expressed as

$$B_z = 2\pi M_z \left[\frac{L/2 + z}{\sqrt{(L/2+z)^2 + a^2}} + \frac{L/2 - z}{\sqrt{(L/2-z)^2 + a^2}} \right].$$

Figure 10.5 shows the variation of the field as a function of z for a magnet of length $L = 10a$.

As we have already noticed in Chap. 6 the field over all space can be obtained starting from the field on the axis, resorting to the expansion in Legendre polynomials, since the magnetostatic field in vacuum is a solution of the equation

$$\nabla^2 \boldsymbol{B} = 0.$$

We leave the calculation as an exercise for the reader.

References

10.1 G. Russakoff: *A Derivation of the Macroscopic Maxwell Equations*, Am. J. Phys. **38**, 1188 (1970).

Further Reading

J.D. Jackson: *Classical Electrodynamics* (Wiley, New York, 1975).

L.D. Landau and E.M. Lifshitz: *Electrodynamics of Cotinuous Media* (Pergamon, New York, 1960).

W.K.H. Panofsky and M. Phillips: *Classical Electricity and Magnetism* (Addison–Wesley, Reading, 1955) Chap. 8.

A.M. Portis: *Electromagnetic Fields: Sources and Media* (Wiley, New York, 1978).

J.R. Reitz, F.J. Milford and R.W. Christy: *Foundations of Electromagnetic Theory* (Addison–Wesley, Reading, 1973).

11. Linear Material Media

In the previous chapter we pointed out that, in order to find the electromagnetic fields in material media starting from Maxwell's laws, it is necessary to specify constitutive equations relating both the polarization and the magnetization of the media to the fields. These equations describe the response of the medium to the electromagnetic excitation.

Some of the most common material media are characterized by a linear response to the applied fields over a wide range of field values. This implies that both the polarization and the magnetization are proportional to the electromagnetic fields. Such materials are called *linear materials.*

In this chapter we treat electrostatic, magnetostatic and conduction phenomena in linear material media. In spite of the conceptual differences between these phenomena, their mathematical treatment is very similar. For this reason it is convenient to present them together. In the case of electrostatic and magnetostatic phenomena this analogy is a new consequence of the covariant character of the theory, which also extends to the case of material media. In this way, it is possible to compare them directly and to discuss their similarities and differences.

11.1 Linear Dielectrics

In linear dielectrics the electric polarization vanishes in the absence of external fields. When an electric field is applied, a separation of positive and negative charges takes place at the atomic level, giving rise to a molecular dipole moment $\boldsymbol{p}_m$. This implies that a dipole moment density $\boldsymbol{P}$ proportional to the electric field,

$$\boldsymbol{P} = \chi_{\mathrm{E}} \boldsymbol{E}, \tag{11.1}$$

is generated. The quantity χ_{E} is always positive and is called the *electric susceptibility.* It is necessary to stress that in this expression $\boldsymbol{E}$ does not represent the external applied field but the average field $\boldsymbol{E} = \langle \boldsymbol{e} \rangle$. As we show later, this field also contains contributions from the induced polarization charges. For isotropic materials χ_{E} is a scalar. In crystalline media, on the other hand, the susceptibility is in general a tensor quantity.

The linear relation between the polarization and the electric field implies that the displacement $\boldsymbol{D}$ is also proportional to $\boldsymbol{E}$:

$$\boldsymbol{D} = \boldsymbol{E} + 4\pi\boldsymbol{P} = (1 + 4\pi\chi_{\mathrm{E}})\boldsymbol{E} = \varepsilon\boldsymbol{E}. \tag{11.2}$$

Here

$$\varepsilon = 1 + 4\pi\chi_{\mathrm{E}} \tag{11.3}$$

is the *dielectric constant* of the material. It is always greater than one. According to Maxwell's equations for material media, in the presence of linear dielectrics the electrostatic field satisfies

$$\begin{aligned} &\nabla \times \boldsymbol{E} = 0, \\ &\nabla \cdot (\varepsilon\boldsymbol{E}) = 4\pi\rho_{\mathrm{F}}. \end{aligned} \tag{11.4}$$

The first equation allows us to define the electrostatic potential as $\boldsymbol{E} = -\nabla\phi$. From the second equation we obtain

$$\nabla^2\phi + \frac{\nabla\varepsilon}{\varepsilon} \cdot \nabla\phi = -4\pi\frac{\rho_{\mathrm{F}}}{\varepsilon}. \tag{11.5}$$

In regions where the dielectric constant does not depend on position this relation reduces to Poisson's equation $\nabla^2\phi = -4\pi\rho_{\mathrm{F}}/\varepsilon$. We see that the sources are given by the density of free charges reduced by a factor $1/\varepsilon$. In regions without free charges, this equation coincides with Laplace's equation.

The proportionality between the displacement and the electric field $\boldsymbol{D} = \varepsilon\boldsymbol{E}$ implies a simple form for the energy density in a linear dielectric. According to (10.50), starting from a field-free configuration and in absence of magnetic fields, the energy per unit volume in the final configuration is

$$w = \frac{1}{8\pi}\boldsymbol{E} \cdot \boldsymbol{D}. \tag{11.6}$$

This reduces to $w = \varepsilon E^2/8\pi$ if the dielectric constant is scalar.

11.2 Polarization of Spherical and Ellipsoidal Bodies

In order to illustrate the difference between the applied field and the macroscopic field $\boldsymbol{E}$ due to the effect of induced polarization charges, we analyze the problem of a sphere of linear dielectric in an uniform external field. This problem will allow us to study in a concrete case the behavior of the electric displacement and of the polarization. Consider a sphere of radius a and dielectric constant ε_1, immersed in an infinite dielectric medium of dielectric constant ε_2 and in a field which at large distances from the sphere is uniform, $\boldsymbol{E}_0 = E_0\hat{\boldsymbol{z}}$ (Fig. 11.1). This problem will also be of interest later, when we study in detail the internal fields in a dielectric.

The problem consists in determining the electric field over all space. Inside each material the potential satisfies Laplace's equation $\nabla^2\phi = 0$. At the separation surface between both materials we must apply the boundary conditions derived in the previous chapter, (10.43) and (10.45).

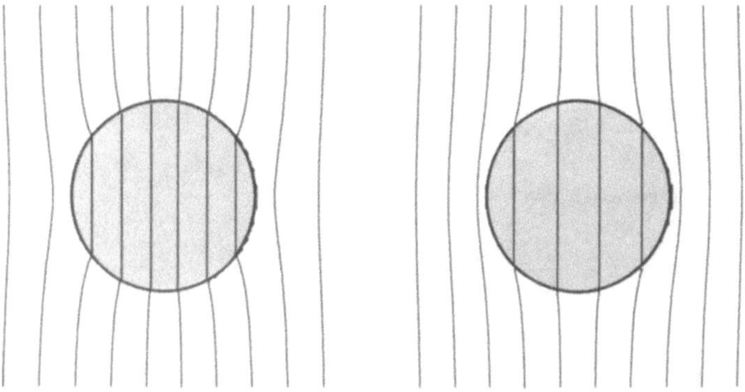

Fig. 11.1. Two dielectric spheres in a uniform field, with different values of the dielectric constants

Given the azimuthal symmetry of the problem, and according to the method of separation of variables in spherical coordinates discussed in Chap. 5, we can assume for the potential inside the sphere the form

$$\phi_1(r, \theta) = \sum_{l=0}^{\infty} A_l r^l P_l(\cos\theta). \tag{11.7}$$

This expression contains positive powers of r only, since ϕ_1 should have no singularities at the origin. For the potential outside the sphere we assume

$$\phi_2(r, \theta) = -E_0 r \cos\theta + \sum_{l=0}^{\infty} B_l r^{-l-1} P_l(\cos\theta). \tag{11.8}$$

Here we have explicitly separated the contributions of the external field and of the fields generated by the dipoles induced in the dielectric.

Continuity of the tangential component of the electric field at the surface of the sphere requires

$$\partial_\theta \phi_1|_{r=a} = \partial_\theta \phi_2|_{r=a}. \tag{11.9}$$

Since there are no free surface charges the continuity of the normal component of the displacement vector implies

$$\varepsilon_1 \partial_r \phi_1|_{r=a} = \varepsilon_2 \partial_r \phi_2|_{r=a}. \tag{11.10}$$

These conditions allow us to establish a system of equations for the coefficients A_l and B_l whose solutions are

$$\begin{aligned} A_1 &= -\frac{3\varepsilon_2}{2\varepsilon_2+\varepsilon_1} E_0 \\ B_1 &= \frac{\varepsilon_1-\varepsilon_2}{2\varepsilon_2+\varepsilon_1} a^3 E_0 \end{aligned} \tag{11.11}$$

and $A_l = B_l = 0$ for $l > 1$.

The potential inside the sphere is then

$$\phi_1(r,\theta) = -\frac{3\varepsilon_2}{2\varepsilon_2+\varepsilon_1} E_0 r\cos\theta, \tag{11.12}$$

while outside the sphere we have

$$\phi_2(r,\theta) = -E_0 r\cos\theta + \frac{\varepsilon_1-\varepsilon_2}{2\varepsilon_2+\varepsilon_1} E_0 \frac{a^3}{r^2} \cos\theta. \tag{11.13}$$

We see that the potential for $r < a$ corresponds to a uniform field in the z direction, parallel to the external field $\boldsymbol{E}_0$ and of modulus $E = 3\varepsilon_2 E_0/(2\varepsilon_2 + \varepsilon_1)$. As a consequence, the polarization inside the sphere is uniform and parallel to the external field. Its modulus is $P = 3\varepsilon_2(\varepsilon_1 - 1)E_0/4\pi(2\varepsilon_2 + \varepsilon_1)$. The field inside the sphere can be written as $\boldsymbol{E} = \boldsymbol{E}_0 - \boldsymbol{E}_{\text{dep}}$, where the *depolarizing field* $\boldsymbol{E}_{\text{dep}}$ is

$$\boldsymbol{E}_{\text{dep}} = \frac{\varepsilon_1-\varepsilon_2}{\varepsilon_1+2\varepsilon_2} \boldsymbol{E}_0 = \frac{4\pi}{3}\frac{\varepsilon_1-\varepsilon_2}{\varepsilon_2(\varepsilon_1-1)} \boldsymbol{P}. \tag{11.14}$$

This field is generated by the charges induced at the interface of both dielectrics, due to the polarization difference between both media. The surface density of these polarization charges is, according to (10.36),

$$\sigma = -(\boldsymbol{P}_2 - \boldsymbol{P}_1)\cdot\boldsymbol{n} = \frac{3}{4\pi}\frac{\varepsilon_1-\varepsilon_2}{\varepsilon_1+2\varepsilon_2} E_0\cos\theta, \tag{11.15}$$

where $\boldsymbol{P}_i$ is the polarization in each medium and $\boldsymbol{n}$ is the unit vector normal to the interface directed outwards from the sphere.

The potential for $r > a$ shows, in addition to the contribution of the external field, a term equivalent to the potential of a dipole placed at the origin. As we see, the induced dipoles in the dielectric modify in a relatively simple form the effect of the applied field.

In this example we see that in the presence of a uniform field a dielectric sphere acquires a homogeneous polarization throughout its volume, so that the resulting field inside is also uniform. Next, we show that this property is shared by all homogeneous dielectric bodies of ellipsoidal shape. This remarkable property, that has its equivalent in magnetic materials, is used experimentally to evaluate the response of a sample subject to an applied uniform external field. To this end, the shape of the sample is approximated by an ellipsoid and the theory to be developed in what follows is applied.

In the first place, we study the electric field inside a uniformly charged ellipsoid. We imagine the ellipsoid divided into concentric ellipsoidal sheets and calculate the contribution of each sheet. From Gauss's theorem, it can be shown that the field inside a spherical shell of radius r is zero, so that the total field at a point at distance R from the centre can be calculated considering only the contribution of the layers with $r < R$.

We show now that the same result is valid for ellipsoidal shells. An ellipsoid can be generated starting from the sphere by making a scale transformation in two orthogonal directions. The charge density, on the other hand, is kept constant in the transformation. Mathematically, such a transformation is called *homotopic* (Fig. 11.2).

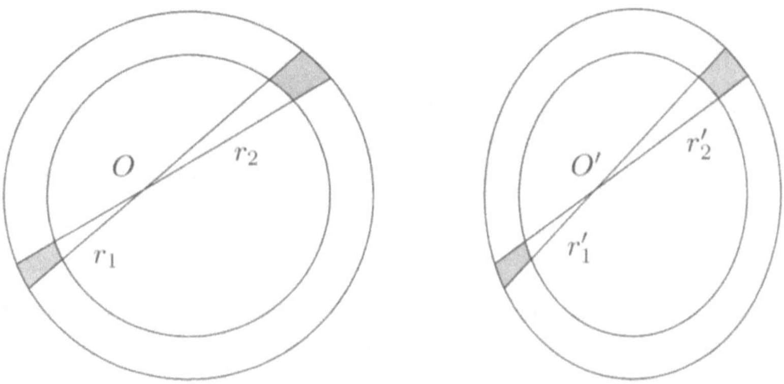

Fig. 11.2. Uniformly charged spherical and ellipsoidal sheets

Considering the change of coordinates that transforms the sphere into the ellipsoid it can be shown that the distances from the points O and O' to the charges satisfy

$$\frac{r'_2}{r_2} = \frac{r'_1}{r_1}. \tag{11.16}$$

As for the spherical shell, the contributions to the electric field produced by the two elements of the shell placed as illustrated in the figure mutually cancel. In fact, the field in O' due to these two elements is

$$\mathrm{d}\boldsymbol{E} = \mathrm{d}q'_1 \frac{\hat{\boldsymbol{r}}'_1}{r'^2_1} + \mathrm{d}q'_2 \frac{\hat{\boldsymbol{r}}'_2}{r'^2_2}. \tag{11.17}$$

Here $\mathrm{d}q'_i$ are the corresponding charge elements. In view of the hypothesis that the charge density is uniform on the ellipsoidal shell, these charge elements are proportional to the corresponding element of solid angle. Thus, (11.16) implies $\mathrm{d}\boldsymbol{E} = 0$. As a consequence, as in the case of the sphere, the total field at a point $\boldsymbol{r}$ inside the ellipsoid is determined only by the charge contained in the volume $V(\boldsymbol{r})$ within the ellipsoidal layer passing through the point $\boldsymbol{r}$. Therefore

$$\boldsymbol{E}(\boldsymbol{r}) = \rho \int_{V(\boldsymbol{r})} \frac{\boldsymbol{r} - \boldsymbol{r}'}{|\boldsymbol{r} - \boldsymbol{r}'|^3} \, \mathrm{d}^3 r', \tag{11.18}$$

where ρ is the charge density. Making in this expression the change of variables $\boldsymbol{r} = \eta \boldsymbol{R}$ and $\boldsymbol{r}' = \eta \boldsymbol{r}''$, where η is an arbitrary constant, we obtain

$$\boldsymbol{E}(\boldsymbol{r}) = \rho \int_{V(\boldsymbol{r})} \frac{\boldsymbol{r} - \boldsymbol{r}'}{|\boldsymbol{r} - \boldsymbol{r}'|^3} \, \mathrm{d}^3 r' = \rho\eta \int_{V(\boldsymbol{R})} \frac{\boldsymbol{R} - \boldsymbol{r}''}{|\boldsymbol{R} - \boldsymbol{r}''|^3} \, \mathrm{d}^3 r''$$

$$= \eta \boldsymbol{E}(\boldsymbol{R}). \tag{11.19}$$

This relation indicates that the fields in two collinear positions $\boldsymbol{r}$ and $\boldsymbol{R}$ inside an evenly charged ellipsoid are parallel, and satisfy

$$\boldsymbol{E}(\boldsymbol{r}) = \frac{r}{R} \boldsymbol{E}(\boldsymbol{R}). \tag{11.20}$$

From this it follows that the field varies linearly with position. In particular, at the center of the ellipsoid we have $\boldsymbol{E}(0) = 0$.

The most general form of $\boldsymbol{E}$ satisfying the linearity relation (11.20) can be written as

$$\boldsymbol{E}(\boldsymbol{r}) = 4\pi\rho \mathcal{N} \cdot \boldsymbol{r}, \tag{11.21}$$

where $\mathcal{N}$ is the so-called *depolarization tensor*, which depends only on geometrical factors. The explicit form of the depolarization tensor can be obtained by calculating the electric field produced by the uniform ellipsoidal charge distribution,

$$\boldsymbol{E}(\boldsymbol{r}) = -\nabla\phi(\boldsymbol{r}) = \rho \int_V \nabla' |\boldsymbol{r} - \boldsymbol{r}'|^{-1} \, \mathrm{d}^3 r', \tag{11.22}$$

where V is the volume of the ellipsoid. Expanding in powers of $\boldsymbol{r}$ around the origin we have

$$\boldsymbol{E}(\boldsymbol{r}) = \boldsymbol{E}(0) + (\boldsymbol{r} \cdot \nabla)\boldsymbol{E}(0) + \cdots. \tag{11.23}$$

As we know, we must have $\boldsymbol{E}(0) = 0$. The second term is given by

$$(\boldsymbol{r} \cdot \nabla)\boldsymbol{E}(\boldsymbol{r}) = \rho(\boldsymbol{r} \cdot \nabla) \int_V \nabla' |\boldsymbol{r} - \boldsymbol{r}'|^{-1} \, \mathrm{d}^3 r' = 4\pi\rho\mathcal{N}(\boldsymbol{r}) \cdot \boldsymbol{r}, \tag{11.24}$$

with

$$\mathcal{N}(\boldsymbol{r}) = \frac{1}{4\pi} \int_V \nabla\nabla' |\boldsymbol{r} - \boldsymbol{r}'|^{-1} \, \mathrm{d}^3 r'. \tag{11.25}$$

This expansion is valid for any uniformly charged body. According to (11.21), for the uniform ellipsoidal charge distribution, $\mathcal{N}$ does not depend on $\boldsymbol{r}$ and the expansion ends at the first-order term. This remarkable property of ellipsoidal bodies explains its relevance in the approximate description of situations found in practice. The components of $\mathcal{N}$ are

$$\mathcal{N}_{ij} = \frac{1}{3}\,\delta_{ij} - \frac{1}{4\pi} \int_V \frac{3n_i n_j - \delta_{ij}}{|\boldsymbol{r} - \boldsymbol{r}'|^3} \, \mathrm{d}^3 r', \tag{11.26}$$

where the indices i, j run over the Cartesian coordinates x, y, z. Here $\boldsymbol{n} = (n_x, n_y, n_z)$ is the unit vector in the direction of $\boldsymbol{r} - \boldsymbol{r}'$, and δ_{ij} is the Kronecker symbol. The tensor $\mathcal{N}$ is diagonal when the coordinate axes coincide with the principal axes of the ellipsoid. For a sphere, the components of the depolarization tensor are $\mathcal{N}_{ij} = \delta_{ij}/3$. For an ellipsoid extremely flattened in the z direction the only non-zero element is $\mathcal{N}_{zz} = 1$. On the other hand, for a very stretched ellipsoid in the same direction, we have $\mathcal{N}_{xx} = \mathcal{N}_{yy} = 1/2$, and $\mathcal{N}_{zz}$ is equal to zero.

In order to find the field inside a uniformly polarized ellipsoid with polarization $\boldsymbol{P}$, we consider first the superposition of two uniformly charged ellipsoids with densities ρ and $-\rho$, slightly displaced relative to each other. Assume that this displacement is given by a vector $\boldsymbol{d}$, in the limit $\boldsymbol{d} \to 0$ and $\rho \to \infty$, with $\rho\boldsymbol{d} = \boldsymbol{P}$. The electric field of such a configuration is

$$\boldsymbol{E}(\boldsymbol{r}) = -4\pi\rho\mathcal{N} \cdot \boldsymbol{r} + 4\pi\rho\mathcal{N} \cdot (\boldsymbol{r} - \boldsymbol{d}) \to -4\pi\mathcal{N} \cdot \boldsymbol{P}. \tag{11.27}$$

We can now tackle the problem of finding $\boldsymbol{E}$ inside an ellipsoid of linear dielectric in the presence of a uniform external field $\boldsymbol{E}_0$. To use the results previously obtained we look for a self-consistent solution: we assume that the external field generates in the initially nonpolarized ellipsoid a uniform polarization $\boldsymbol{P}$. The internal field is then

$$\boldsymbol{E} = \boldsymbol{E}_0 - 4\pi\mathcal{N} \cdot \boldsymbol{P}. \tag{11.28}$$

Since $\boldsymbol{P} = \chi_{\mathrm{E}}\boldsymbol{E} = (\varepsilon - 1)\boldsymbol{E}/4\pi$, we have

$$\boldsymbol{E} = \boldsymbol{E}_0 - (\varepsilon - 1)\mathcal{N} \cdot \boldsymbol{E}. \tag{11.29}$$

This relation indicates that the field inside the ellipsoid is effectively constant and allows us to obtain $\boldsymbol{E}$ starting from $\boldsymbol{E}_0$, as $\boldsymbol{E} = [1 + (\varepsilon - 1)\mathcal{N}]^{-1} \cdot \boldsymbol{E}_0$. For the case of a sphere, where $\mathcal{N}_{ij} = \delta_{ij}/3$, we obtain again the result derived at the beginning of this section, $\boldsymbol{E} = 3\boldsymbol{E}_0/(2 + \varepsilon)$.

We notice finally that, due to the tensor character of $\mathcal{N}$, the polarization induced in an ellipsoid in the presence of an external field is parallel to the field when this is applied along one of the main axes of the ellipsoid only.

11.3 Local Field in a Dielectric

In several applications of the electrostatics of material media at the microscopic level, it is important to evaluate the electric field $\boldsymbol{E}_{\mathrm{loc}}$ acting on a particular atom or molecule, called the *local field*. This is necessary, for example, to relate macroscopic quantities like the dielectric constant ε to the molecular polarizability α defined by

$$\boldsymbol{p} = \alpha\boldsymbol{E}_{\mathrm{loc}}, \tag{11.30}$$

where $\boldsymbol{p}$ is the dipole moment induced in the molecule by the electric field.

The local field $\boldsymbol{E}_{\text{loc}}$ is the field at the position of the atom or molecule, measured in the absence of the particle in question. Let us assume, for concreteness, that we are dealing with a molecular material. The field $\boldsymbol{E}_{\text{loc}}$ is measured at the center of the molecule, assuming that the environment is not altered when we remove the molecule. The local field does not coincide with the macroscopic field, since this is an average over a finite volume of material.

In principle, the local field inside a dielectric can be calculated by adding up the contributions of the charges that give rise to the external field and those of all the dipoles in the material medium. This leads us to consider the effect of the shape of the material body in question. In order to apply the results obtained in the previous section we assume the body to be ellipsoidal. The problem of a body of arbitrary shape cannot be treated analytically in a general way. The local field in a dielectric is made up of various contributions which we write

$$\boldsymbol{E}_{\text{loc}} = \boldsymbol{E}_0 + \boldsymbol{E}_1 + \boldsymbol{E}_2 + \boldsymbol{E}_3. \tag{11.31}$$

Here, $\boldsymbol{E}_0$ is the external field applied to the material and $\boldsymbol{E}_1$ is the depolarization field, produced by the polarization charges on the surface of the dielectric. The contributions $\boldsymbol{E}_2$ and $\boldsymbol{E}_3$ correspond to the field generated by the charges within the material.

The field $\boldsymbol{E}_3$ is produced by the molecules in the immediate vicinity of the molecule in question. In order to evaluate it, we must add the contributions of the dipoles placed inside a sphere (*Lorentz sphere*) centered at the molecule. The field $\boldsymbol{E}_2$ is produced by the molecules placed beyond the limiting surface of that sphere. To determine $\boldsymbol{E}_2$ we consider that it is produced by a continuous medium (Fig. 11.3). The result for the total local field should not depend on the radius of the Lorentz sphere considered.

According to the discussion of the previous section, the macroscopic field is given by $\boldsymbol{E} = \boldsymbol{E}_0 + \boldsymbol{E}_1$, with $\boldsymbol{E}_1 = -4\pi\mathcal{N} \cdot \boldsymbol{P}$. The Lorentz field $\boldsymbol{E}_2$ can be calculated as the field produced in a spherical cavity carved in a dielectric whose polarization **stays uniform and constant**. In fact, we should not confuse this field with that produced in a dielectric with a cavity in the presence of a uniform field. In the latter case, the field and, therefore, the polarization are distorted near the cavity. Integrating the surface charge density we obtain

$$\boldsymbol{E}_2 = \frac{4\pi}{3}\,\boldsymbol{P}. \tag{11.32}$$

The field $\boldsymbol{E}_3$ produced by the molecules within the sphere depends on the detailed structure of their distribution. Its general form is

$$\boldsymbol{E}_3 = \sum_i \frac{3(\boldsymbol{p}_i \cdot \boldsymbol{r}_i)\boldsymbol{r}_i - r_i^2 \boldsymbol{p}_i}{r_i^5}. \tag{11.33}$$

The sum runs over the molecules within the cavity, $\boldsymbol{p}_i$ being the dipole moment of each molecule. When the dielectric has a cubic structure, it can be shown that $\boldsymbol{E}_3 = 0$. The same result is obtained with spherical symmetry as in an amorphous material, a liquid or a gas. In general, crystal symmetries make $\boldsymbol{E}_3$ relatively small with respect to the other contributions to the local field.

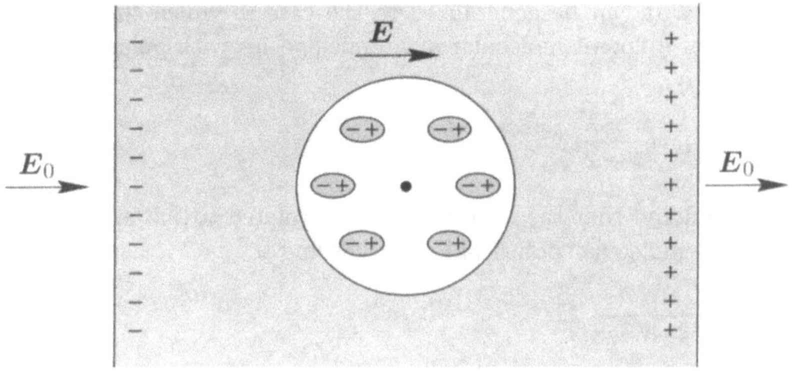

Fig. 11.3. Determination of the local field inside a dielectric

As an application of the local field theory to a particular case let us consider a piece of dielectric placed between the plates of a plane capacitor. Assume the plates are separated by a distance d and kept at constant potential V. Under these circumstances the fields point in the direction normal to the plates so they can all be treated as scalars. As we mentioned above, the macroscopic field $E = V/d$ is the sum of the external field E_0 produced by the plates and the polarization field E_1. The latter is calculated using the depolarization tensor. Approximating the dielectric portion by a flattened ellipsoid, we obtain $E_1 = -4\pi P$. Note that the macroscopic field E is not modified by the presence of the dielectric since the potential is constant. Indeed when the polarization is induced in the dielectric there is a rearrangement of charges in the plates to keep V fixed. Neglecting the dipole field E_3 we obtain

$$E_{\text{loc}} = E_0 + E_1 + E_2 = E + \frac{4\pi}{3}P. \tag{11.34}$$

This identity, called the *Lorentz relation*, allows us to link the molecular polarizability α of the material to its dielectric constant ε. Indeed, the dipole moment density $\boldsymbol{P}$ is given by

$$P = Np = N\alpha E_{\text{loc}}, \tag{11.35}$$

where N is the number of molecules per unit volume and p is the dipole moment of each molecule. Using the expression for the local field found above, we obtain

$$(1 - 4\pi N\alpha/3)P = N\alpha E. \tag{11.36}$$

Recalling that $P = (\varepsilon - 1)E/4\pi$ we finally get

$$\frac{4\pi}{3} N\alpha = \frac{\varepsilon - 1}{\varepsilon + 2}. \tag{11.37}$$

This relation, which links a microscopic property, the polarizability α, to a macroscopic property, the dielectric constant ε, is called the *Clausius–Mossotti equation.* It can be generalized to the case in which the material is made of several different molecular or atomic species with polarizabilities α_i and densities N_i:

$$\frac{4\pi}{3} \sum_i N_i \alpha_i = \frac{\varepsilon - 1}{\varepsilon + 2}. \tag{11.38}$$

It should be noted that the local field can be related to the macroscopic field through the molecular polarizability, according to

$$E_{\text{loc}} = \frac{1}{1 - 4\pi N\alpha/3} E. \tag{11.39}$$

Expanding E_{loc} for small values of α, we get

$$E_{\text{loc}} = \left(1 + \frac{4\pi}{3} N\alpha + \cdots\right) E. \tag{11.40}$$

This expression shows that the difference between the local field and the macroscopic field is due to the contribution of the material dipoles described by the density N and the polarizability α.

In order to complete the analysis of the microscopic fields in a dielectric we must consider that, in general, the polarizability consists of several contributions. The electronic polarizability originates in the relative displacement of positive and negative atomic charges when a field is applied. This displacement takes place until the internal forces in the atoms and molecules balance the effect of the external field, giving rise to a well-defined dipole moment for each of them.

In a simple model, we can consider that an atom consists of a spherical electronic cloud of radius a and total charge $-q$ and a point-like nucleus of positive charge q placed at the center of the cloud, where the electric field is zero. When an external field is applied, the electronic cloud moves relative to the nucleus by a distance x, giving rise to a force of modulus $q^2x/4\pi a^3$ between both distributions. In equilibrium this force equals that produced by the local field qE_{loc}. The dipole moment of the atom is given by $p = qx$, so that, according to (11.30), the atomic polarizability in this model turns out to be $\alpha = 4\pi a^3$. We see that it is of the order of the atomic volume, namely, $\alpha \sim 10^{-23}\,\text{cm}^3$. More complex models, taking into account the quantum mechanical aspects of atomic structure, give results of the same order of magnitude.

The ionic polarizability in a crystal is of similar character, since it corresponds to the relative displacement of ions of different sign. Note that the electronic and ionic contributions to the polarizability, and therefore to the electric susceptibility, are positive.

In polar liquids such as water and alcohol each molecule has a permanent dipole moment. At finite temperature and in the absence of an external field, these dipole moments are randomly oriented. When a field is applied the dipoles are aligned with the field, giving rise to an additional effective polarization. This polarization depends on temperature, since thermal disorder competes with the field in aligning the molecules. Keeping in mind that the potential energy of a permanent dipole in an electric field is $U = -\boldsymbol{P} \cdot \boldsymbol{E} = -pE\cos\theta$, the polarization of a material with N molecules per unit volume in thermodynamical equilibrium can be calculated as

$$\begin{aligned} P = Np\,\langle\cos\theta\rangle_T &= Np\,\frac{\int \exp(-U/k_\mathrm{B}T)\cos\theta\;\mathrm{d}\Omega}{\int \exp(-U/k_\mathrm{B}T)\;\mathrm{d}\Omega} \\ &= Np\,[\coth(pE/k_\mathrm{B}T) - kT/pE]. \end{aligned} \tag{11.41}$$

Here k_B is Boltzmann constant. As a function of temperature the polarization reaches its maximum value for $T = 0$ and falls monotonically, approaching zero as $T \to \infty$, where thermal fluctuations dominate over the effect of the applied field.

At room temperature and for a relatively strong electric field, $E \sim 1000\,\mathrm{V/cm}$, the ratio $pE/k_\mathrm{B}T$ is very small for almost all materials. Under these conditions we can approximate

$$P \approx Np\,\frac{pE}{3k_\mathrm{B}T}. \tag{11.42}$$

From this expression it is possible to define a temperature-dependent effective polarizability for the polar medium, $\alpha_T = p^2/3k_\mathrm{B}T$. The susceptibility of the ensemble of polar molecules is $\chi_T = Np^2/3k_\mathrm{B}T$. The linear dependence of χ_T on the inverse of temperature, which also applies to systems whose atoms have a permanent magnetic dipole moment is known as Curie's law.

Taking into account the contributions due to the electronic polarizability and the permanent dipole moment, we have in general

$$\frac{\varepsilon - 1}{\varepsilon + 2} = \frac{4\pi}{3}N\left(\alpha + \frac{p^2}{3k_\mathrm{B}T}\right). \tag{11.43}$$

Since α and p are molecular properties, the ratio $(\varepsilon-1)/(\varepsilon+2)$ is proportional to the density and varies linearly with the inverse of the temperature. For $1/T \to 0$, it is proportional to α.

In polar liquids the typical values of α_T are of the order of $10^{-23}\,\mathrm{cm}^3$, which coincides with the order of magnitude of atomic and molecular polarizabilities. This is also the approximate value of the molecular volume. Consequently the ratio $(\varepsilon-1)/(\varepsilon+2)$ is close to unity, indicating that $\varepsilon \gg 1$. In fact for water $\varepsilon \approx 80$. The high value of ε for polar liquids is related to their properties as solvents. Indeed the electrostatic force of cohesion between the charges of a substance diluted in the liquid diminishes by a factor ε with respect to its value in vacuum.

Once the total molecular polarizability α is determined according to (11.39) the dipole moment density in a dielectric can be written as

$$\boldsymbol{P} = \frac{N\alpha}{1 - 4\pi N\alpha/3}\,\boldsymbol{E}. \tag{11.44}$$

This expression indicates that when $N\alpha$ approaches $3/4\pi$ the electric susceptibility increases indefinitely, and can even become singular. Beyond this point, the dielectric material can show a spontaneous polarization in the absence of an external field, and the linear theory ceases to be valid. This phenomenon is called *ferroelectricity* and is present in several materials, in particular, in the perovskite family. An example of these materials is barium titanate $BaTiO_3$, with a critical temperature of 381 K.

11.4 Linear Magnetic Media

In a linear magnetic medium the net magnetization in the absence of an external magnetic field vanishes. When a field is applied, two effects occur. One is the appearance of induced currents at the atomic level which, according to the Lenz's law, give rise to a magnetic moment opposed to the field. This induced moment is called the *diamagnetic moment* and is present in all material media. A second effect is the alignment with the external field of preexistent microscopic magnetic moments associated with the spin and the orbital moment of the electrons. If present, this *paramagnetic* effect is usually larger than the diamagnetic effect, and dominates over it. As a consequence of both processes a density of dipole magnetic moment $\boldsymbol{M}$ proportional to the average magnetic field,

$$\boldsymbol{M} = \chi_{\mathrm{M}}\boldsymbol{B}, \tag{11.45}$$

appears at the macroscopic level. Here χ_{M} is the *magnetic susceptibility.* In general, the magnetic susceptibility is a tensor, but for isotropic materials it reduces to a scalar. In contrast to what happens with the electric susceptibility, χ_{M} can be positive or negative according to whether the paramagnetic or the diamagnetic effect dominates.

The linear relation between magnetization $\boldsymbol{M}$ and magnetic field $\boldsymbol{B}$ makes the field $\boldsymbol{H}$ also proportional to $\boldsymbol{B}$:

$$\boldsymbol{H} = \boldsymbol{B} - 4\pi\boldsymbol{M} = (1 - 4\pi\chi_{\mathrm{M}})\boldsymbol{B} = \frac{1}{\mu}\boldsymbol{B}, \tag{11.46}$$

where

$$\mu = \frac{1}{1 - 4\pi\chi_{\mathrm{M}}} \tag{11.47}$$

is the *permeability* of the magnetic material. This quantity can be greater or smaller than unity.

In static situations and for linear magnetic media, Maxwell's equations for $\boldsymbol{B}$ reduce to

$$\nabla \cdot \boldsymbol{B} = 0, \qquad \nabla \times (\boldsymbol{B}/\mu) = \frac{4\pi}{c}\boldsymbol{j}_{\mathrm{F}}. \tag{11.48}$$

Introducing the vector potential $\boldsymbol{A}$, such that $\boldsymbol{B} = \nabla \times \boldsymbol{A}$, the second of these equations implies

$$\nabla \times \nabla \times \boldsymbol{A} - \frac{\nabla\mu}{\mu} \times \nabla \times \boldsymbol{A} = \frac{4\pi}{c}\mu\boldsymbol{j}_{\mathrm{F}}. \tag{11.49}$$

In regions of space where the material is homogeneous this equation reduces to $\nabla \times \nabla \times \boldsymbol{A} = 4\pi\mu\boldsymbol{j}_{\mathrm{F}}/c$. This is similar to the equation for the vector potential in the absence of material media. In fact, it reduces to a vector Poisson equation choosing $\nabla \cdot \boldsymbol{A} = 0$. The sources of this equation appear now affected by a factor μ.

According to (11.46), in a linear magnetic medium the energy density in the absence of electric fields can be written as

$$w = \frac{1}{8\pi}\boldsymbol{B} \cdot \boldsymbol{H} = \frac{1}{8\pi\mu}\, B^2. \tag{11.50}$$

In regions where no currents flow, equations (11.48) imply that the field $\boldsymbol{H}$ can be derived from a scalar magnetic potential ϕ_{M} such that

$$\boldsymbol{H} = -\nabla\phi_{\mathrm{M}}. \tag{11.51}$$

As we have already mentioned in Chap. 6, ϕ_{M} is not a single-valued quantity. Except for this difficulty, the magnetic scalar potential is useful for the solution of problems in linear material media, since in regions where μ is constant ϕ_{M} satisfies Laplace's equation. As a consequence, it is possible to establish a correspondence between problems with magnetic materials and dielectrics with the same geometry. So, for instance, the solution for a magnetic sphere of permeability μ_1 immersed in a medium of permeability μ_2 in the presence of a uniform external magnetic field can be obtained from the solution found in Sect. 11.2, through the substitution $\varepsilon_i \to \mu_i$. This substitution is justified by comparing the boundary conditions in both problems. The magnetic potential inside the sphere is

$$\phi_{\mathrm{M}1}(r,\theta) = -\frac{3}{2\mu_2 + \mu_1}\, B_0 r\cos\theta. \tag{11.52}$$

Outside the sphere we have

$$\phi_{\mathrm{M}2}(r,\theta) = -\frac{1}{\mu_2}B_0 r\cos\theta + \frac{\mu_1/\mu_2 - 1}{2\mu_2 + \mu_1}\, B_0\frac{a^3}{r^2}\,\cos\theta. \tag{11.53}$$

Inside the sphere, the magnetic field is uniform and parallel to the external field. Its modulus is $B = 3\mu_1 B_0/(2\mu_2 + \mu_1)$. Correspondingly, the magnetization is also homogeneous. By analogy with the dielectric case, it is possible to introduce a demagnetizing field by writing the magnetic field inside the sphere as $\boldsymbol{B} = \boldsymbol{B}_0 - \boldsymbol{B}_{\text{dem}}$. This demagnetizing field, generated by the magnetic moments induced in the material, reads

$$\boldsymbol{B}_{\text{dem}} = 2\frac{\mu_2 - \mu_1}{(2\mu_2 + \mu_1)}\,\boldsymbol{B}_0. \tag{11.54}$$

When $\mu_1 > \mu_2$ the demagnetizing field is negative and therefore the field inside the sphere is larger than $\boldsymbol{B}_0$. On the other hand, if $\mu_1 < \mu_2$ the internal field is smaller than the applied field. In contrast with the dielectric case, these two alternatives remain even when the sphere is in vacuum, i.e. when $\mu_2 = 1$. In this situation, if the sphere is paramagnetic the demagnetizing field is negative and $B > B_0$. If the sphere is diamagnetic, the opposite effect appears.

In the presence of a uniform magnetic field, linear magnetic materials with ellipsoidal shape present a homogeneous magnetization. As a consequence, the magnetic field inside is also uniform, in correspondence with the result obtained for dielectrics. When the ellipsoid is in vacuum this field can be calculated starting from its vector potential. When $\boldsymbol{M}$ is constant and we are in the presence of an external field with potential $\boldsymbol{A}_0$, (7.25) implies

$$\boldsymbol{A} = \boldsymbol{A}_0(\boldsymbol{r}) + \boldsymbol{M} \times \int_V \nabla'|\boldsymbol{r} - \boldsymbol{r}'|^{-1}\mathrm{d}^3r', \tag{11.55}$$

where V is the volume of the ellipsoid. Using vector identities the field derived from this potential can be written as

$$\boldsymbol{B}(\boldsymbol{r}) = \boldsymbol{B}_0 - \int_V \left[\boldsymbol{M}\cdot\nabla^2|\boldsymbol{r} - \boldsymbol{r}'|^{-1} + (\boldsymbol{M}\cdot\nabla)\nabla'|\boldsymbol{r} - \boldsymbol{r}'|^{-1}\right] \mathrm{d}^3r'. \tag{11.56}$$

In the first term in the integral we have a contribution proportional to $\delta(\boldsymbol{r} - \boldsymbol{r}')$. The second term can be identified with the depolarization tensor introduced in Sect. 10.2. As a consequence, the magnetic field inside the ellipsoid turns out to be $\boldsymbol{B} = \boldsymbol{B}_0 + 4\pi\boldsymbol{M} - 4\pi\mathcal{N}\cdot\boldsymbol{M}$. For the field $\boldsymbol{H}$ we have thus

$$\boldsymbol{H} = \boldsymbol{H}_0 - 4\pi\mathcal{N}\cdot\boldsymbol{M}, \tag{11.57}$$

since $\boldsymbol{H}_0 = \boldsymbol{B}_0$.

Regarding the microscopic fields in a linear magnetic medium, it would be possible to evaluate the contribution of the magnetic dipoles of the medium to the local magnetic field acting on an atom or molecule, as in the case of dielectrics. This calculation, however, cannot reproduce a very well-known effect in magnetic materials, namely, the spontaneous magnetization present in magnets, called *ferromagnetism.* The explanation of this nonlinear effect can only be given in the framework of quantum mechanics.

The molecular field that tends to align the dipoles giving rise to ferromagnetism is not a magnetic field, but rather is due to the Coulomb interactions enhanced by quantum effects. In ferromagnetic materials the linear effects discussed above are a correction of little relevance.

11.5 Linear Conducting Media

Conducting media, i.e. media which in the presence of an electric field generate an electric current, constitute a case of great interest in the applications. In a linear conductor, the relationship between the electric field and the density of generated current is given by *Ohm's law*:

$$\boldsymbol{j} = \sigma \boldsymbol{E}. \tag{11.58}$$

The quantity σ is the *electric conductivity* of the material. In the cgs system it is measured in s^{-1}. In the SI system the conductivity is measured in $\mathrm{m}^{-1}\,\mathrm{S}$, where S is the siemens, also called the "mho," or in $\mathrm{m}^{-1}\,\Omega^{-1}$, where Ω is the ohm. These units of conductivity are equivalent to $9 \times 10^9\,\mathrm{s}^{-1}$.

Ordinary material media are characterized by a wide range of values of the electrical conductivity. Good metallic conductors such as gold or silver have conductivities of the order of $10^8\,\mathrm{m}^{-1}\,\mathrm{S}$, whereas the so-called good insulators have conductivities of the order of $10^{-18}\,\mathrm{m}^{-1}\,\mathrm{S}$. At this end of the conductivity scale are the insulating plastics. Other materials cover the intermediate range between these values. Table 11.1 illustrates qualitatively the values of the conductivity of different materials, according to the conduction mechanism involved. Given the wide range of variation of the conductivity, it is usual to indicate the logarithm of the conductivity referred to an arbitrary constant value σ_0.

Table 11.1. Electrical conductivity of some materials

Type of material	Carriers	$\log(\sigma/\sigma_0)$	Examples
metals and alloys	free electrons	8	Cu, Ag
		6	Hg
		4	
electrolytes	ions in solution	2	NaCl (solution)
		0	
semiconductors	excited electrons/holes	−2	Ge, Si
		−4	H_2O
ionic solids	ion diffusion	−6	NaCl (crystalline)
		−8	
insulators	electron diffusion or other	−10	glasses
		−12	mica
		−14	polyestyrene
		−16	thermoplastics
		−18	PTFE

The value of σ is determined in different materials by different transport mechanisms. From the electromagnetic point of view, however, it is only the existence of the linear relationship implied by Ohm's law that interests us.

In Sect. 12.5 we shall study the relation between the conductivity σ and the dielectric constant within the framework of the classical Lorentz model.

In a metal, the charges participating in the conduction process are almost-free electrons accelerated by the applied electric field. These electrons collide with ions of the metallic lattice suffering on average an energy loss, which is then dissipated as heat. In this way, the average velocity of the conduction electrons is constant and, therefore, σ is well defined. In the case of ions in solution, the conductivity is given by the motion of the ions, due both to the applied field and to diffusion. In a solution of NaCl in water, for instance, *ionic conduction* is due to the motion of the Na and Cl ions within the liquid. A special class of solid ionic conductors are the *superionic conductors*, such as AgS. In spite of being in the solid state, in these materials the ions of a given species can easily move through the crystal lattice due to their reduced size. In this way, relatively high conductivities, comparable to those of ionic solutions, occur.

In the stationary state, once the transient in which the current develops vanishes, charge transport is governed by the corresponding continuity equation, which takes the form

$$\nabla \cdot \boldsymbol{j} = 0. \tag{11.59}$$

If the regions where the electromotive force originates are excluded from the volume under consideration, we have

$$\boldsymbol{E} = -\nabla\phi. \tag{11.60}$$

Combining this with (11.59) it follows that

$$\nabla \cdot \boldsymbol{j} = -\nabla \cdot (\sigma\nabla\phi) = 0. \tag{11.61}$$

If the medium is homogeneous, this equation implies

$$\nabla^2\phi = 0. \tag{11.62}$$

That is to say, ϕ satisfies the same equation as in electrostatics. Comparing with the results of Sects. 11.1 and 11.2, we see that σ plays the same role as ε in the case of dielectric media. For this reason, many of the results of electrostatics can be adapted to the solution of conduction problems in linear media and *vice versa.*

The appropriate boundary conditions are derived from the fact that the potential and the normal component of the current must be continuous at the interface between two media:

$$\begin{aligned} \sigma_1\frac{\partial\phi_1}{\partial n} &= \sigma_2\frac{\partial\phi_2}{\partial n}, \\ \phi_1 &= \phi_2. \end{aligned} \tag{11.63}$$

The boundary conditions for the electric displacement imply that, if a current flows through the interface, a density σ_{F} of free charge accumulates at the surface:

$$4\pi\sigma_{\mathrm{F}} = \left(\frac{\varepsilon_1}{\sigma_1} - \frac{\varepsilon_2}{\sigma_2}\right) \boldsymbol{j}\cdot\boldsymbol{n}. \tag{11.64}$$

In a practical case, the charge accumulated at the interface takes a certain time to reach the stationary value given by (11.64). The boundary condition for the tangential component of $\boldsymbol{E}$ follows from the relationship $\nabla \times \boldsymbol{E} = 0$, which is satisfied in the regions where (11.60) holds.

In Chap. 4 we defined the capacity coefficients of a system of conductors. If the space between the conductors is filled with a linear dielectric material of dielectric constant ε and the potentials on the conductors are $V_1, V_2, \ldots, V_n$, the charges in each of them are given by

$$Q_i = -\int_{S_i} \varepsilon\nabla\phi\cdot \mathrm{d}\boldsymbol{s}. \tag{11.65}$$

This implies that in presence of a dielectric medium the capacity coefficients C_{ij} are increased, in relation to the values in vacuum, by a factor ε. It is interesting to compare this situation with that in which the space between the conductors is occupied by a material of electric conductivity σ so that, if each conductor is kept at constant potential V_i, a current I_i flows in or out each of them. Since the medium surrounding the conductors is linear, we can assume that the currents and the potentials are linked through a relation of the type

$$I_i = \sum_j G_{ij} V_j. \tag{11.66}$$

The constants G_{ij} are called the *coefficients of conductance* of the conductors. The inverse relation allows us to express the potential on each conductor in terms of the currents flowing through each one of them

$$V_i = \sum_j R_{ij} I_j. \tag{11.67}$$

Here, R_{ij} are the *coefficients of resistance* of the system.

The similarity between the problems we are considering results from the fact that for a dielectric medium the potential ϕ satisfies Laplace's equation with the boundary conditions

$$\begin{aligned} \phi &= \phi_n, \\ Q_n &= -\int_S \varepsilon\,\frac{\partial\phi}{\partial n}\,\mathrm{d}s, \end{aligned} \tag{11.68}$$

while for a conducting medium ϕ also satisfies Laplace's equation but with the boundary conditions

$$\begin{aligned} \phi &= \phi_n, \\ I_n &= -\int_S \sigma\nabla\phi\cdot \mathrm{d}\boldsymbol{s}. \end{aligned} \tag{11.69}$$

From this analogy, we conclude that the conductance coefficients for one case and the coefficients of capacity for the other are related according to

$$G_{ij}/\sigma = C_{ij}/\varepsilon. \tag{11.70}$$

Thus, if we have determined the capacities of a system of conductors in the presence of a linear dielectric, we also have the conductivities of the same system in the presence of a linear conducting material.

11.6 Variational Principle for Conducting Media

Similarly to the case of electrostatics and of magnetostatics, we can show that the potential distribution and the currents in a linear conducting medium respond to a variational principle. In this case, because the conduction of current is a dissipative process, the principle does not involve the energy, but rather the dissipated power. The principle implies that the distribution of potentials and currents involved in conduction is such that the dissipated power is a minimum.

The total power dissipated in a volume V by the Joule effect is given by the work done by the electric field on the conduction charges, which is

$$\begin{aligned}\frac{\mathrm{d}W}{\mathrm{d}t} &= \int_V \boldsymbol{j}\cdot\boldsymbol{E}\,\mathrm{d}^3r = \int_V \sigma E^2\mathrm{d}^3r = \int_V \sigma(\nabla\phi)^2\mathrm{d}^3r \\ &= \frac{1}{\sigma}\int j^2\mathrm{d}^3r.\end{aligned} \tag{11.71}$$

Let us now assume that the current $\boldsymbol{j}$ violates Ohm's law by an amount $\boldsymbol{j}'$, more precisely, $\boldsymbol{j} = \sigma\boldsymbol{E} + \boldsymbol{j}'$, but such that the total current $\boldsymbol{j}$ still satisfies the same boundary conditions. In that case we have

$$\begin{aligned}\frac{\mathrm{d}W}{\mathrm{d}t} &= \frac{1}{\sigma}\int(\sigma\boldsymbol{E}+\boldsymbol{j}')^2\mathrm{d}^3r = \frac{1}{\sigma}\int(-\sigma\nabla\phi+\boldsymbol{j}')^2\mathrm{d}^3r \\ &= \frac{1}{\sigma}\int\left[\sigma^2(\nabla\phi)^2 - 2\sigma\,\boldsymbol{j}'\cdot\nabla\phi + j'^2\right]\mathrm{d}^3r.\end{aligned} \tag{11.72}$$

Now, from the divergence theorem, we can prove that

$$\int_V (\boldsymbol{j}'\cdot\nabla\phi + \phi\nabla\cdot\boldsymbol{j}')\,\mathrm{d}^3r = \phi\int_S \boldsymbol{j}'\cdot\mathrm{d}\boldsymbol{s}. \tag{11.73}$$

The right-hand side of this identity is zero because the integral equals the total flow of the current $\boldsymbol{j}'$. The second term on the left-hand side is also zero, since $\boldsymbol{j}'$ also satisfies the continuity of the total current $\nabla\cdot\boldsymbol{j}' = 0$. In this way we obtain

$$\int_V \boldsymbol{j}'\cdot\nabla\phi\,\mathrm{d}^3r = 0. \tag{11.74}$$

Consequently, the dissipated power is

$$\frac{\mathrm{d}W}{\mathrm{d}t} = \frac{1}{\sigma} \int \left[\sigma^2 (\nabla\phi)^2 + j'^2\right] \mathrm{d}^3 r. \tag{11.75}$$

Any deviation $\boldsymbol{j}'$ of the current distribution thus produces an increment in the generated power. Therefore, this power is a minimum for the potential ϕ which satisfies Laplace's equation.

Problems

Most problems in this chapter can also be analyzed with PhysicSolver.

11.1 Find the angle determined by the lines of $\boldsymbol{E}$ and $\boldsymbol{D}$ at each side of the interface of two dielectrics of constants ε_1 and ε_2.

11.2 Use the method of images to find the electric field produced over all space by a charge inside a sphere of linear dielectric. Find the polarization charge density induced in the dielectric.

11.3 A large portion of linear dielectric with a spherical cavity is immersed in a constant electric field. Find the field over all space. Show that outside the cavity the field corresponds to that produced by a dipole placed at the center. Find the charge induced on the surface of the cavity.

Hint: Use the axially symmetric expansion of the potential and apply boundary conditions to determine the coefficients.

11.4 A spherical shell of dielectric material of radii a and b is immersed in a uniform electric field. What is the effect of the shell on the field in the inner cavity? Analyze the limit in which the dielectric constant tends to infinity.

11.5 A charged metallic sphere is half immersed in a linear dielectric filling half-space. Find the field over all space and the charges induced on the surface of the dielectric.

11.6 A tiny sphere of linear dielectric material is placed at a certain distance from a uniformly charged wire. Calculate the force on the sphere.

Hint: Assume the radius of the sphere to be much smaller than the distance to the wire.

11.7 A spherical shell of permeability μ and radii a and b is placed in a uniform magnetic field. Determine $\boldsymbol{B}$ over all space.

11.8 A very long cylindrical capacitor of coaxial plates is partially immersed in a vertical position into a non-conducting dielectric liquid of density ρ and dielectric constant ε. The plates of the capacitor are kept at constant potential. Find the height reached by the liquid between the capacitor plates.

Hint: Neglect edge effects and assume the field to be radial inside the capacitor.

11.9 Express the energy of a material ellipsoid of linear dielectric in the presence of a uniform electric field in terms of the depolarization tensor $\mathcal{N}$. Show that the ellipsoid tends to align its longer axis with the applied field.

11.10 The parallel plates of a plane capacitor are covered with a material of dielectric constant ε_1, conductivity σ_1 and thickness d_1. This material does not fill all the space between the plates, since a second material of dielectric constant ε_2 and conductivity σ_2 occupies the central region of thickness d_2 (Fig. 11.4). The potential difference between the plates is constant and equal to V. Determine the field inside the capacitor. Assume different values for the conductivities σ_1 and σ_2 and establish the different regimes that may appear.

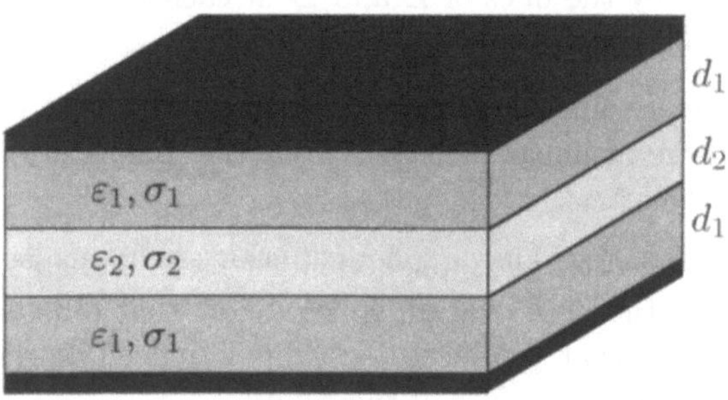

Fig. 11.4. Dielectrics inside a plane capacitor

Solution: Assuming that the distance between the plates is much smaller than any of the transverse dimensions, we can suppose that both the potential and the fields depend on the coordinate normal to the plates only, which we call x. Let I label the relevant quantities in the region $0 \le x < d_1$, II in $d_1 \le x < d_1 + d_2$, and III in $d_1 + d_2 \le x < d$. The electrostatic potential $\phi(x)$ satisfies Laplace's equation, which in this case reduces to

$$\frac{\mathrm{d}^2\phi(x)}{\mathrm{d}x^2} = 0,$$

in each region. The solutions are of the type

$$\phi_i(x) = a_i + b_i\, x,$$

with $i = \mathrm{I, II, III}$, and where the constants a_i and b_i vary from one region to another and are determined by the boundary conditions. Due to the finite conductivity of both media, a current normal to the plates flows. Once the stationary regime is established, the condition of charge conservation requires that this current is the same through all regions. In each one of them Ohm's law must be satisfied:

$$j = \sigma_1 E_{\mathrm{I}} = \sigma_2 E_{\mathrm{II}} = \sigma_1 E_{\mathrm{III}}.$$

As, in turn, $E_i = -b_i$, we have $\sigma_1 b_{\mathrm{I}} = \sigma_2 b_{\mathrm{II}} = \sigma_1 b_{\mathrm{II}}$, from which $b_{\mathrm{III}} = b_{\mathrm{I}}$. Moreover, $b_{\mathrm{II}} = \sigma_1 b_{\mathrm{I}}/\sigma_2$. Making use of the boundary conditions at $x = 0$,

$x = d_1$, $x = d_1 + d_2$, and $x = 2d_1 + d_2 = d$, we obtain the values of the coefficients in terms of the resistivities $\rho_1 = 1/\sigma_1$ and $\rho_2 = 1/\sigma_2$. They are $b_{\rm I} = \rho_1 V/R$, $a_{\rm I} = 0$, $a_{\rm II} = -(\rho_2 - \rho_1)d_1 V/R$, and $a_{\rm III} = (\rho_2 - \rho_1)d_2 V/R$, where $R = (2d_1\rho_1 + d_2\rho_2)$ represents the total resistance per unit area of the set of materials filling the capacitor. The three segments of the potential can be written as

$$\phi_{\rm I} = \frac{V}{R}\rho_1 x,$$

$$\phi_{\rm II} = \frac{V}{R}\left[\rho_1 d_1 + \rho_2 (x - d_1)\right],$$

$$\phi_{\rm III} = \frac{V}{R}\left[\rho_2 d_2 + \rho_1 (x - d_2)\right].$$

We see that for dielectrics with finite conductivity, the values of the potential, and therefore of the electric field, are determined by the conductivity and not by the dielectric constant. This latter quantity determines the values of the displacement vector and the amount of charge accumulated on the dielectric interfaces. The values of the potential at the points $x = d_1$ and $x = d_1 + d_2$ are

$$\phi_{\rm I}(d_1) = \phi_{\rm II}(d_1) = \frac{V}{R}\rho_1 d_1,$$

and

$$\phi_{\rm II}(d_1 + d_2) = \phi_{\rm III}(d_1 + d_2) = \frac{V}{R}\left(\rho_2 d_2 + \rho_1 d_1\right).$$

According to (11.64), the charge density accumulated at the interfaces between the dielectrics is

$$\sigma_{\rm F} = \frac{1}{4\pi}\left(\frac{\varepsilon_1}{\sigma_1} - \frac{\varepsilon_2}{\sigma_2}\right)\frac{V}{R}.$$

The displacement vector $\boldsymbol{D}$ can be evaluated from this relation.

Further Reading

R. Becker and F. Sauter: *Electromagnetic Fields and Interactions* (Blaisdell, New York, 1964).

J.D. Jackson: *Classical Electrodynamics* (Wiley, New York, 1975).

L.D. Landau and E.M. Lifshitz: *Electrodynamics of Continuous Media* (Pergamon, New York, 1960).

W.K.H. Panofsky and M. Phillips: *Classical Electricity and Magnetism* (Addison–Wesley, Reading, 1955) Chap. 8.

Further Reading

12. Waves in Material Media

In this chapter we study the problem of wave propagation in linear material media. This is one of the most important applications of the electromagnetic theory to these media.

We obtain the main properties of waves in linear media and compare them with the case of waves in vacuum discussed in Chap. 8. The effect of an interface between two media is also considered and the electromagnetic foundations of Snell laws are presented. Total internal reflection, an interesting phenomenon with several practical applications of current interest, is discussed in some detail. The propagation of waves in conducting media, where energy dissipation plays an important role, is also analyzed.

The propagation of electromagnetic waves, as for all other electromagnetic phenomena in material media, takes place in the empty space surrounding atoms and molecules. Observable phenomena are the result of the cooperative effect of the interaction of atoms and molecules with the propagating electromagnetic fields. In order to analyze this interaction, we briefly discuss Maxwell's equations in a material medium from the point of view of the field sources.

To illustrate the influence of the microscopic properties on the behavior of dielectrics under the effect of dynamical fields the Lorentz model is presented. This model describes the response of material media on the basis of a classical model of molecular dynamics. Finally, we present the general properties of the linear response of the medium. This is an important concept in many physical systems which can be well exemplified through the properties of the electromagnetic response.

12.1 Wave Equations in Linear Media

Starting from Maxwell's equations in material media in the absence of free charges and currents

$$\begin{cases} \nabla \cdot \boldsymbol{D} = 0, \\ \nabla \times \boldsymbol{H} - \frac{1}{c}\partial_t \boldsymbol{D} = 0, \\ \nabla \cdot \boldsymbol{B} = 0, \\ \nabla \times \boldsymbol{E} + \frac{1}{c}\partial_t \boldsymbol{B} = 0, \end{cases} \tag{12.1}$$

it is possible to introduce the electromagnetic potentials ϕ and $\boldsymbol{A}$ which allow us to express the fields as $\boldsymbol{E} = -\nabla\phi - c^{-1}\partial_t \boldsymbol{A}$ and $\boldsymbol{B} = \nabla \times \boldsymbol{A}$.

We have already mentioned that, for electromagnetic potentials in vacuum, the Lorentz gauge is a convenient choice because of the symmetric form in which scalar and vector potentials appear in this gauge. This symmetry is, as we have shown in Chap. 9, a manifestation of the underlying covariance of the theory. On the other hand, when dealing with the electrodynamics of material media, there is a privileged reference frame – in which the material medium is at rest. In this case, it is convenient to choose a gauge that also privileges the properties of the medium. For linear media, we impose

$$\nabla \cdot \boldsymbol{A} + \frac{\varepsilon\mu}{c}\frac{\partial\phi}{\partial t} = 0. \tag{12.2}$$

Certainly, this choice does not mean that the electrodynamics of material media cannot be formulated in covariant form. In fact, this problem, which was faced by Einstein in his work *Zur Electrodynamik bewegter Koerper*, can be completely solved.

Through a derivation similar to that presented in Chap. 7 for vacuum, and using the constitutive laws for linear media $\boldsymbol{P} = \chi_{\mathrm{E}}\boldsymbol{E}$ and $\boldsymbol{M} = \chi_{\mathrm{M}}\boldsymbol{B}$ which imply $\boldsymbol{D} = \varepsilon\boldsymbol{E}$ and $\boldsymbol{H} = \boldsymbol{B}/\mu$, it can be shown that in the gauge given by (12.2) the potentials satisfy the following equations

$$\begin{aligned} \left(\nabla^2 - \tfrac{\varepsilon\mu}{c^2}\partial_{tt}^2\right)\phi = 0, \\ \left(\nabla^2 - \tfrac{\varepsilon\mu}{c^2}\partial_{tt}^2\right)\boldsymbol{A} = 0. \end{aligned} \tag{12.3}$$

We leave the reader the task of showing that the electromagnetic fields $\boldsymbol{E}$ and $\boldsymbol{B}$, as well as the electric and magnetic polarizations $\boldsymbol{P}$ and $\boldsymbol{M}$, satisfy the same wave equation as the potentials. This implies that all fields of interest in a material medium propagate with velocity c/n, where $n = \sqrt{\varepsilon\mu}$ is the *refractive index* of the medium. Since in the most common situations $n > 1$, the phase velocity of the wave in the medium – that is to say, the speed at which the surfaces of equal phase move – is smaller than c.

As in the case of vacuum analyzed in Chap. 8, plane waves are an interesting particular solution to the wave equations in material media (12.3). Introducing complex fields, a plane wave solution of the wave equations can be written as

$$\begin{aligned}\phi(\boldsymbol{r},t) &= \phi_0 \exp[\mathrm{i}(\boldsymbol{k}\cdot\boldsymbol{r}-\omega t)],\\ \boldsymbol{A}(\boldsymbol{r},t) &= \boldsymbol{A}_0 \exp[\mathrm{i}(\boldsymbol{k}\cdot\boldsymbol{r}-\omega t)].\end{aligned} \tag{12.4}$$

Here the amplitudes ϕ_0 and $\boldsymbol{A}_0$ are, in general, complex constants. The wave equations imply the following relation between wave vector and frequency

$$k^2 = \varepsilon\mu\,\frac{\omega^2}{c^2} = n^2\,\frac{\omega^2}{c^2}. \tag{12.5}$$

This determines the dispersion relation $\omega(k) = ck/n$. The gauge given by (12.2) together with the radiation gauge (see Chap. 7) requires that $\phi_0 = 0$ and $\boldsymbol{A}_0\cdot\boldsymbol{k} = 0$. The resulting electromagnetic fields are

$$\begin{aligned}\boldsymbol{E}(\boldsymbol{r},t) &= \mathrm{i}(\omega/c)\boldsymbol{A}_0 \exp[\mathrm{i}(\boldsymbol{k}\cdot\boldsymbol{r}-\omega t)] = \boldsymbol{E}_0 \exp[\mathrm{i}(\boldsymbol{k}\cdot\boldsymbol{r}-\omega t)],\\ \boldsymbol{B}(\boldsymbol{r},t) &= \mathrm{i}\boldsymbol{k}\times\boldsymbol{A}_0 \exp[\mathrm{i}(\boldsymbol{k}\cdot\boldsymbol{r}-\omega t)] = \boldsymbol{B}_0 \exp[\mathrm{i}(\boldsymbol{k}\cdot\boldsymbol{r}-\omega t)].\end{aligned} \tag{12.6}$$

Both $\boldsymbol{E}_0$ and $\boldsymbol{B}_0$ are then perpendicular to $\boldsymbol{k}$, so that, if the wave vector is real, electromagnetic waves in a linear material medium are transverse. The following relation is also relevant:

$$\boldsymbol{B}_0 = n\,\hat{\boldsymbol{k}}\times\boldsymbol{E}_0. \tag{12.7}$$

As in vacuum, the magnetic field is perpendicular to the electric field and to the propagation direction $\hat{\boldsymbol{k}}$ and both fields are in phase if $\boldsymbol{k}$ is real. On the other hand, their moduli are not equal, $B_0 = nE_0 > E_0$. The polarization properties of waves in material media are similar to those in vacuum, and are independent of the gauge choice.

The fact that the electric and magnetic polarization $\boldsymbol{P}$ and $\boldsymbol{M}$ obey the same wave equations as the fields imply that the propagation of the latter is accompanied by the propagation of *polarization waves.* In Sect. 12.4 we analyze in more detail the role of the polarizations as sources of the electromagnetic fields and their relevance in the formation of the resulting wave, which propagates with velocity c/n.

We have shown in Chap. 10 that the Poynting vector in a material medium is defined as

$$\boldsymbol{S} = \frac{c}{4\pi}\,\boldsymbol{E}\times\boldsymbol{H}. \tag{12.8}$$

In the special case of a plane wave, the time average of $\boldsymbol{S}$ is

$$\langle\boldsymbol{S}\rangle = \frac{c}{8\pi}\sqrt{\frac{\varepsilon}{\mu}}E_0^2\,\hat{\boldsymbol{k}}. \tag{12.9}$$

As we see, it has the direction of propagation of the fields. The energy density associated with the wave can be written as

$$\langle w\rangle = \frac{\varepsilon}{8\pi}E_0^2. \tag{12.10}$$

This implies that the velocity of energy flow is

$$\boldsymbol{v} = \frac{\langle \boldsymbol{S} \rangle}{\langle w \rangle} = \frac{c}{n}\,\hat{\boldsymbol{k}}. \tag{12.11}$$

As in vacuum, the energy of a monochromatic plane wave in a material medium flows with the phase velocity.

12.2 Waves at an Interface

Let us consider now a situation of particular importance in practice, i.e. the transmission of a wave from one material medium to another through the interface. Consider two media which are linear materials of constants (ε, μ) and (ε', μ') separated by a plane surface. If a plane wave is incident from the first medium onto this interface, we expect to have in general a reflected wave propagating back into the first medium and a transmitted wave propagating into the second medium (Fig. 12.1). We write the fields associated with the incident wave as

$$\begin{aligned} \boldsymbol{E} &= \boldsymbol{E}_0 \exp[\mathrm{i}(\boldsymbol{k}\cdot\boldsymbol{r} - \omega t)], \\ \boldsymbol{B} &= \sqrt{\varepsilon\mu}\,\hat{\boldsymbol{k}} \times \boldsymbol{E}, \end{aligned} \tag{12.12}$$

and for the transmitted and reflected waves we assume

$$\begin{aligned} \boldsymbol{E}' &= \boldsymbol{E}'_0 \exp[\mathrm{i}(\boldsymbol{k}'\cdot\boldsymbol{r} - \omega' t)], \\ \boldsymbol{B}' &= \sqrt{\varepsilon'\mu'}\,\hat{\boldsymbol{k}}' \times \boldsymbol{E}', \end{aligned} \tag{12.13}$$

and

$$\begin{aligned} \boldsymbol{E}'' &= \boldsymbol{E}''_0 \exp[\mathrm{i}(\boldsymbol{k}''\cdot\boldsymbol{r} - \omega'' t)], \\ \boldsymbol{B}'' &= \sqrt{\varepsilon\mu}\,\hat{\boldsymbol{k}}'' \times \boldsymbol{E}'', \end{aligned} \tag{12.14}$$

respectively. The dispersion relation requires $k = \omega n/c$ and $k'' = \omega'' n/c$, where $n = \sqrt{\varepsilon\mu}$ is the refractive index of the first medium. Similarly $k' = \omega n'/c$.

In Chap. 10 we have analyzed the boundary conditions to be satisfied by the fields at the interface of two media. In the present case, two basic constraints for these conditions to hold at any time and at any point in the interface are

$$\begin{aligned} &\omega = \omega' = \omega'', \\ &\boldsymbol{k}\cdot\boldsymbol{r}|_{z=0} = \boldsymbol{k}'\cdot\boldsymbol{r}|_{z=0} = \boldsymbol{k}''\cdot\boldsymbol{r}|_{z=0}, \end{aligned} \tag{12.15}$$

where $z = 0$ is the plane of the interface (Fig. 12.1). The second of these conditions implies $k_x x + k_y y = k'_x x + k'_y y = k''_x x + k''_y y$.

Without loss of generality let us consider the incident wave vector $\boldsymbol{k}$ to be in the xz plane, so that $k_y = 0$. For the previous condition to be satisfied

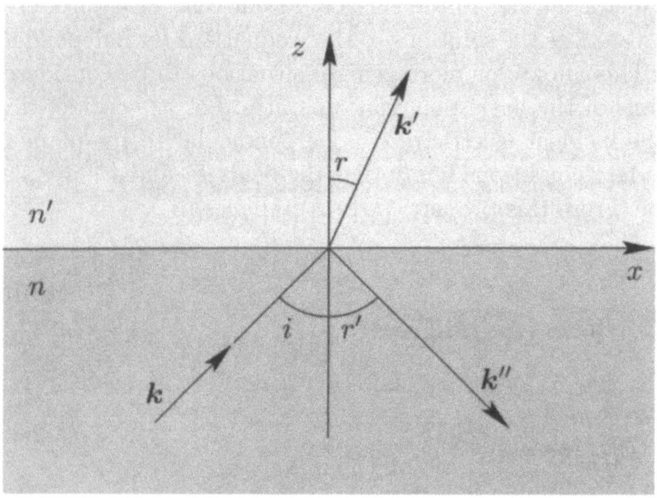

Fig. 12.1. A plane wave incident on the interface of two dielectrics

at all points of the interface the relation $k'_y = k''_y = 0$ must hold. This result implies that the incident, reflected and transmitted waves are coplanar. The common plane containing them all is defined by the wave vector $\boldsymbol{k}$ and the unit vector normal to the interface, and is called the *plane of incidence.*

The constraint on the wave vectors is equivalent to the condition that their components along x should be equal, $k_x = k'_x = k''_x$. Introducing the angles of incidence i, of refraction r, and of reflection r', this implies

$$\begin{aligned} \sin i &= \sin r', \\ k\ \sin i &= k' \sin r. \end{aligned} \tag{12.16}$$

From this, the *kinematic* or *Snell laws* for reflection and refraction are readily derived:

$$\begin{aligned} i &= r', \\ n\ \sin i &= n' \sin r. \end{aligned} \tag{12.17}$$

The continuity of the normal components of the displacement $\boldsymbol{D}$ and of the magnetic field $\boldsymbol{B}$, as well as of the tangential components of $\boldsymbol{E}$ and $\boldsymbol{H}$ through the interface imply the following conditions:

$$\begin{aligned} &\varepsilon(\boldsymbol{E}_0 + \boldsymbol{E}''_0) \cdot \boldsymbol{n} = \varepsilon' \boldsymbol{E}'_0 \cdot \boldsymbol{n}, \\ &(\boldsymbol{k} \times \boldsymbol{E}_0 + \boldsymbol{k}'' \times \boldsymbol{E}''_0) \cdot \boldsymbol{n} = (\boldsymbol{k}' \times \boldsymbol{E}'_0) \cdot \boldsymbol{n}, \\ &(\boldsymbol{E}_0 + \boldsymbol{E}''_0) \times \boldsymbol{n} = \boldsymbol{E}'_0 \times \boldsymbol{n}, \\ &\mu^{-1}(\boldsymbol{k} \times \boldsymbol{E}_0 + \boldsymbol{k}'' \times \boldsymbol{E}''_0) \times \boldsymbol{n} = \mu'^{-1}(\boldsymbol{k}' \times \boldsymbol{E}'_0) \times \boldsymbol{n}. \end{aligned} \tag{12.18}$$

Here, $\boldsymbol{n}$ is the unit vector normal to the surface. In our case, $\boldsymbol{n} = (0, 0, 1)$.

For simplicity, we analyze the solutions to these equations for independent polarization states. This allows for more general situations to be considered as linear combinations of the basic polarization states. Let us consider first the case in which the incident electric field is perpendicular to the plane of incidence. The boundary conditions imply that the fields $\boldsymbol{E}'$ and $\boldsymbol{E}''$ are also normal to that plane. From this property, (12.18) reduces to

$$\begin{aligned} &E_0 + E_0'' = E_0', \\ &\sqrt{\varepsilon/\mu}(E_0 - E_0'')\cos i = \sqrt{\varepsilon'/\mu'}E_0'\cos r. \end{aligned} \tag{12.19}$$

Their solutions are

$$E_0' = \frac{2n\cos i}{n\cos i + (\mu/\mu')n'\cos r}E_0, \tag{12.20}$$

and

$$E_0'' = \frac{n\cos i - (\mu/\mu')n'\cos r}{n\cos i + (\mu/\mu')n'\cos r}E_0. \tag{12.21}$$

Both quantities are given in terms of the incident amplitude E_0.

The transmissivity and reflectivity of the interface are usually defined by the transmission and reflection coefficients T and R. These are defined as the ratios between the averages of the normal component of the transmitted and reflected energy flow to the corresponding component of the incident flow

$$\begin{aligned} &R = \left|\langle \boldsymbol{S}'' \cdot \boldsymbol{n}\rangle / \langle \boldsymbol{S} \cdot \boldsymbol{n}\rangle\right|, \\ &T = \left|\langle \boldsymbol{S}' \cdot \boldsymbol{n}\rangle / \langle \boldsymbol{S} \cdot \boldsymbol{n}\rangle\right|. \end{aligned} \tag{12.22}$$

Here $\boldsymbol{S}$, $\boldsymbol{S}'$ and $\boldsymbol{S}''$ are the Poynting vectors corresponding to the incident, transmitted and reflected fields respectively. These quantities satisfy $R+T = 1$, as required by energy conservation.

When the electric field is perpendicular to the plane of incidence, we obtain

$$\begin{aligned} &R_\perp = \left[\frac{\sin r\cos i - (\mu/\mu')\sin r\cos i}{\sin r\cos i + (\mu/\mu')\sin r\cos i}\right]^2, \\ &T_\perp = \frac{4(\mu/\mu')\sin r\cos i\sin r\cos r}{[\sin r\cos i + (\mu/\mu')\sin r\cos i]^2}. \end{aligned} \tag{12.23}$$

The expressions for R and T are considerably simplified if we assume that $\mu/\mu' \approx 1$, as is true for most dielectrics. Thus, we obtain

$$R_\perp = \frac{\sin^2(i-r)}{\sin^2(i+r)}, \qquad T_\perp = \frac{\sin 2i\ \sin 2r}{\sin^2(i+r)}. \tag{12.24}$$

In the second case, the incident electric field is parallel to the plane of incidence. We have now

$$E_0' = \frac{2n(\mu'/\mu)\cos i}{n'\cos i + (\mu'/\mu)n\cos r}\, E_0, \tag{12.25}$$

and

$$E_0'' = \frac{n'\cos i - (\mu'/\mu)n\cos r}{n'\cos i + (\mu'/\mu)n\cos r}\, E_0. \tag{12.26}$$

The reflection and transmission coefficients, in the limit $\mu/\mu' = 1$, take the form

$$R_{\|} = \frac{\tan^2(i-r)}{\tan^2(i+r)}, \qquad T_{\|} = \frac{\sin 2i\ \sin 2r}{\sin^2(i+r)\ \sin^2(i-r)}. \tag{12.27}$$

For normal incidence, $i = 0$, there is no distinction between parallel and perpendicular polarization, and the reflection and transmission coefficients reduce to

$$R = \left(\frac{n-n'}{n+n'}\right)^2, \qquad T = \frac{4nn'}{(n+n')^2}. \tag{12.28}$$

In Fig. 12.2 the reflection coefficients $R_\perp$ and $R_{\|}$ are shown as functions of the angle of incidence in two qualitatively different situations. In the first plot $n < n'$. We see that $R_\perp$ increases monotonically with i and reaches the value $R_\perp = 1$ at $i = \pi/2$. On the other hand, $R_{\|}$ – which is equal to $R_\perp$ at $i = 0$ and $i = \pi/2$ – vanishes at an intermediate angle i_B, called the *Brewster angle*. Analyzing the expression for $R_{\|}$ we find that the Brewster angle satisfies

$$i_\mathrm{B} + r_\mathrm{B} = \pi/2, \tag{12.29}$$

where $r_\mathrm{B} = \arcsin(n\ \sin i_\mathrm{B}\ /n')$. At this angle of incidence, the dipoles induced in the material in which the transmitted wave propagates oscillate in the direction of the reflected wave. As we have shown in Chap. 8, there is no radiation emitted in that direction. In this case, if the incident wave is unpolarized – namely, if it has polarization components both parallel and perpendicular to the plane of incidence – the parallel polarization component is completely transmitted and, therefore, the reflected wave is polarized in the perpendicular direction.

Reflection at a dielectric interface is therefore a method of generating polarized light. The polarizing lenses used in photographic cameras and sun glasses are designed to eliminate the reflections on dielectric surfaces (glass, water) by cancelling the polarized component that remains after reflection. The same principle applies in telecommunications, where the use of polarization-sensitive antennas can reinforce or eliminate the reflected waves.

For $n > n'$ the reflection coefficient has a similar behavior to that in the previous case. However, both $R_\perp$ and $R_{\|}$ become now equal to unity at a critical angle $i_\mathrm{C} < \pi/2$. This phenomenon is known as *total internal reflection*. Indeed, for angles of incidence greater than i_C all energy flow normal to the surface is reflected, the transmission coefficient vanishes, and

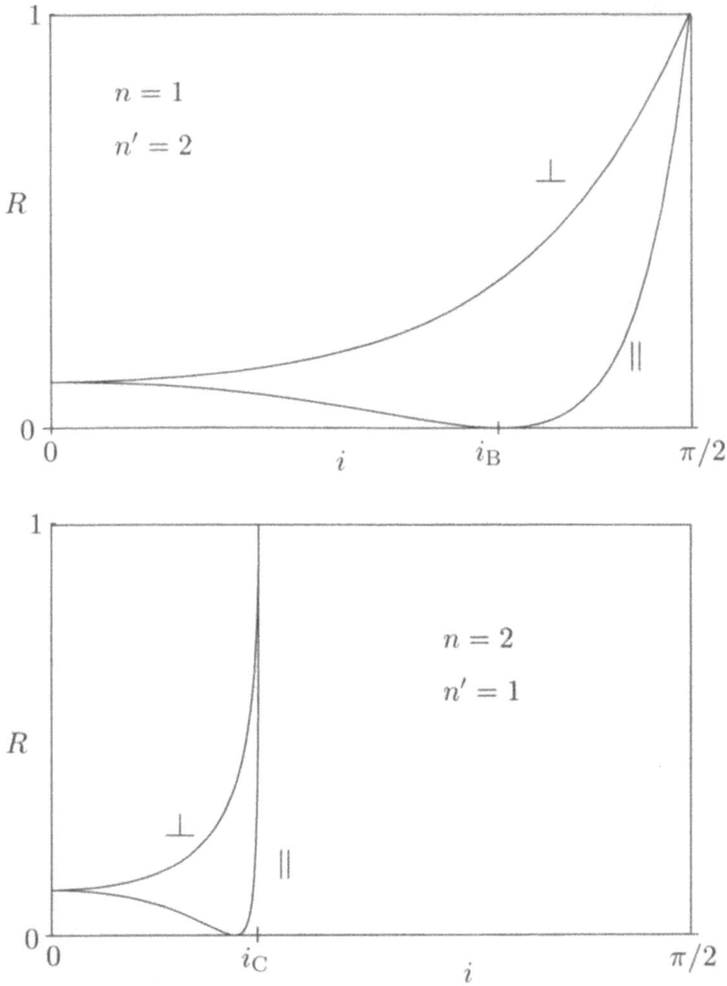

Fig. 12.2. Reflection coefficients as functions of the angle of incidence

the interface acts as a perfect mirror. The angle of total reflection is reached when $r = \pi/2$, i.e. when

$$\sin i_{\mathrm{C}} = \frac{n'}{n}. \tag{12.30}$$

From this we see that this phenomenon requires $n' < n$.

It is interesting to analyze in some detail the transmitted fields in the situation of total reflection. When $i > i_{\mathrm{C}}$ the angle of refraction r becomes imaginary, $\sin r = \sin i/\sin i_{\mathrm{C}}$ is real and larger than unity, and $\cos r = i\beta$, where $\beta = \sqrt{(\sin i/\sin i_{\mathrm{C}})^2 - 1}$, is imaginary. The space dependence of the transmitted field $\boldsymbol{E}'$ is given by

$$\boldsymbol{k}' \cdot \boldsymbol{r} = k' \sin r \; x + k' \cos r \; z = \alpha x + \mathrm{i}\gamma z. \tag{12.31}$$

Here we have used the same system of coordinates as in Fig. 12.1. The transmitted electric field is thus

$$\boldsymbol{E}' = \boldsymbol{E}_0' \exp[\mathrm{i}(\alpha x - \omega t)] \; \exp(-\gamma z). \tag{12.32}$$

Therefore, the transmitted field falls off exponentially in the direction normal to the interface, while it propagates parallel to the interface in the x direction. The wave associated with this field is called the *evanescent wave*. It represents a *bound state* associated with the interface, since it is a solution to Maxwell's equations whose existence is possible only thanks to the presence of the surface of separation between both media. In Fig. 12.3 we show schematically the phase planes of the incident, reflected and transmitted waves in the case of total internal reflection.

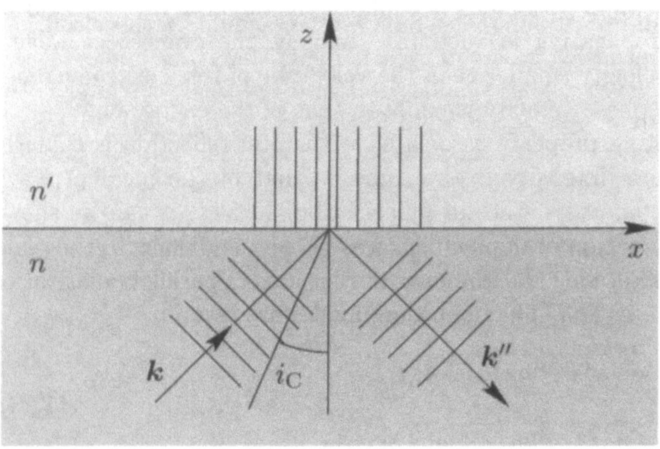

Fig. 12.3. Phase planes in total internal reflection

To analyze the flow of energy associated with the evanescent wave, we consider the particular case of an electric field normal to the plane of incidence. The real components of the transmitted electromagnetic fields are

$$\begin{aligned} E_y' &= E_0' \exp(-\gamma z) \cos(\alpha x - \omega t), \\ B_x' &= \tfrac{c\gamma}{\omega} E_0' \exp(-\gamma z) \sin(\alpha x - \omega t), \\ B_z' &= \tfrac{c\alpha}{\omega} E_0' \exp(-\gamma z) \cos(\alpha x - \omega t), \\ E_x' &= E_z' = B_y' = 0. \end{aligned} \tag{12.33}$$

Note that, in contrast to what happens with waves propagating in material media far from the interfaces, the component B_x' is now dephased by $\delta = \pi/2$ relative to E_y'.

The corresponding Poynting vector has components in the x and z directions. Due to the dephasing between B'_x and E'_y, the component S_z vanishes after time averaging. The only net flow of energy associated with the evanescent wave therefore propagates in the x direction, and is given by

$$\langle S_x \rangle = \frac{c^2 \alpha}{8\pi\omega} E_0'^{\,2} \exp(-2\gamma z). \tag{12.34}$$

The energy density is

$$\langle w \rangle = \frac{c^2 \alpha^2}{8\pi\omega^2} E_0'^2 \exp(-2\gamma z). \tag{12.35}$$

From here, we can evaluate the velocity of energy propagation to be

$$v_x = \frac{\langle S_x \rangle}{\langle w \rangle} = \frac{\sin i_\mathrm{C}}{\sin i}\, c. \tag{12.36}$$

Note that the medium with index n' could be the vacuum. In this situation we have a wave propagating in vacuum – namely, the evanescent wave – which moves at a velocity smaller than the velocity c of free electromagnetic perturbations. The phase velocity is equal to that of the energy flow.

Another interesting property associated with total reflection is that the reflected wave, whose intensity is the same as that of the incident wave, differs from the latter in its state of polarization. In fact, it can be shown that when the polarization of the incident wave is perpendicular to the plane of incidence, $E_0'' = \exp(-\mathrm{i}\delta_\perp)E_0$. Similarly, in the case of parallel polarization we have $E_0'' = \exp(-\mathrm{i}\delta_\parallel)E_0$. The dephasing angles are given by

$$\begin{aligned} \tan(\delta_\perp/2) &= \sqrt{\sin^2 i - \sin^2 i_\mathrm{C}}/\cos i, \\ \tan(\delta_\parallel/2) &= \sqrt{\sin^2 i - \sin^2 i_\mathrm{C}}/\sin^2 i_\mathrm{C} \cos i. \end{aligned} \tag{12.37}$$

Because of this property, total reflection can be used to produce special polarization states starting from linearly polarized light. The *Fresnel prism* is a device that produces circularly polarized light through multiple total internal reflections.

Internal reflection presents an additional property, connected with other undulatory phenomena. If a piece of dielectric with refractive index n is placed at a certain distance from the interface, in the region occupied by the medium with index n', the "exponential tail" of the evanescent wave can excite an ordinary wave in the second medium. This phenomenon constitutes an analog of the *tunnel effect* of quantum mechanics. It demonstrates the similarity between both theories based on their undulatory character.

If plane waves are replaced by a collimated beam or a ray of light, interesting new phenomena associated with total internal reflection can be observed. In this case, the point of incidence on the interface is well defined. In an experiment carried out in 1947, F.Goos and H. Hänchen demonstrated that the reflected ray **does not** start from the point of incidence [12.1]. It appears that the light, after reaching the interface, travels a certain distance along the interface before reappearing in the first medium. This shift is of the order of the wavelength of the incident wave. J. Agudín and A.M. Platzeck showed that the *Goos–Hänchen effect* can be explained by means of Fermat's *principle of minimum time* [12.2]. In this explanation, one must keep in mind that the propagation velocity between the point of incidence and that of reflection is that indicated in (12.36). In 1970 Ch. Imbert showed that if the incident light is circularly polarized the displacement has a component transverse to the propagation direction of the beam [12.3]. The sign of this displacement depends on the sense of the polarization. The *transverse Goos–Hänchen phenomenon* was studied theoretically by Agudín [12.4].

The phenomenon of total internal reflection has important applications in the design of optical fibers used in communications and in endoscopic techniques in medicine.

12.3 Waves in Conducting Media

As we have shown in the previous chapter, conducting media are characterized by an internal distribution of free charges. In fact, the presence of an electric field in a linear conductor generates a current density proportional to the field. We have already indicated that the basic equation describing these materials is Ohm's law,

$$\boldsymbol{j} = \sigma \boldsymbol{E}. \tag{12.38}$$

Here σ is the conductivity of the medium, which is in general a tensor quantity.

In the presence of the current generated by the electric field the wave equation for $\boldsymbol{E}$ reads

$$\nabla \times \nabla \times \boldsymbol{E} + \frac{\varepsilon\mu}{c^2}\, \partial_{tt}^2 \boldsymbol{E} + \frac{4\pi\sigma\mu}{c^2}\, \partial_t \boldsymbol{E} = 0. \tag{12.39}$$

Assuming a harmonic solution propagating in the z direction,

$$\boldsymbol{E} = \boldsymbol{E}_0\ \exp[\mathrm{i}(kz - \omega t)], \tag{12.40}$$

we find that the wave number k must satisfy

$$k = \frac{\omega}{c}\ \sqrt{\varepsilon\mu + 4\pi\mathrm{i}\ \sigma\mu/\omega}. \tag{12.41}$$

The wavenumber, and therefore the refractive index $n = ck/\omega$, is complex. Writing the wavenumber in the form $k = \alpha + \mathrm{i}\beta$ we have for the electric field

$$\boldsymbol{E} = \boldsymbol{E}_0\ \exp[\mathrm{i}(\alpha z - \omega t)]\ \exp(-\beta z). \tag{12.42}$$

In a conductor, therefore, the electric field decays exponentially as it propagates.

The magnetic field is given, as in dielectric media, by $\boldsymbol{B} = c\boldsymbol{k} \times \boldsymbol{E}/\omega$. However, as $\boldsymbol{k}$ is here a complex vector, $\boldsymbol{B}$ is dephased relative to $\boldsymbol{E}$. The dephasing φ is given by the phase of the complex number k

$$\tan\varphi = \frac{\beta}{\alpha} = \frac{1}{2}\arctan\left(\frac{4\pi\sigma}{\omega\varepsilon}\right). \tag{12.43}$$

The amplitude ratio of the electric and magnetic perturbations is $B/E = c|k|/\omega$. In the limit of very high conductivity, $\sigma \gg \omega\varepsilon$ and the magnetic field is therefore much larger than the electric field.

The time average of the Poynting vector associated with the fields propagating in the conductor is given by

$$\langle \boldsymbol{S} \rangle = \frac{c^2\alpha}{8\pi\mu\omega}\, E_0^2 \exp(-2\beta x)\; \hat{\boldsymbol{k}}. \tag{12.44}$$

Its modulus falls off exponentially in the propagation direction, indicating that part of the energy of the electromagnetic wave is dissipated in the material. Indeed, this energy is transferred to the free charges of the medium – the electrons – thus generating the current density $\boldsymbol{j}$. The electrons in turn dissipate the energy in inelastic collisions with the other particles of the medium. In a metal these collisions take place with the ions of the crystal lattice where, as mentioned in the previous chapter, the energy is eventually converted into heat.

12.4 Polarization as Source of the Wave Fields

We have already pointed out that electromagnetic phenomena in material media actually take place in the interatomic space, and that the fields measured in the laboratory are averages of the corresponding microscopic quantities. It is thus interesting to review wave propagation phenomena from this point of view. We consider the electromagnetic fields in a material medium, in the absence of free charges and currents, and consider the role of the polarization and the magnetization. Maxwell equations (10.33) can be written in terms of the fields $\boldsymbol{E}$ and $\boldsymbol{B}$ as

$$\begin{cases} \nabla\cdot\boldsymbol{E} = -4\pi\nabla\cdot\boldsymbol{P}, \\ \nabla\times\boldsymbol{B} - \frac{1}{c}\partial_t\boldsymbol{E} = \frac{4\pi}{c}(c\nabla\times\boldsymbol{M} + \partial_t\boldsymbol{P}). \\ \nabla\cdot\boldsymbol{B} = 0, \\ \nabla\times\boldsymbol{E} + \frac{1}{c}\partial_t\boldsymbol{B} = 0, \end{cases} \tag{12.45}$$

Here we have separated the quantities $\rho_{\mathrm{P}} = -\nabla\cdot\boldsymbol{P}$ and $\boldsymbol{j}_{\mathrm{P}} = \partial_t\boldsymbol{P} + c\nabla\times\boldsymbol{M}$ which act as charge and current densities, respectively, and therefore as sources for the fields $\boldsymbol{E}$ and $\boldsymbol{B}$.

The left-hand side of (12.45) have the same form as Maxwell's equations in vacuum. Writing now the equations for the potentials $\boldsymbol{A}$ and ϕ in the Lorentz gauge, we obtain

$$\begin{aligned} \nabla^2\phi - \tfrac{1}{c^2}\partial^2_{tt}\phi &= -4\pi\left(-\nabla\cdot\boldsymbol{P}\right), \\ \nabla^2\boldsymbol{A} - \tfrac{1}{c^2}\partial^2_{tt}\boldsymbol{A} &= -\tfrac{4\pi}{c}\left(\tfrac{\partial\boldsymbol{P}}{\partial t} + c\nabla\times\boldsymbol{M}\right). \end{aligned} \tag{12.46}$$

According to these equations the fields propagate as waves of velocity c perturbed by sources associated with $\rho_{\rm P} = -\nabla\cdot\boldsymbol{P}$ and $\boldsymbol{j}_{\rm P} = \partial\boldsymbol{P}/\partial t + c\nabla\times\boldsymbol{M}$. In most cases of interest the magnetization current $\nabla\times\boldsymbol{M}$ can be neglected.

How do we reconcile the wave equations (12.46) with the result obtained in the previous sections, according to which electromagnetic waves in a material medium propagate with velocities smaller than c? To answer this question we must analyze the role of molecular dipoles as radiation sources.

When an electromagnetic wave is incident on the surface of a material medium, it excites dipoles of the medium which in turn reemit radiation. This situation can be described, in the dipole approximation, by adding to the incident wave the wave produced by the sources $\rho_{\rm P}$ and $\boldsymbol{j}_{\rm P}$ given by the general solution of (12.46) in that approximation. The solution for the vector potential has the form

$$\boldsymbol{A}(\boldsymbol{r},t) = \boldsymbol{A}_{\rm inc}(\boldsymbol{r},t) + \boldsymbol{A}_{\rm dip}(\boldsymbol{r},t), \tag{12.47}$$

where $\boldsymbol{A}_{\rm inc}(\boldsymbol{r},t)$ represents the incident wave and $\boldsymbol{A}_{\rm dip}(\boldsymbol{r},t)$ is given by

$$\boldsymbol{A}_{\rm dip}(\boldsymbol{r},t) = \frac{1}{c}\partial_t \int \frac{1}{|\boldsymbol{r}-\boldsymbol{r}'|}\boldsymbol{P}\left(\boldsymbol{r}', t - \frac{|\boldsymbol{r}-\boldsymbol{r}'|}{c}\right) \mathrm{d}^3r. \tag{12.48}$$

This represents the field created by the dipoles. The integral extends over the region occupied by the material medium.

According to (11.35), the polarization is given by $\boldsymbol{P} = N\boldsymbol{p} = N\alpha\boldsymbol{E}_{\rm loc}$ where $\boldsymbol{E}_{\rm loc}$ is the local field introduced in Chap. 11. Expressing this field in terms of the total field as $\boldsymbol{E}_{\rm loc} = \boldsymbol{E}+4\pi\boldsymbol{P}/3$, (12.47) becomes an integral equation for the fields. The solution describes all the properties of the fields in the material medium and also the reflection and refraction phenomena at interfaces.

For a monochromatic incident wave the solution (12.47) can be interpreted as follows. It is assumed that the field $\boldsymbol{A}_{\rm dip}(\boldsymbol{r}, t)$ consists of a part propagating with velocity c and another part that propagates with velocity c/n. The part propagating with velocity c *extinguishes* the incident wave in the material medium in a transition region of the order of the wavelength of the fields, and gives rise to the wave reflected in the external region. The part of the solution propagating with velocity c/n survives in the material sufficiently far away from the surface, and gives rise to the refracted wave. This result is called the *extinction theorem* of Ewald and Oseen and explains how the transmitted wave becomes the perturbation finally propagating in the second medium.

12.5 General Properties of the Linear Response

In our study of material media we have emphasized the relevance, both formal and practical, of linear media, in which the effect induced by an applied field is proportional to that field. For example, we have shown that in linear dielectrics the polarization is related to the electric field through the electric susceptibility $\boldsymbol{P} = \chi_{\mathrm{E}}\boldsymbol{E}$. In a similar way, in linear magnetic media the magnetization per unit volume $\boldsymbol{M}$ is proportional to the magnetic field $\boldsymbol{B}$.

Although the approximation of a linear response to an external excitation is valid only within certain limits, it turns out to be characteristic of a wide variety of physical systems. Examples are:

- scattering phenomena of particles or waves by a potential, where the outgoing flow is proportional to the incoming flow;
- diffusion problems, where the particle current depends linearly on the concentration gradient;
- thermal diffusion, where the heat flow is proportional to the temperature gradient.

In view of this general applicability, it is useful to derive a general formalism for the linear response of a physical system, independently of the particular problem under study. We present here the theory as it applies to electromagnetic phenomena in linear materials, but the reader should note that the main ideas introduced here can be generalized to the other systems mentioned above.

A general form of a linear relation between two quantities which, like $\boldsymbol{P}$ and $\boldsymbol{E}$, depend on position and time is

$$\boldsymbol{P}(\boldsymbol{r},t) = \int \mathrm{d}^3r' \int_{-\infty}^{\infty} \mathrm{d}t' \; \chi_{\mathrm{E}}(\boldsymbol{r}-\boldsymbol{r}', t-t') \; \boldsymbol{E}(\boldsymbol{r}',t'). \tag{12.49}$$

This relationship is *nonlocal* both in space and in time, and is characterized by the susceptibility $\chi_{\mathrm{E}}(\boldsymbol{r},t)$. The local relationship which we have used up to now is obtained with $\chi_{\mathrm{E}}(\boldsymbol{r},t) = \chi_{\mathrm{E}}\delta(\boldsymbol{r})\delta(t)$.

When the characteristic length associated with the electromagnetic perturbation, namely the wavelength, is much larger than the distances between the polarization charges, nonlocal effects in space are not important. As a consequence we can limit ourselves to the study of the nonlocal form in time, by writing

$$\boldsymbol{P}(\boldsymbol{r},t) = \int_{-\infty}^{\infty} \mathrm{d}t' \; \chi_{\mathrm{E}}(t-t') \; \boldsymbol{E}(\boldsymbol{r},t'). \tag{12.50}$$

For the electric displacement $\boldsymbol{D} = \boldsymbol{E} + 4\pi\boldsymbol{P}$, we find that

$$\boldsymbol{D}(\boldsymbol{r},t) = \int_{-\infty}^{\infty} \mathrm{d}t' \; \varepsilon(t-t') \; \boldsymbol{E}(\boldsymbol{r},t'), \tag{12.51}$$

where $\varepsilon(t) = \delta(t) + 4\pi\chi_{\mathrm{E}}(t)$.

These expressions are considerably simplified in the Fourier representation. We define the Fourier transform of $\chi_{\mathrm{E}}(t)$ and its inverse as

$$\begin{aligned} \chi_{\mathrm{E}}(\omega) &= \tfrac{1}{2\pi}\int_{-\infty}^{\infty} \mathrm{d}t \ \exp(\mathrm{i}\omega t)\ \chi_{\mathrm{E}}(t), \\ \chi_{\mathrm{E}}(t) &= \int_{-\infty}^{\infty} \mathrm{d}\omega \ \exp(-\mathrm{i}\omega t)\ \chi_{\mathrm{E}}(\omega). \end{aligned} \tag{12.52}$$

Similar expressions for $\boldsymbol{P}$, $\boldsymbol{E}$ and $\boldsymbol{D}$ can be introduced. The relation between electric field and polarization can be written as

$$\boldsymbol{P}(\omega) = \chi_{\mathrm{E}}(\omega)\boldsymbol{E}(\omega), \tag{12.53}$$

while we have

$$\boldsymbol{D}(\omega) = \varepsilon(\omega)\boldsymbol{E}(\omega), \tag{12.54}$$

where $\varepsilon(\omega) = 1 + 4\pi\chi_{\mathrm{E}}(\omega)$.

The requirement of *causality* implies that $\chi_{\mathrm{E}}(t)$ must vanish for $t < 0$, since the polarization at a given time can depend on the fields at previous times only. For the Fourier transform, this condition implies that the singularities of $\chi_{\mathrm{E}}(\omega)$ must lie in the lower half of the complex ω plane, and that $\chi_{\mathrm{E}}(\omega)$ must be analytic in the upper halfplane. Indeed, to calculate the antitransform $\chi_{\mathrm{E}}(t)$ we must analytically extend the function $\chi_{\mathrm{E}}(\omega)$ to the whole complex plane by defining $\omega = \omega_1 + \mathrm{i}\omega_2$. Causality implies that the integral defining $\chi_{\mathrm{E}}(\omega)$ in (12.52) must be restricted to $t > 0$ so that

$$\chi_{\mathrm{E}}(\omega) = \chi_{\mathrm{E}}(\omega_1 + \mathrm{i}\omega_2) = \frac{1}{2\pi}\int_0^{\infty} \mathrm{d}t \ \mathrm{e}^{-\omega_2 t}\exp(\mathrm{i}\omega_1 t)\chi_{\mathrm{E}}(t). \tag{12.55}$$

If this integral converges for real ω ($\omega_2 = 0$), due to the exponential factor the convergence improves for $\omega_2 > 0$. Therefore $\chi_{\mathrm{E}}(\omega)$ is well defined and holomorphic in the upper complex halfplane. The singularities of this function thus have a negative imaginary part (Fig. 12.4).

These properties allow us to prove that, according to the Cauchy theorem, the Fourier transform of the susceptibility satisfies

$$\chi_{\mathrm{E}}(\omega) = \frac{1}{\pi i}\ \mathrm{p.v.}\int_{-\infty}^{\infty} \frac{\chi_{\mathrm{E}}(\omega')}{\omega' - \omega}\ \mathrm{d}\omega'. \tag{12.56}$$

Here p.v. stands for the *principal value* of the integral. Taking real and imaginary parts of this equation we obtain the *Plemelj relations*:

$$\begin{aligned} \mathrm{Re}[\chi_{\mathrm{E}}(\omega)] &= \frac{1}{\pi}\ \mathrm{p.v.}\int_{-\infty}^{\infty} \frac{\mathrm{Im}[\chi_{\mathrm{E}}(\omega')]}{\omega' - \omega}\ \mathrm{d}\omega', \\ \mathrm{Im}[\chi_{\mathrm{E}}(\omega)] &= -\frac{1}{\pi}\ \mathrm{p.v.}\int_{-\infty}^{\infty} \frac{\mathrm{Re}[\chi_{\mathrm{E}}(\omega')]}{\omega' - \omega}\ \mathrm{d}\omega'. \end{aligned} \tag{12.57}$$

These equations imply that $\mathrm{Im}[\chi_{\mathrm{E}}]$ and $\mathrm{Re}[\chi_{\mathrm{E}}]$ are linked together, and are not independent quantities. This link is important because $\chi_{\mathrm{E}}(\omega)$ determines

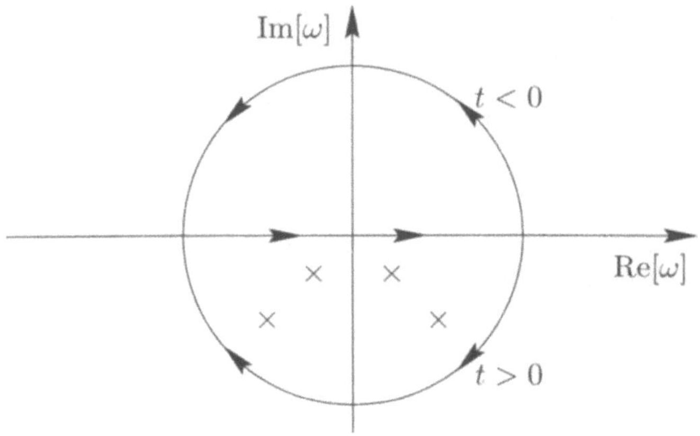

Fig. 12.4. Integration path for the calculation of susceptibility

the dielectric constant of the material and therefore the refractive index. As we have shown when studying wave propagation in metals, the imaginary part of the refractive index describes the absorption of radiation in the material, while its real part determines the propagation characteristics or *dispersion* of the waves. The Plemelj relations indicate that dispersion phenomena and radiation absorption in a material are not independent, and that their link follows from the principle of causality.

In order to show that the physical origin of the Plemelj relations is directly related to causality, let us imagine a light pulse (wave train) incident on a dielectric slab of finite width. In Fig. 12.5 we show schematically the electric field associated with the pulse as a function of position. Since it is limited in space, the incident electric field $\boldsymbol{E}$ is a superposition of an infinite number of components of well-defined frequency, $\boldsymbol{E}_\omega(x,t) = \boldsymbol{E}^0_\omega \exp[\mathrm{i}(kx - \omega t)]$. Each of these components extends over all space at all times.

Let us now assume that the only effect of the dielectric slab is to act as a filter for a given frequency ω_0. The transmitted light pulse $\boldsymbol{E}'(x,t)$ would then be of the form

$$\boldsymbol{E}' = \boldsymbol{E} - \boldsymbol{E}_{\omega_0}. \tag{12.58}$$

Due to the presence of the term $\boldsymbol{E}_{\omega_0}$, this field extends over the whole of space at all times. In particular, we should be able to find it on the second side of the slab at times prior to the arrival of the light pulse on the first side, thus violating the principle of causality. Therefore the absorption process in a filter must be associated with a dispersion process. Different frequency components of the incoming pulse should interfere destructively after leaving the slab, preventing any signal from appearing prior to the arrival of the pulse.

The fundamental consequence of the Plemelj relations is that a constitutive equation of the form $\boldsymbol{P} = \chi_\mathrm{E}\boldsymbol{E}$ with χ_E real, like the equation used in the

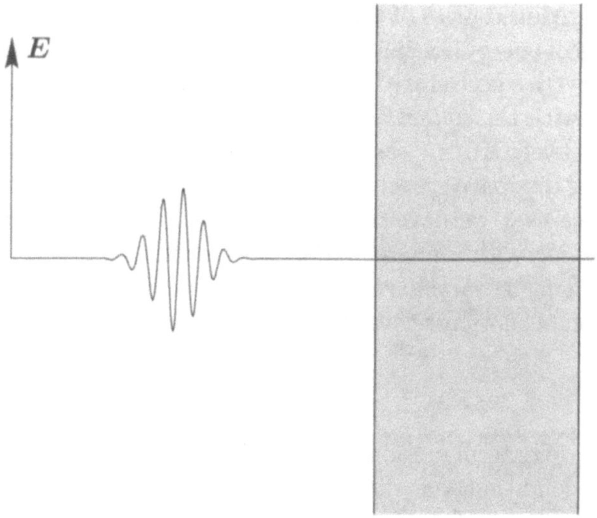

Fig. 12.5. A light pulse incident on a dielectric slab

analysis of wave propagation in material media, is only valid within certain approximations. In the general case the susceptibility should be a complex number. Its detailed form can be derived starting from a microscopic model for the dynamics of charges in the material media. In next section we study one of these models.

12.6 Lorentz Model for the Electric Susceptibility

The detailed analysis of the interaction of radiation with matter can only be carried out within the framework of quantum theory. Such an approach makes it possible to determine the possible modes of excitation of atoms and molecules in the presence of electromagnetic fields. However, the most relevant results of a quantum-mechanical model concerning the response of materials to radiation fields are reproduced by a classical model introduced by H.A. Lorentz in 1907.

In the Lorentz model each atom or molecule of the material is represented as a harmonic oscillator of frequency ω_0. This representation is based on approximating the relative motion of the internal charges of the molecule by small oscillations. Such an approach is valid in the limit of small deviations from equilibrium whatever the true interaction potential is. The equation of motion for a coordinate $\boldsymbol{d}$ measuring these deviations in the presence of an electromagnetic field can be written as

$$m(\ddot{\boldsymbol{d}} + \omega_0^2 \boldsymbol{d} + \gamma \dot{\boldsymbol{d}}) = q\boldsymbol{E}, \tag{12.59}$$

where m is the effective or reduced mass of the system of charges. The coefficient γ represents a process of energy dissipation and can be linked classically to the radiation emitted by the accelerated system of charges. In quantum mechanics it is associated with the *natural width* of the spectral lines of the system. Its value varies between $10^7\,\mathrm{s}^{-1}$ and $10^9\,\mathrm{s}^{-1}$. Additional contributions to γ come from other dissipative processes, such as collisions between atoms and electrons. Notice that we have neglected magnetic effects in the equation of motion, in keeping with a nonrelativistic approach.

Assuming the applied field to vary harmonically as $\boldsymbol{E} = \boldsymbol{E}_0 \exp(-\mathrm{i}\omega t)$, the solution of the equation of motion relevant to our problem is

$$\boldsymbol{d} = \frac{q/m}{\omega_0^2 - \omega^2 - \mathrm{i}\gamma\omega}\,\boldsymbol{E}. \tag{12.60}$$

This solution represents the long term motion of the charge, once the transient associated with the initial conditions has disappeared because of dissipation effects. Defining the dipole moment of the molecule as $\boldsymbol{p}_m = q\boldsymbol{d}$ the dynamical molecular polarizability $\alpha(\omega)$ is given by

$$\alpha(\omega) \doteq \frac{q^2/m}{\omega_0^2 - \omega^2 - \mathrm{i}\gamma\omega}, \tag{12.61}$$

If the distribution of molecules is uniform, the polarization density can be written as $\boldsymbol{p}(r) = N\boldsymbol{p}_m$, where N is the number of molecules per unit volume. Within this assumption, the polarization of the medium is $\boldsymbol{P} = N\,\boldsymbol{p}_m$ and the susceptibility therefore is $\chi_{\mathrm{E}}(\omega) = N\alpha(\omega)$.

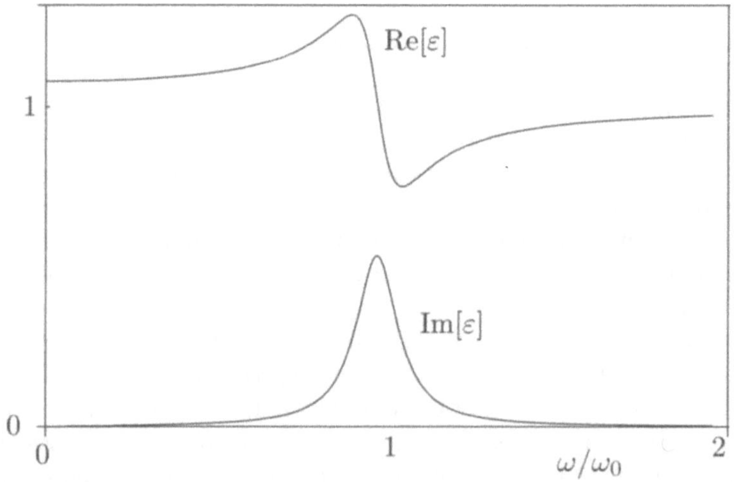

Fig. 12.6. Dielectric constant as a function of frequency, in the vicinity of a resonance

The dielectric constant $\varepsilon(\omega) = 1 + 4\pi\chi_E(\omega)$ and, therefore, the refractive index $n(\omega) = \sqrt{\varepsilon(\omega)}$ are complex numbers. In particular, we can write

$$\mathrm{Re}[\varepsilon(\omega)] = 1 + \frac{4\pi q^2 N}{m} \frac{\omega_0^2 - \omega^2}{(\omega_0^2 - \omega^2)^2 + \gamma^2\omega^2}, \tag{12.62}$$

and

$$\mathrm{Im}[\varepsilon(\omega)] = \frac{4\pi q^2 N}{m} \frac{\gamma\omega}{(\omega_0^2 - \omega^2)^2 + \gamma^2\omega^2}. \tag{12.63}$$

In Fig. 12.6 the real and imaginary parts of the dielectric constant are plotted as functions of the frequency, near the resonance frequency ω_0. At $\omega \approx \omega_0$, $\mathrm{Im}[\varepsilon(\omega)]$ has a maximum, corresponding to a maximum in the absorption. In the same region $\mathrm{Re}[\varepsilon(\omega)]$ changes its slope, which is positive for all other frequencies, giving rise to the phenomenon of *anomalous dispersion.*

The Lorentz model can be generalized to the case in which each molecule contains several electrons, each one with a characteristic frequency ω_i and an associated dissipative constant γ_i. In this case, the electric dipole moment of the molecule turns out to be

$$\boldsymbol{p}_m = \sum_{i(m)} q\boldsymbol{d}_i = \sum_i \frac{q^2/m}{\omega_i^2 - \omega^2 - \mathrm{i}\gamma_i\omega} \boldsymbol{E}. \tag{12.64}$$

If a fraction f_k of the electrons contributes to one mode of oscillation of frequency ω_k and dissipation constant γ_k, the dipole moment of the molecule can be written as a sum over such modes:

$$\boldsymbol{p}_m = \frac{q^2}{m} \sum_k \frac{f_k}{\omega_k^2 - \omega^2 - \mathrm{i}\gamma_k\omega} \boldsymbol{E}. \tag{12.65}$$

The coefficients f_k, called *oscillator strengths*, satisfy $\sum_k f_k = Z$, where Z is the total electronic charge of the molecule. The proportionality factor in (12.65) is the molecular polarizability

$$\alpha(\omega) = \frac{q^2}{m} \sum_k \frac{f_k}{\omega_k^2 - \omega^2 - \mathrm{i}\gamma_k\omega}. \tag{12.66}$$

The dielectric constant is

$$\varepsilon(\omega) = 1 + \frac{4\pi N q^2}{m} \sum_k \frac{f_k}{\omega_k^2 - \omega^2 - \mathrm{i}\gamma_k\omega}. \tag{12.67}$$

In the limit $\omega \to \infty$ the dielectric constant reduces to

$$\varepsilon(\omega) = 1 - \frac{\omega_\mathrm{p}^2}{\omega^2}, \tag{12.68}$$

where the *plasma frequency* ω_p is

$$\omega_\mathrm{p} = \sqrt{\frac{4\pi N q^2}{m} \sum_k f_k} = \sqrt{\frac{4\pi N Z q^2}{m}}. \tag{12.69}$$

This frequency corresponds to the collective oscillations of the negative charges with respect to the positive ones. In this limit ω is much larger than all frequencies in the problem, and the molecular charges behave as free. When $\omega \ll \omega_{\mathrm{p}}$ the dielectric constant is $\varepsilon \sim 1$. In these circumstances, the material becomes transparent to radiation.

For this model we can also consider the case in which one of the oscillation modes has a vanishing characteristic frequency, $\omega_0 = 0$, corresponding to the fact that a certain fraction f_0 of electrons in a molecule can move freely inside the material. This is the case for conduction electrons in a metal. Taking explicitly into account the contribution of this mode, the dielectric constant $\varepsilon(\omega)$ can be written, according to (12.67), as

$$\varepsilon(\omega) = \varepsilon_{\mathrm{P}}(\omega) + \mathrm{i}\,\frac{4\pi N f_0 q^2}{m\omega(\gamma_0 - \mathrm{i}\omega)} \tag{12.70}$$

Here $\varepsilon_{\mathrm{P}}(\omega)$ represents the contribution of the rest of the oscillation modes to the polarizability. The product $N f_0 = n_{\mathrm{c}}$ represents the number of free electrons per unit volume. In order to interpret the contribution of these electrons to the dielectric constant we observe that, according to the Ampère law, the displacement current is

$$\boldsymbol{j} = \frac{1}{4\pi}\frac{\partial \boldsymbol{D}}{\partial t}. \tag{12.71}$$

In the Fourier representation we have

$$\begin{aligned} \boldsymbol{j}(\omega) &= -\frac{\mathrm{i}\omega}{4\pi}\boldsymbol{D}(\omega) = -\frac{\mathrm{i}\omega}{4\pi}\varepsilon_{\mathrm{P}}(\omega)\boldsymbol{E}(\omega) + \sigma(\omega)\boldsymbol{E}(\omega) \\ &= \boldsymbol{j}_{\mathrm{D}}(\omega) + \boldsymbol{j}_{\mathrm{C}}(\omega), \end{aligned} \tag{12.72}$$

where $\boldsymbol{j}_{\mathrm{D}} = -\mathrm{i}\omega\varepsilon_{\mathrm{P}}(\omega)\boldsymbol{E}(\omega)/4\pi$ is the displacement current proper, and $\boldsymbol{j}_{\mathrm{C}} = \sigma(\omega)\boldsymbol{E}(\omega)$ is a conduction current generated by the free charges. From this expression, the relationship between dielectric constant and conductivity is

$$\varepsilon(\omega) = \varepsilon_{\mathrm{P}}(\omega) + \frac{4\pi\mathrm{i}}{\omega}\sigma(\omega). \tag{12.73}$$

The conductivity of the material can thus be written as

$$\sigma(\omega) = \frac{n_{\mathrm{c}} q^2}{m(\gamma_0 - \mathrm{i}\omega)}. \tag{12.74}$$

For $\omega = 0$, this expression gives us the value of the DC conductivity

$$\sigma_{\mathrm{DC}} = \sigma(0) = \frac{n_{\mathrm{c}} q^2}{m\gamma_0} = \frac{n_{\mathrm{c}} q^2}{m}\tau. \tag{12.75}$$

Here $\tau = 1/\gamma_0$ is called the *relaxation time* of conduction electrons.

The quantum result for the molecular polarizability has exactly the same form as (12.66), except that now ω_k are the characteristic frequencies of the allowed energy transitions, and γ_k are the widths of the corresponding spectral lines. The coefficients f_k are the oscillator strengths for each transition, and are related to quantum mechanical properties.

It is interesting to evaluate the energy loss in the medium and relate it to the quantities introduced in this section. This aspect is discussed in one of the problems at the end of this chapter.

Until this point we have neglected in the Lorentz model the action of the field generated by the dipoles induced in other molecules of the medium. This approximation corresponds to a dilute medium. For a denser medium the electric field in the equations of motion should be replaced by the local field $\boldsymbol{E}_{\text{loc}} = \boldsymbol{E} + 4\pi\boldsymbol{P}/3$. Taking into account this local field, we find that the dielectric constant is related to the dynamical molecular polarizability according to

$$\frac{\varepsilon(\omega)-1}{\varepsilon(\omega)+2} = \frac{4\pi}{3} N\alpha(\omega). \tag{12.76}$$

This equation, similar to the Clausius–Mossotti equation, is called the *Lorentz–Lorenz relation.*

Problems

12.1 A light ray is incident perpendicularly to one of the faces of a dielectric wedge of angle β and refractive index n (Fig. 12.7). Determine how many times is the ray reflected inside the wedge before the radiation becomes trapped.

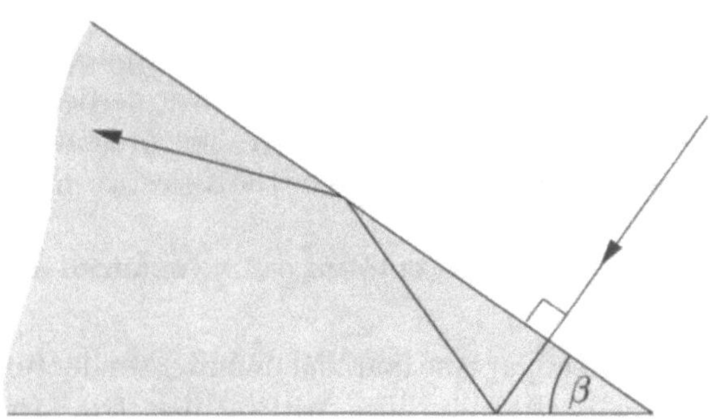

Fig. 12.7. Light ray incident on a dielectric wedge. Only the reflected rays are shown

12.2 Two semiinfinite dielectric layers of refractive indices n_1 and n_3 enclose another layer of refractive index n_2 and thickness d. A plane wave is incident from the first layer perpendicularly to the interfaces. (a) Calculate the reflection and transmission coefficients of the system. Study its behavior as function of frequency for (n_1, n_2, n_3) equal to (1,2,3), (3,2,1) and (2,4,1).

(b) Assume the first medium to be a lens, and the third one to be air. Determine what thickness the intermediate layer should have to optimize the transparency of the whole system by eliminating the reflected wave.

Hint: Notice that there is an incident and a reflected wave in the first medium, a transmitted and a reflected wave in the second, and only a transmitted wave in the third. Using the *impedance* for medium j, $Z_j = E_j/B_j$, the resulting relations can be written in a simple form.

12.3 Two semi-infinite dielectric media are separated by a vacuum layer of thickness d. Calculate the transmission and reflection coefficients for a wave incident from one of the media at an angle of incidence i, polarized parallel to the interfaces.

12.4 Consider a plane wave, polarized perpendicularly to the plane of incidence, under conditions of total reflection at the interface between a dielectric of refractive index n and vacuum. Study the dynamics of a charge q initially at rest placed in vacuum a distance d from the interface.

Hint: Consider only the nonrelativistic approximation, and therefore neglect the effect of the magnetic field on the dynamics of the particle.

12.5 (a) Show that the boundary conditions at the interface between vacuum and a conductor of finite conductivity are the same as for vacuum and a dielectric. (b) Study the reflected and transmitted fields when a plane wave is incident normal to the surface of the conductor.

12.6 A mirror is formed by a dielectric layer of thickness d with dielectric constant ε and $\mu = 1$ with a perfect conducting metallic film behind it. A linearly polarized plane wave is incident from the vacuum, perpendicularly to the surface of the dielectric. (a) Find the electric fields in vacuum and in the dielectric. (b) Determine the thickness of the dielectric layer needed for the fields in vacuum to be the same as if the dielectric were not present. (c) Discuss the boundary conditions for the magnetic field at the dielectric–metal interface.

Hint: Discuss whether the solution to this problem can be obtained as a limiting case of Problem 12.2.

12.7 The space between two perfectly plane parallel infinite conducting plates separated by a distance a is partially filled with a dielectric layer of dielectric constant ε and thickness b in contact with one of the plates. It is known that inside this system plane waves propagate normal to the interfaces. Through measurements it is established that the electric field at the dielectric–vacuum interface is zero. Find the conditions a, b and ε should satisfy for the electric field to be different from zero in the space between the plates. Calculate the field over all space.

12.8 Starting from the wave equations (12.47) obtain the corresponding expressions for the electric and magnetic fields. Expressing the polarization and magnetization in terms of $\boldsymbol{E}$ and $\boldsymbol{B}$, integral equations for the fields in material media are obtained. Obtain these equations in the static case and in dynamic situations. Outline the starting steps for a proof of the extinction theorem of Ewald and Oseen.

Hint: Use Green's function for the wave equation.

12.9 Determine the total energy loss in a dielectric in terms of the polarizability $\chi(\omega)$. Consider the case where the medium has finite conductivity.

Solution: According to Section 10.4, if there are no free currents in the medium the total work done by the electric field reduces to the work carried out on the polarization current, which is given by

$$\frac{\mathrm{d}W}{\mathrm{d}t} = \int_{-\infty}^{\infty} \mathrm{d}t \int \boldsymbol{E} \cdot \partial_t \boldsymbol{P} \, \mathrm{d}^3 r.$$

Using the expressions

$$\boldsymbol{E}(t) = \textstyle\int_{-\infty}^{\infty} \mathrm{d}\omega \exp(-\mathrm{i}\omega t) \boldsymbol{E}(\omega),$$

$$\boldsymbol{P}(t) = \textstyle\int_{-\infty}^{\infty} \mathrm{d}\omega \exp(-\mathrm{i}\omega t) \boldsymbol{P}(\omega),$$

$$\boldsymbol{P}(t) = \textstyle\int_{-\infty}^{\infty} \mathrm{d}t' \chi(t-t') \boldsymbol{E}(t'),$$

the dissipated power can be written as

$$\frac{\mathrm{d}W}{\mathrm{d}t} = \int_{-\infty}^{\infty} \mathrm{d}\omega \; \mathrm{i}\omega \chi(\omega) \left|\boldsymbol{E}(\omega)\right|^2 .$$

Since $\chi'(\omega)$ is an even function of ω, its integral vanishes because of the presence of the factor $\mathrm{i}\omega$. The power is then

$$\frac{\mathrm{d}W}{\mathrm{d}t} = -\int_{-\infty}^{\infty} \mathrm{d}\omega \; \omega \, \chi''(\omega) \left|\boldsymbol{E}(\omega)\right|^2$$

and, in terms of the dielectric constant, we have

$$\frac{\mathrm{d}W}{\mathrm{d}t} = -\frac{1}{4\pi} \int_{-\infty}^{\infty} \mathrm{d}\omega \; \omega \varepsilon''(\omega) \left|\boldsymbol{E}(\omega)\right|^2 .$$

If the medium contains free charges, the dielectric constant can be divided into components of the form (12.73). It can be seen from this that the conductivity contributes to the dissipation in the form

$$\frac{\mathrm{d}W}{\mathrm{d}t} = \frac{1}{4\pi} \int_{-\infty}^{\infty} \mathrm{d}\omega \; \left[-\omega \varepsilon_{\mathrm{P}}''(\omega) + \sigma(\omega)\right] \left|\boldsymbol{E}(\omega)\right|^2 .$$

References

12.1 F. Goos and H. Hänchen: *Ein neuer und fundamentaler Versuch zur Totalreflexion*, [1] Ann. Physik **1**, 333 (1947).

[1] *A New and Fundamental Experiment on Total Reflection*

12.2 J.L. Agudín and A.M. Platzeck: *Fermat's Principle and Evanescent Waves*, J. Optics (Paris) **9**, 101 (1977).
12.3 Ch. Imbert: *Experimental Proof of the Photon's Translational Inertial Spin Effect*, Phys. Lett. **31 A**, 337 (1970).
12.4 J.L. Agudín: *Mean Velocity of Energy Transport in the Transverse Goos and Hänchen Effect*, Phys. Lett. **35 A**, 107 (1971).

Further Reading

R. Becker and F. Sauter: *Electromagnetic Fields and Interactions* (Blaisdell, New York, 1964) Chap. DII.
R.P. Feynman, R.B. Leighton and M. Sands: *The Feynman Lectures on Physics, Vol. II* (Addison–Wesley, New York, 1964) Chaps. 32 and 33.
J.D. Jackson: *Classical Electrodynamics* (Wiley, New York, 1975) Chap. 7.
L.D. Landau and E.M. Lifshitz: *Electrodynamics of Continuous Media* (Pergamon, New York, 1960).
W.K.H. Panofsky and M. Phillips: *Classical Electricity and Magnetism* (Addison–Wesley, Massachusetts, 1955) Chaps. 11 and 12.
J.M. Stone: *Radiation and Optics* (McGraw–Hill, New York, 1963) Chaps. 16 and 17.

13. Electromagnetic Theory of Superconductivity

Superconductivity has been known since 1911, when the Dutch physicist Heike Kamerlingh Onnes discovered that the electrical resistivity of mercury vanishes completely below a temperature of 4.2 K [13.1]. Kamerlingh Onnes had been the first physicist to achieve, three years earlier, the liquefaction of helium, the gas of lowest boiling point. This gave him the opportunity to carry out experiments at very low temperatures. As in other instances in history, access to a given technology – in this case, cryogenics – brought with it important scientific progress.

Kamerlingh Onnes took on the study of the electrical resistivity of different metals. At that time two different theories were in dispute. On the one hand the Lorentz model, discussed in the previous chapter, claimed that the conductivity is proportional to the number of conduction electrons and to the average time between electron–ion collisions. On the other hand, a hypothesis due to Lord Kelvin maintained that, at low temperatures, conduction electrons should return to the atoms to which they originally belonged.

To diminish the effect of impurities, Kamerlingh Onnes worked with mercury, because it can be easily purified by distillation. The results showed that the resistivity decreases with temperature, but they produced a surprise: *at a temperature of 4.2 K, the resistivity of mercury vanishes suddenly* (Fig. 13.1).

Today we know that this is a common phenomenon, exhibited by a great variety of pure metals and alloys. The absence of electrical resistivity allows superconductors to carry a persistent electric current. This makes them useful in the construction of electromagnets occupying less space than if made of a good conductor like copper, and less expensive to operate. As mentioned in Chap. 6, these magnets are in general use today in research laboratories and in medical equipment designed for nuclear magnetic resonance tomography.

13.1 Phenomenology

The transition to the superconducting state takes place at a well defined temperature T_c, called the *critical temperature*, which is characteristic of each material. The phenomenon has all the properties of a phase transition, and in the superconducting state the material is in thermodynamic equilibrium. This phase transition does not involve changes in the crystalline lattice or

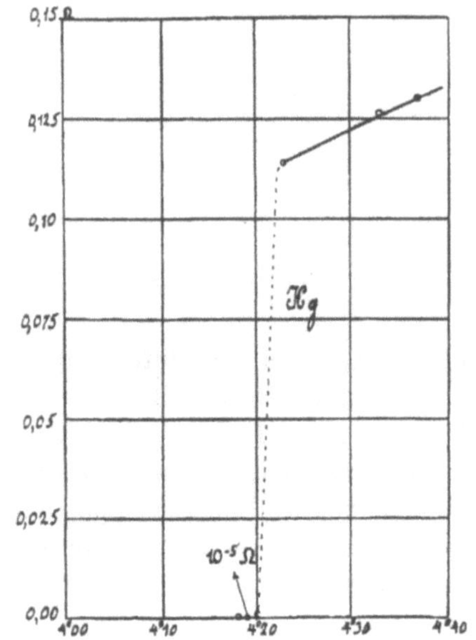

Fig. 13.1. Dependence of resistivity on temperature in a superconducting material (Hg). Original picture by Kamerlingh Onnes (Leyden Commun. **122b** (1911))

in the structure of the material and, therefore, must be interpreted as an electronic transformation.

Besides the loss of electrical resistance, superconductors are characterized by the so-called *Meissner effect*, discovered by Walter Meissner and Rudolf Ochsenfeld in 1933 [13.2]. These researchers found that below T_c the magnetic flux is expelled from the sample, except in a small region close to the surface. This means that, independently of the previous history, the thermodynamically stable superconducting state corresponds to the situation in which there is no magnetic field inside the material.

In this respect, superconductors differ from a perfect ideal conductor. In such a material, if it existed, the pre-existent flux would tend to persist by virtue of Lenz's law. In a transition to the state of perfect conduction, the inner flux would depend on the previous history of the sample. The behavior of a superconductor immersed in a magnetic field is described in Fig. 13.2:

- in (1) the material is at $T > T_c$ and the magnetic field is zero;
- in (2) $T > T_c$ and a magnetic field is applied;
- in (3) the temperature is lower than the critical value, $T < T_c$, and the field is zero; finally,
- in (4) the temperature is $T < T_c$ and the field is different from zero.

The final state of flux expulsion is independent of the path followed to reach state (4). For a perfect conductor, when going from (2) to (4) the magnetic flux would remain unaffected.

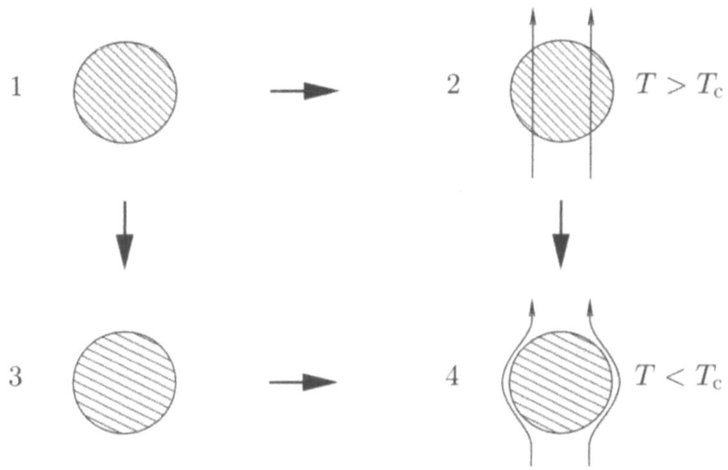

Fig. 13.2. Expulsion of the magnetic field in a superconductor. The final state is independent of the path, (*1,2,4*) or (*1,3,4*)

Another important property of superconducting materials is the existence of a critical magnetic field, characteristic of each material, above which superconductivity disappears. This field is temperature dependent.

From the point of view of the magnetic properties, two types of superconducting material, called type I and type II, exist. In those of type I, flux expulsion takes place at any field smaller than the critical value. There are relatively few materials of this type. Among them we find some pure metals such as Cd, In, Sn, Hg, Tl and Pb. In superconductors of type II the flux expulsion occurs only at fields smaller than a certain critical value, called H_{c1} (the field values correspond to a cylindrical sample in a longitudinal field)[1].

Above H_{c1} the magnetic field penetrates as flux tubes, within which the material is normal, surrounded by vortices of superconducting current. The energy needed to create the interfaces of these tubes is negative, which explains the formation of the vortices as an energetically convenient configuration. Superconductivity disappears totally above a second critical field called H_{c2}.

[1] We use the symbol $\boldsymbol{H}_0$ to indicate the applied field, assuming that in vacuum we have $\boldsymbol{B}_0 = \boldsymbol{H}_0$. Due to the deformation produced by the presence of the superconductor, the field $\boldsymbol{H}$ outside the sample differs from $\boldsymbol{H}_0$, except for a very long cylinder in a longitudinal uniform field. The indicated critical fields refer to values of $\boldsymbol{H}_0$.

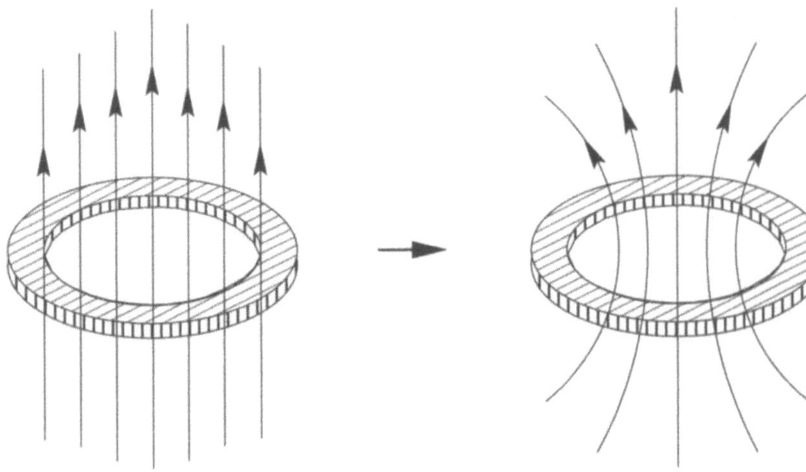

Fig. 13.3. Flux quantization in a superconductor

A third characteristic of superconductors is the property of flux quantization, present in multiply-connected samples. Let us consider the ring of superconducting material shown in Fig. 13.3, cooled below T_c in the presence of a magnetic field. When the field is removed, a certain amount of magnetic flux remains trapped in the ring, quantized in multiples of a magnetic flux unit. This phenomenon was predicted by London and verified experimentally by B.S. Deaver and W. M. Fairbank [13.3], and also by R. Doll and M. Naebauer [13.4]. Superconductivity is a macroscopic manifestation of quantum effects, and flux quantization is directly related to angular momentum quantization in microscopic systems.

In spite of its early discovery in 1911 and many years of research, a satisfactory explanation for the origin of superconductivity was not available until after the complete development of quantum mechanics. However, a phenomenological model which describes the behavior of superconductors on a semiclassical basis was developed by the brothers Fritz and Heinz London in 1933.

In 1956 Lev Davidovich Landau and Vitali Ginzburg proposed a new theory which takes into account the quantum nature of the phenomenon in a more precise way. The microscopic theory of George Bardeen, Leon Cooper and Robert Schrieffer (BCS) was introduced in 1959. Each of these successive advances contains the previous one as a particular limiting case. In the BCS theory it is essential to assume the existence at the microscopic level of an attractive force between the electrons of the material medium. This phenomenon – impossible in vacuum, where the only force between electrons is the Coulomb repulsion – can take place inside a metal due to the presence of positive charges. An electron moving in a metal affects the dynamics of the positive charges and induces local variations of charge density that attract a second electron. The forces of attraction lead to the formation of bound electron pairs whose dynamics determines the characteristics of the superconducting state.

In 1986 the study of superconductors acquired momentum due to the discovery by the Swiss physicists J. Gernod Bednorz and K. Alex Müller of materials with critical temperatures markedly higher than those of all superconductors known at the time [13.5]. This revived interest in the phenomenon because of the possibility of extending its applications. From the point of view of basic research, it is a real challenge to devise a theory explaining the behavior of the new materials. This is due to the fact that, although the microscopic mechanisms responsible for traditional superconductivity seem to be absent, many characteristics common to the classic superconductors are present in the new materials. In particular, in flux quantization experiments the presence of pairs of electrons has been verified in these materials. Table 13.1 shows critical temperatures of some pure elements and alloys, together with the year when it was possible to identify the critical temperature or to prepare the corresponding compound.

Table 13.1. Critical temperatures of some superconductors

Material	**Critical temperature (K)**	**Year**
Hg	4.153	1911
Pb	7.193	1912
Nb	9.1	1956
V_3Si	17.1	1964
Nb_3Sn	18.05	1965
Nb_3Ge	23.2	1966
$La_{2-x}Ba_xCuO_4$	33	1986
$YBa_2Cu_3O_x$	93	1987
$Tl_2Ba_2Ca_2Cu_3O_{10}$	125	1988
$HgBa_2Ca_2Cu_3O_x$ (under pressure)	133-160	1993

The table shows that in the twenty years between 1966 and 1986 no superconductors with critical temperatures higher than 23.3 K were found. This is one of the reasons why the discovery of Bednorz and Müller had such repercussions. In contrast to traditional superconductors, which are metals, the new materials are not good electrical conductors in the normal state.

In what follows we develop the London phenomenological theory, which allows for a description of the Meissner state and for an understanding of the origin of magnetic field expulsion within the framework of the electrodynamics of material media.

13.2 The London Theory

This theory assumes that the electrons participating in the phenomenon, the superconducting electrons, are only a fraction of the total number of conduction electrons in the material. It is further assumed that they form a charged perfect fluid of density n_s. For simplicity, London considered this fluid to be uniform, which is one of the limitations of the theory. More elaborate theories assume n_s to depend on position, a more realistic assumption.

The average density of superconducting current is

$$\boldsymbol{j}_s(\boldsymbol{r},t) = q n_s \boldsymbol{v}(\boldsymbol{r},t). \tag{13.1}$$

The velocity $\boldsymbol{v}(\boldsymbol{r},t)$ is a macroscopic quantity, averaged over microscopic variables, of the same type as those defined when studying material media. The dynamics of $\boldsymbol{v}(\boldsymbol{r},t)$ and of $\boldsymbol{j}_s(\boldsymbol{r},t)$ is determined by the macroscopic fields

$$\boldsymbol{E}(\boldsymbol{r},t) = \langle \boldsymbol{e}(\boldsymbol{r},t)\rangle; \qquad \boldsymbol{B}(\boldsymbol{r},t) = \langle \boldsymbol{b}(\boldsymbol{r},t)\rangle. \tag{13.2}$$

In the stationary state, the current satisfies

$$\nabla \cdot \boldsymbol{j}_s(\boldsymbol{r}) = 0. \tag{13.3}$$

The dynamics of the velocity $\boldsymbol{v}$ is given by Newton's law:

$$m\frac{\mathrm{d}\boldsymbol{v}}{\mathrm{d}t} = q\left(\boldsymbol{E} + \frac{1}{c}\boldsymbol{v}\times\boldsymbol{B}\right). \tag{13.4}$$

For an ordinary conductor the equivalent equation should contain a friction term, accounting for the resistivity of the medium. In this phenomenological description, the superconductor is characterized by the absence of this term. A detailed theory taking into account the fundamental aspects of the phenomenon should explain this assumption starting from microscopic properties.

Equation (13.4) is known as the London first equation. The time derivative in (13.4) is a total derivative. It includes the local variation of $\boldsymbol{v}$ at point $\boldsymbol{r}$ as well as the change due to the motion of the fluid. This total variation is expressed as

$$\frac{\mathrm{d}\boldsymbol{v}}{\mathrm{d}t} = \partial_t \boldsymbol{v} + (\boldsymbol{v}\cdot\nabla)\boldsymbol{v}, \tag{13.5}$$

which can also be written as

$$\frac{\mathrm{d}\boldsymbol{v}}{\mathrm{d}t} = \partial_t \boldsymbol{v} + \frac{1}{2}\nabla v^2 - \boldsymbol{v}\times(\nabla\times\boldsymbol{v}). \tag{13.6}$$

Combining this equation with (13.4), we obtain

$$\partial_t \boldsymbol{v} - \frac{q}{m}\boldsymbol{E} + \frac{1}{2}\nabla v^2 = \boldsymbol{v}\times\left(\nabla\times\boldsymbol{v} + \frac{q}{mc}\boldsymbol{B}\right). \tag{13.7}$$

Taking into account the Faraday equation$\nabla\times\boldsymbol{E} = -\partial_t\boldsymbol{B}/c$, we can write

$$\partial_t \boldsymbol{Q} = \nabla\times(\boldsymbol{v}\times\boldsymbol{Q}), \tag{13.8}$$

where the quantity $\boldsymbol{Q}$ is given by

$$\boldsymbol{Q} = \nabla \times (m\boldsymbol{v}) + \frac{q}{c}\,\boldsymbol{B} = \nabla \times \left(m\boldsymbol{v} + \frac{q}{c}\boldsymbol{A}\right) = \nabla \times \boldsymbol{p}. \tag{13.9}$$

In this equation, $\boldsymbol{p}$ is the canonical momentum of the charged fluid,

$$\boldsymbol{p} = m\boldsymbol{v} + \frac{q}{c}\boldsymbol{A}, \tag{13.10}$$

where $m\boldsymbol{v}$ is the kinetic momentum. As we mentioned in Chap. 3 when dealing with relativistic dynamics, the presence of the vector potential in the momentum of particles – or, in this case, of the fluid – is, from the viewpoint of the covariance of electromagnetism, completely equivalent to the presence of the scalar potential in the total energy.

In hydrodynamics, (13.8) is called the *Helmholtz equation*. The quantity $\boldsymbol{Q}$ represents, in that framework, the *vorticity* of the fluid. Because we have here a charged fluid in the presence of a magnetic field, vorticity is defined using the canonical momentum $\boldsymbol{p}$. Equation (13.8) is of general validity and can therefore be applied to any charged fluid.

As we show later, the London theory predicts $\boldsymbol{Q} \equiv 0$ in the superconducting state. This means that the charged fluid must generate a vorticity field to compensate for contributions to $\boldsymbol{Q}$ due to the vorticity of the field $\boldsymbol{A}$, $\nabla \times \boldsymbol{A} = \boldsymbol{B}$. For this reason a current appears around the sample, whose effect is to shield the material from the external field (Meissner effect).

F. and H. London put forward their theory in 1935. At that time the applications of quantum mechanics to complex material systems had just began. Although in many cases it was possible to identify the quantum origin of a given phenomenon, few calculation methods were available – and many concepts developed later were lacking – to attack problems involving systems with many degrees of freedom. However, just as Heisenberg in 1933 identified the origin of ferromagnetism in the Coulomb interaction combined with the symmetry properties of the wave function under particle exchange, so the Londons hypothesised that the characteristics of superconductors were a consequence of the properties of a *macroscopic quantum wave function.* This means that a finite fraction of the conduction electrons, those contributing to superconductivity, behave as part of a "giant atom," to which the laws of quantum mechanics apply.

According to the London hypothesis, this wave function has a constant value throughout the material and is not affected by the magnetic field. In the London terminology this corresponds to a *rigid wave function.* Since the square of the modulus of the wave function is a measure of the number of superconducting electrons per unit volume, the assumption of rigidity is directly related to the hypothesis of a uniform density n_{s}. This hypothesis, which is a consequence of the properties of the wave function, implies that the angular momentum of the superconducting fluid $\boldsymbol{L} = \boldsymbol{r} \times \boldsymbol{p}$ is also a rigid variable or, in other words, is independent of coordinates. Accordingly

$$\nabla \cdot \boldsymbol{L} = \nabla \cdot \left[\boldsymbol{r} \times \left(m\boldsymbol{v} + \frac{q}{c}\boldsymbol{A} \right) \right] = \boldsymbol{r} \cdot \boldsymbol{Q} = 0. \tag{13.11}$$

Since $\boldsymbol{r}$ is arbitrary we conclude that $\boldsymbol{Q} = 0$. In some circumstances there can also be solutions in which (13.11) is satisfied with $\boldsymbol{Q} \neq 0$, and $\boldsymbol{Q} \perp \boldsymbol{r}$. This is the case for type II superconductors in the so-called mixed state (Problem 13.9). If we limit ourselves to the case $\boldsymbol{Q} = 0$, we have

$$\boldsymbol{Q} = \nabla \times \left(m\boldsymbol{v} + \frac{q}{c}\boldsymbol{A} \right) = 0, \tag{13.12}$$

or, equivalently,

$$\nabla \times \boldsymbol{j}_{\mathrm{s}} + \frac{q^2 n_{\mathrm{s}}}{mc}\boldsymbol{B} = 0. \tag{13.13}$$

This is the so-called second London equation. It is a relation between a property of the medium, the superconducting current $\boldsymbol{j}_{\mathrm{s}}$, and the magnetic field $\boldsymbol{B}$, and it should be interpreted as a constitutive equation. In particular, it replaces the relation between electric field and current in an ordinary conductor, i.e. Ohm's law. Combining it with $\nabla \times \boldsymbol{B} = 4\pi \boldsymbol{j}_{\mathrm{s}}/c$ we obtain the law governing magnetic field behavior in a superconductor, which is another form of the second London equation,

$$\nabla \times \nabla \times \boldsymbol{B} = -\frac{1}{\lambda_{\mathrm{L}}^2}\,\boldsymbol{B}. \tag{13.14}$$

Here λ_{L} is the London *penetration depth*, defined as

$$\lambda_{\mathrm{L}}^2 = \frac{mc^2}{4\pi n_{\mathrm{s}} q^2}. \tag{13.15}$$

This length is characteristic of each material and, due to the presence of the density n_{s}, is a function of temperature. Near the critical temperature it is singular and, as the temperature decreases, it reaches finite values between 100 and 2000 Å for classical superconductors, and between 1400 and 2500 Å for high-T_{c} superconductors.

Combining the condition $\boldsymbol{Q} = 0$ with (13.7) we obtain

$$\partial_t \boldsymbol{v} + \nabla \frac{v^2}{2} = -\frac{q}{m}\boldsymbol{E}, \tag{13.16}$$

which describes the dynamics of the superconducting electrons. For moderate velocities the quadratic term in the velocity can be neglected.

The London equation (13.14) allows us to explicitly calculate the current associated with the Meissner effect. For instance, imagine a plain superconducting slab of width $2d$ placed in a longitudinal uniform magnetic field $\boldsymbol{B}_0$, such as that of a solenoid much longer than the sample (Fig. 13.4). Under these conditions, far from the ends of the sample we can assume translational symmetry along the z axis. The London equation for the field reduces to

$$\partial_x^2 B_z = \frac{1}{\lambda_{\mathrm{L}}^2}\,B_z. \tag{13.17}$$

For the boundary condition that the external field has the same value on both faces of the slab, the solution is

$$B_z(x) = B_0 \frac{\cosh(z/\lambda_L)}{\cosh(d/\lambda_L)}. \tag{13.18}$$

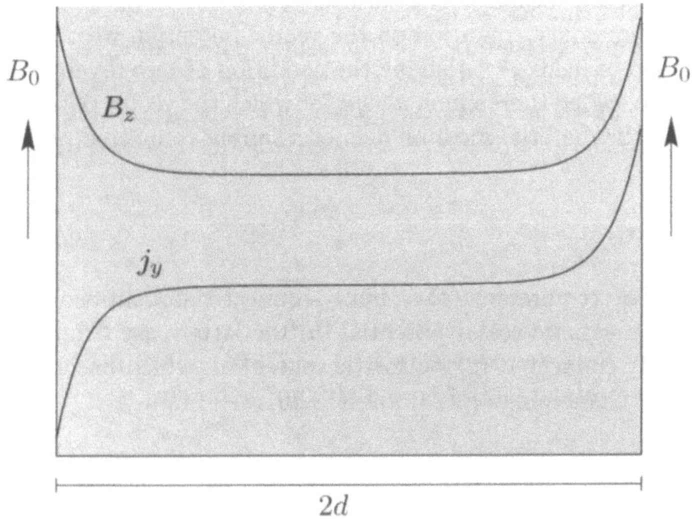

Fig. 13.4. Distribution of current and field in a superconducting sheet

The magnetic field inside the slab thus decays with a characteristic length λ_L. Since in general $\lambda_L \ll d$, the field $\boldsymbol{B}$ is practically zero inside the material.

The relation between field and current, (13.13), implies that a superconducting current in the direction y is established within the slab

$$j_y(z) = j_0 \frac{\sinh(z/\lambda_L)}{\sinh(d/\lambda_L)}. \tag{13.19}$$

The relationship between B_0 and j_0 is given by $j_0 = -(c/4\pi)B_0 \tanh(d/\lambda_L)$. Figure 13.4 shows the current and field distribution inside the slab. From this figure it can be qualitatively seen how the Meissner current j_y produces the field needed to shield the external field B_0.

By an appropriate choice of gauge it is possible to rewrite the London equation to directly link the current to the vector potential. In a stationary state the current satisfies the relationship $\nabla \cdot \boldsymbol{j}_s = 0$. For the vector potential we can always choose a gauge such that $\nabla \cdot \boldsymbol{A} = 0$. Combining both relations we have

$$\nabla \cdot \left(\boldsymbol{j}_s + \frac{q^2 n_s}{mc} \boldsymbol{A} \right) = 0. \tag{13.20}$$

On the other hand, (13.13) implies

$$\nabla \times \left(\boldsymbol{j}_{\mathrm{s}} + \frac{q^2 n_{\mathrm{s}}}{mc} \boldsymbol{A} \right) = 0. \tag{13.21}$$

Since there is no current flowing out of the superconductor, the current $\boldsymbol{j}_{\mathrm{s}}$ at the material surface satisfies the boundary condition

$$\boldsymbol{n} \cdot \boldsymbol{j}_{\mathrm{s}} = 0, \tag{13.22}$$

where $\boldsymbol{n}$ is the unit vector normal to the surface.

In order to set a boundary condition on the vector potential, we observe that, within the group of fields $\boldsymbol{A}$ satisfying the condition of zero divergence, we can choose a gauge, called the *London gauge*, for which the vector potential satisfies $\boldsymbol{A} \cdot \boldsymbol{n} = 0$. If the material medium defines a simply-connected region, (13.21) implies

$$\boldsymbol{j}_{\mathrm{s}} + \frac{q^2 n_{\mathrm{s}}}{mc} \boldsymbol{A} = \nabla \eta, \tag{13.23}$$

where η is a scalar. The requirement of a simply-connected domain is equivalent to that for the magnetic scalar potential. In the latter case, Ampère's law defines a multiply-connected region for the potential, which becomes a multivalued function. Combining (13.23) and (13.20) we obtain

$$\nabla^2 \eta = 0. \tag{13.24}$$

The function η is a solution of Laplace's equation in the volume of the material. The boundary conditions discussed above imply that $\boldsymbol{n} \cdot \nabla \eta = 0$ on the boundary, thus defining a Neumann problem with zero normal derivative (Chap. 5). The solution for η is a constant, and therefore $\nabla \eta = 0$. According to (13.23) it follows that

$$\boldsymbol{j}_{\mathrm{s}} + \frac{q^2 n_{\mathrm{s}}}{mc} \boldsymbol{A} = 0. \tag{13.25}$$

This is another form of the second London equation, valid in simply-connected domains for the gauge defined above.

13.3 Magnetization and H-Field in a Superconductor

According to Maxwell's equations in stationary situations, the magnetic field $\boldsymbol{B}$ is determined by the relation

$$\nabla \times \boldsymbol{B} = \frac{4\pi}{c} \boldsymbol{j} = \frac{4\pi}{c} (\boldsymbol{j}_{\mathrm{F}} + \boldsymbol{j}_{\mathrm{s}}), \tag{13.26}$$

where the free currents $\boldsymbol{j}_{\mathrm{F}}$ are the sources of the external field and $\boldsymbol{j}_{\mathrm{s}}$ are the superconducting currents. In the theory of material media we found it convenient to define an auxiliary vector $\boldsymbol{H}$ whose sources are the free currents circulating in the laboratory coils and in the conducting medium. Both have the property of being measurable with an ammeter.

In an isolated superconductor the currents due to the Meissner effect cannot be extracted out of the material and therefore are not susceptible of being measured directly. The presence of the currents $\boldsymbol{j}_\mathrm{s}$ determines the existence of a magnetic moment per unit volume similar to the one observed in a magnetic medium due to atomic or Ampère currents. It is interesting to point out here that, when the relationship between magnetism and electric currents was established and Ampère advanced the idea that material magnetism could be due to the existence of microscopic currents, Fresnel commented that these should be of atomic origin, since they do not dissipate as ordinary currents do. In the case of superconductivity, Meissner currents are macroscopic currents that do not die away due to the quantum character of this phenomenon. As stated above, the material behaves as a giant atom: the absence of dissipation in a superconductor in the Meissner state is a consequence of the fact that it represents the *ground state of a quantum system.* The same argument explains the absence of radiation in an atom in its ground state.

Superconducting currents produce a total magnetic moment that can be associated with a magnetization given by $c\nabla \times \boldsymbol{M} = \boldsymbol{j}_\mathrm{s}$. Similarly to what we did for magnetic media, we can introduce the auxiliary field $\boldsymbol{H}$ defined as $\boldsymbol{H} = \boldsymbol{B} - 4\pi\boldsymbol{M}$, which satisfies the relation

$$\nabla \times \boldsymbol{H} = \frac{4\pi}{c}\, \boldsymbol{j}_\mathrm{F}. \tag{13.27}$$

In the absence of a material medium, this field $\boldsymbol{H}$ is identical to $\boldsymbol{B}$. The average superconductor current alters the distribution of $\boldsymbol{B}$ over all space and, as we show later, the boundary conditions make $\boldsymbol{H}$ different from the case in vacuum. Only in the case of an infinite cylinder placed in a magnetic field parallel to its axis, do the average superconducting currents not alter $\boldsymbol{H}$ outside the material and $\boldsymbol{H}$ equals the applied field.

As in the case of a magnetic medium, the total magnetic moment of the material is calculated from $\boldsymbol{M}$ as $\boldsymbol{m} = \int \boldsymbol{M}\, \mathrm{d}^3 r$. It should be noted that the same value of $\boldsymbol{m}$ would be obtained if the magnetization were defined as $\boldsymbol{M} = \boldsymbol{r} \times \boldsymbol{j}_\mathrm{s}/2c$. However, the two definitions are not locally equivalent. In a superconductor, in fact, the only physically meaningful quantity is the total magnetic moment. In spite of this lack of uniqueness, it is useful to analyze in more detail the relation between $\boldsymbol{M}$ and $\boldsymbol{B}$, namely, the constitutive equation. Combining the identity $c\nabla \times \boldsymbol{M} = \boldsymbol{j}_\mathrm{s}$ with the London equation (13.13), it is possible to obtain the relation $\nabla \times \nabla \times \boldsymbol{M} = -\boldsymbol{B}/4\pi\lambda_\mathrm{L}^2$. This represents the constitutive equation for the magnetization in a superconductor according to the London theory. In contrast to the case of an ordinary material this is a differential relation. This fact and the expulsion of the magnetic field $\boldsymbol{B}$ pose some problems in the definition of the susceptibility as introduced in Chap. 11. For this reason it is more appropriate to define the *susceptance* in the limit $\lambda_\mathrm{L} \ll a$, when the magnetization is uniform, as the relationship between $\boldsymbol{M}$ and the applied field $\boldsymbol{B}_0 = \boldsymbol{H}_0$:

$$\boldsymbol{M} = \xi_\mathrm{M} \boldsymbol{B}_0. \tag{13.28}$$

13.4 Application of the London Theory: Sphere in a Uniform Field

As an example of the application of the London theory, we consider now a sphere of radius a immersed in an initially uniform field. This example illustrates in a concrete case many elements of the electromagnetic theory of superconductors. Combining (13.25) with Maxwell's equations, the London equation for the vector potential $\boldsymbol{A}$ can be written as

$$\nabla^2 \boldsymbol{A} = \frac{1}{\lambda_{\mathrm{L}}^2} \boldsymbol{A}. \tag{13.29}$$

To deal with the case of a sphere, it is convenient to use spherical coordinates. In this system of coordinates the vector potential has components $\boldsymbol{A} = A_r \hat{\boldsymbol{r}} + A_\theta \hat{\boldsymbol{\theta}} + A_\varphi \hat{\boldsymbol{\varphi}}$. If the sphere is placed in an initially uniform field, this is deformed due to the induced superconducting currents that compensate the applied field inside the sphere. It is expected that the Meissner currents circulate around the sphere preserving the axial symmetry of the problem, so that the only non-vanishing component is j_φ. In view of (13.25), in the London gauge $\boldsymbol{A}$ also has a φ component only. For this reason, we choose to describe the uniform magnetic field $\boldsymbol{B}_0 = \boldsymbol{H}_0 = B_0 \hat{\boldsymbol{z}}$ in the absence of the sphere, or when it is in the normal state, through a vector potential $\boldsymbol{A}_0 = A_\varphi \hat{\boldsymbol{\varphi}}$ such that

$$A_\varphi = \frac{r B_0}{2} \sin\theta. \tag{13.30}$$

To determine A_φ in the presence of the sphere we have to solve the London equation with appropriate boundary conditions. Eventually we may have to modify the solution outside the material to take into account the effect of the superconducting current.

According to (13.29), A_φ satisfies the equation

$$\nabla^2 A_\varphi - \frac{1}{r^2 \sin\theta} A_\varphi = \frac{1}{\lambda_{\mathrm{L}}^2} A_\varphi. \tag{13.31}$$

To solve it we assume a solution of the form

$$A_\varphi = \frac{U(r)}{r} P(\theta). \tag{13.32}$$

Separating variables we obtain an equation for the functions $P(\theta)$ and $U(r)$. For the first function we get

$$\frac{1}{\sin\theta} \frac{\mathrm{d}}{\mathrm{d}\theta} \left(\sin\theta \frac{\mathrm{d}P}{\mathrm{d}\theta} \right) - \frac{P}{\sin^2\theta} + \Lambda P = 0, \tag{13.33}$$

where Λ is the separation constant. We recognize in (13.33) the differential equation of the associated Legendre functions. For $P(\theta)$ to be a regular function at $\theta = 0$ and $\theta = \pi$ we must have $\Lambda = l(l+1)$, with l a non-negative integer. The appropriate solution to the problem turns out to be $P_1^0(\cos\theta) = \sin\theta$. The equation for $U(r)$ is

$$\frac{\mathrm{d}^2U}{\mathrm{d}x^2} - \left(1 + \frac{2}{x^2}\right) U = 0. \tag{13.34}$$

This can be identified with the equation for the modified Bessel functions of order 3/2. Its regular solution in the interval of interest, $0 < r < a$, is

$$U(x) = \sqrt{x}\, I_{3/2}(x), \tag{13.35}$$

where $x = r/\lambda_{\mathrm{L}}$. This Bessel function can be expressed in terms of elementary functions, and the general solution is then

$$A_\varphi(r, \theta, \phi) = Af(x) \sin\theta, \tag{13.36}$$

where

$$f(x) = \frac{\cosh x}{x} - \frac{\sinh x}{x^2}. \tag{13.37}$$

The constant A is determined later. According to the London equation, the vector potential inside the superconductor is directly proportional to the circulating Meissner current j_φ, so that we have

$$j_\varphi(r, \theta, \phi) = -\frac{c}{4\pi\lambda_{\mathrm{L}}^2}\, A_\varphi(r, \theta, \phi). \tag{13.38}$$

The magnetic field inside the sphere is given by

$$\begin{aligned} B_r &= \frac{2A}{\lambda_{\mathrm{L}} x} f(x) \cos\theta, \\ B_\theta &= \frac{2A}{\lambda_{\mathrm{L}} x} \left[\frac{\sinh x}{x} - f(x)\right], \\ B_\varphi &= 0. \end{aligned} \tag{13.39}$$

Equation (13.30) gives the form of A_φ in free space, before immersing the superconducting sphere in the magnetic field. In the presence of the sphere the field outside it is also modified so that the general solution, with the same symmetry as the solution in the absence of superconductivity, is

$$A_\varphi = \left(\frac{B_0}{2}\, r + \frac{m}{r^2}\right) \sin\theta. \tag{13.40}$$

Here m is a constant to be determined, representing the net magnetic moment of the sphere. At sufficiently large distances we obtain the field in the absence of the sphere, and so determine the coefficient B_0.

The magnetic field outside the sphere, given by $\boldsymbol{B} = \nabla \times \boldsymbol{A}$, has components

$$\begin{aligned} B_r &= 2\cos\theta \left(B_0/2 + m/r^3\right), \\ B_\theta &= \sin\theta \left(-B_0 + m/r^3\right), \\ B_\varphi &= 0. \end{aligned} \tag{13.41}$$

The boundary conditions can be outlined more clearly if we consider a finite penetration length λ_L. In fact, this length is always greater than several interatomic distances, i.e. it is a macroscopic quantity from the viewpoint of our definition of the average fields. Some authors assume from the outset $\lambda_L \to 0$, which implies taking a coarse average which neglects details of the magnetization current distribution.

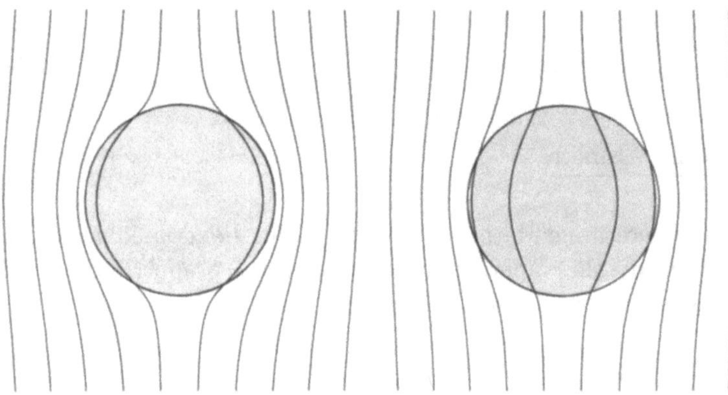

Fig. 13.5. Superconducting sphere in a magnetic field; (**a**) $\lambda_L = 0.2a$, (**b**) $\lambda_L = 0.3a$

The magnetization $\boldsymbol{M}$ should vanish near the surface in a distance of the order of λ_L. Accordingly, the boundary conditions at the surface of the sphere, derived in Sect. 10.3, imply that $B_{n1} = B_{n2}$ and $B_{t1} = B_{t2}$. These equations allow us to determine the constants A and m, which are

$$\begin{aligned} A &= -\tfrac{3}{2} a B_0 / \sinh(a/\lambda_L), \\ m &= -\tfrac{1}{2} a^3 B_0 \left[1 - \tfrac{3\lambda_L}{a} \coth(a/\lambda_L) + 3(\lambda_L/a)^2\right]. \end{aligned} \tag{13.42}$$

Figure 13.5 shows fields inside and outside the sphere in two particular cases. From (13.42) we note that, in the limit $\lambda_L \to 0$, the total magnetic moment is $m = -a^3 B_0/2$. In the exercise solved at the end of this chapter, this result is related to the demagnetization coefficient of an evenly-magnetized sphere.

13.5 Flux Quantization

We have mentioned above that in multiply connected samples the London theory predicts the quantization of the magnetic flux trapped in a hole. We

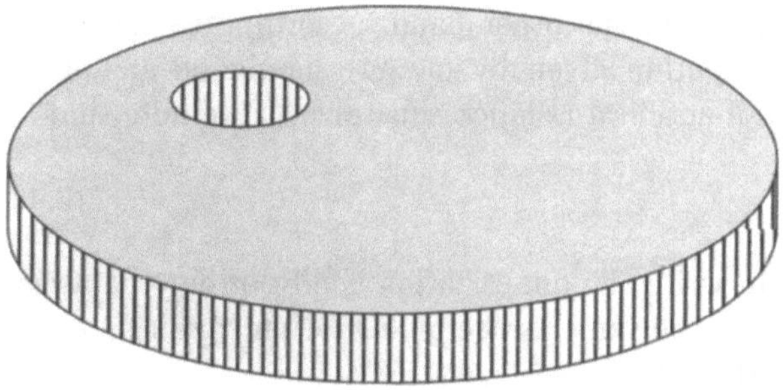

Fig. 13.6. Multiply-connected piece of superconductor

show now that this result is a consequence of the London equations and of the angular momentum quantization for a quantum system.

Imagine a multiply-connected body as in Fig. 13.6. Assuming that the problem has axial symmetry, let us consider the angular momentum of the superconducting fluid around the z axis. Replacing in $\boldsymbol{L} = \boldsymbol{r} \times \boldsymbol{p}$ the value of $\boldsymbol{p}$ derived in previous sections, the z component of the angular momentum is

$$L_z = \frac{q}{c}\left(\frac{4\pi}{c}\lambda_{\mathrm{L}}^2 j_\varphi + A_\varphi\right) r \sin\theta, \tag{13.43}$$

where θ is the polar angle and $r \sin\theta$ is the projection of the normal radius onto the z axis.

In 1913 Niels Bohr and Arnold Sommerfeld introduced the hypothesis of the quantization of the action, as a way of explaining the energy level structure of atoms. In modern quantum theory quantization conditions take a different form, but the Bohr–Sommerfeld condition on the action is useful as a means of introducing a quantum principle into a classical theory. In our case it allows us to set up a semiclassical theory of superconductors. For the present case we define the action associated with the angular variable φ as

$$J_\varphi = \oint p_\varphi \mathrm{d}\varphi, \tag{13.44}$$

where the integral extends over a closed curve. It can be shown that the canonical momentum p_φ is equal to the angular momentum L_z, and therefore

$$J_\varphi = \frac{q}{c}\oint \left(\frac{4\pi}{c}\lambda_{\mathrm{L}}^2 j_\varphi + A_\varphi\right) \mathrm{d}l, \tag{13.45}$$

where $\mathrm{d}l$ is the differential element of arc. The expression in brackets in the integral corresponds to the quantity defined in (13.20), whose curl vanishes. As mentioned above, this quantity can be written as the gradient of a scalar function η only when the condition (13.20) is satisfied in a simply-connected region. Considering a multiply connected material, it is not possible to write (13.22) as a gradient, and therefore it should not be expected that the integral over a closed circuit vanishes. On the other hand, according to the Bohr–Sommerfeld hypothesis, the action given by (13.44) should be an integer multiple n of the quantum of action h (Planck constant). If we note that

$$\oint A_{\varphi} \mathrm{d}\varphi = \int \boldsymbol{B} \cdot \mathrm{d}\boldsymbol{s} = \Phi \tag{13.46}$$

is the magnetic flux, we can write the quantization condition for the action $J_{\varphi} = nh$, which in this case is the same as the quantization condition of the angular momentum L_z, in the form

$$\Phi + \frac{4\pi}{c} \lambda_{\mathrm{L}}^2 \oint j_{\varphi} \mathrm{d}l = n \frac{hc}{q} = n\Phi_0 . \tag{13.47}$$

Here $\Phi_0 = hc/q$ is the so-called *flux quantum*. The combination appearing on the left-hand side of this equation is called a *fluxoid*. Equation (13.47) implies the quantization of the fluxoid.

By onveniently choosing the circuit of integration, the integral on the left-hand side can become zero, which implies *flux quantization*. Experiments demonstrate that the quantization condition is fulfilled with $q = 2e$, where e is the charge of the electron. This turns out to be a direct confirmation of the fact that charge carriers in a superconductor are pairs of electrons. This condition has also been verified in high temperature superconductors. The value of Φ_0 is $2.10^{-7}\,\mathrm{G\,cm^{-2}}$.

13.6 Energy of a Superconductor in a Magnetic Field

To analyze the energy balance of a superconductor in a magnetic field we begin by recalling the general expression for the energy in a material medium. In Chap. 10, we have shown that when the magnetic field is increased by an amount $\delta \boldsymbol{B}$, the energy change is given by

$$\delta W = \int \frac{1}{4\pi} \boldsymbol{H} \cdot \delta \boldsymbol{B} \, \mathrm{d}^3 r . \tag{13.48}$$

As shown previously, starting from the relation $\boldsymbol{H} = \boldsymbol{B} - 4\pi \boldsymbol{M}$ we can divide this energy variation into a purely magnetic contribution,

$$\delta W_1 = \int \frac{1}{4\pi} \boldsymbol{B} \cdot \delta \boldsymbol{B} \, \mathrm{d}^3 r , \tag{13.49}$$

and a contribution that can be identified as the magnetization work,

$$\delta W_2 = - \int \boldsymbol{M} \cdot \delta \boldsymbol{B} \, \mathrm{d}^3 r . \tag{13.50}$$

When the magnetic field varies from zero to its final value $\boldsymbol{B}$, the magnetic energy can be integrated and we obtain $W_1 = \int B^2 \mathrm{d}^3 r / 8\pi$.

The calculation of the magnetization work requires knowing the relation between magnetization $\boldsymbol{M}$ and magnetic field $\boldsymbol{B}$, that is to say, the constitutive equation. In a superconductor this should be a relationship between the magnetic field and the current, or between the vector potential and the current, as given by the London equation. As discussed above, the magnetization density is not uniquely defined. To evaluate the term δW_2 we can use the London equation, obtaining

$$\delta W_2 = -\int \boldsymbol{M} \cdot \delta \boldsymbol{B} \, \mathrm{d}^3 r = \int \frac{4\pi\lambda_\mathrm{L}^2}{c} \boldsymbol{M} \cdot (\nabla \times \delta \boldsymbol{j}_\mathrm{s}) \, \mathrm{d}^3 r. \tag{13.51}$$

For the total magnetization work we get, using a vector transformation,

$$\begin{aligned} W_2 &= \int_0^{\boldsymbol{j}_\mathrm{s}} \int_{V_0} \delta W_2 \mathrm{d}^3 r \\ &= \frac{4\pi\lambda_\mathrm{L}^2}{c} \int_0^{\boldsymbol{j}_\mathrm{s}} \int_{V_0} [\delta \boldsymbol{j}_\mathrm{s} \cdot \nabla \times \boldsymbol{M} + \nabla \cdot (\delta \boldsymbol{j}_\mathrm{s} \times \boldsymbol{M})] \, \mathrm{d}^3 r. \end{aligned} \tag{13.52}$$

Here V_0 is the volume of the superconductor. The second integral can be transformed into a surface integral of $\delta \boldsymbol{j}_\mathrm{s} \times \boldsymbol{M}$. This vanishes because the magnetization and the current $\boldsymbol{j}_\mathrm{s}$ are zero at the surface of the superconductor. In the previous sections we have defined the magnetization density through the relation $c\nabla \times \boldsymbol{M} = \boldsymbol{j}_\mathrm{s}$. This allows us to write

$$\begin{aligned} W_2 &= \frac{4\pi\lambda_\mathrm{L}^2}{c^2} \int_0^{j_\mathrm{s}} \int_{V_0} \boldsymbol{j}_\mathrm{s} \cdot \delta \boldsymbol{j}_\mathrm{s} \mathrm{d}^3 r = \frac{2\pi\lambda_\mathrm{L}^2}{c^2} \int_{V_0} j_\mathrm{s}^2 \mathrm{d}^3 r \\ &= \int_{V_0} \frac{m v_\mathrm{s}^2}{2} \, \mathrm{d}^3 r. \end{aligned} \tag{13.53}$$

This relation indicates that in a superconductor the magnetization work is identical to the kinetic energy of the superconducting currents. This means that the work done in establishing the magnetization of a superconductor is reversibly accumulated as kinetic energy of the superconducting electrons. The total energy can thus be expressed as

$$W = \frac{1}{8\pi} \int_V \left(B^2 + \frac{m v_\mathrm{s}^2}{2} \right) \mathrm{d}^3 r. \tag{13.54}$$

It is interesting to calculate the energy difference between the configuration of the superconductor in the field, and the field configuration in the absence of the superconductor. This is equivalent, if we neglect magnetic effects in the normal state, to calculating the energy difference between the material in the superconducting and the normal state when the sources producing the field are held fixed. Denoting the total volume of the system by V this energy difference is

$$\Delta W = \frac{1}{8\pi} \int_V \boldsymbol{H} \cdot \boldsymbol{B} \, \mathrm{d}^3 r - \frac{1}{8\pi} \int_V B_0^2 \, \mathrm{d}^3 r. \tag{13.55}$$

This can be transformed into

$$\begin{aligned}\Delta W &= \frac{1}{8\pi}\int_V (B^2 - B_0^2)\ \mathrm{d}^3r - \frac{1}{2}\int_V \boldsymbol{M}\cdot\boldsymbol{B}\ \mathrm{d}^3r \\ &= \Delta W_1 + \Delta W_2. \end{aligned} \tag{13.56}$$

From this it follows that

$$\Delta W = \frac{1}{8\pi}\int_V (\boldsymbol{B} - \boldsymbol{B}_0)\cdot(\boldsymbol{B} + \boldsymbol{B}_0)\ \mathrm{d}^3r + \Delta W_2. \tag{13.57}$$

Since $\boldsymbol{B} + \boldsymbol{B}_0 = \nabla \times (\boldsymbol{A} + \boldsymbol{A}_0)$ and, since we keep the sources fixed, $\nabla \times (\boldsymbol{B} - \boldsymbol{B}_0) = 4\pi \boldsymbol{j}_\mathrm{s}/c$, the magnetic energy difference in the superconducting material is

$$\begin{aligned}\Delta W_1 = &\ \frac{1}{8\pi}\int_S (\boldsymbol{A} + \boldsymbol{A}_0)\times(\boldsymbol{B} - \boldsymbol{B}_0)\cdot \mathrm{d}\boldsymbol{s} \\ &+ \frac{1}{2c}\int_V (\boldsymbol{A} + \boldsymbol{A}_0)\cdot \boldsymbol{j}_\mathrm{s}\ \mathrm{d}^3r. \end{aligned} \tag{13.58}$$

Here S is the surface enclosing the volume V. If we extend this volume beyond the region where the sources are located, the surface integral vanishes because the integrand falls to zero faster than R^{-2}. The constitutive equation (13.26) relates the vector potential linearly to the superconducting current, which in turn is proportional to the velocity, according to (13.1). As a consequence, the first term in the volume integral of (13.55) corresponds, except for the sign, to the kinetic energy density associated with the superconducting currents ΔW_2. The energy difference that we look for is therefore

$$\Delta W = \frac{1}{2c}\int_{V_0} \boldsymbol{A}_0 \cdot \boldsymbol{j}_\mathrm{s} \mathrm{d}^3r. \tag{13.59}$$

Note that this expression involves the vector potential $\boldsymbol{A}_0$ associated with the external field and that the integral extends only to the volume of the superconductor, since $\boldsymbol{j}_\mathrm{s}$ vanishes outside the material. We leave it as an exercise for the reader to show that, making use of the relation between $\boldsymbol{j}_\mathrm{s}$ and $\boldsymbol{M}$, this last integral can be written as

$$\Delta W = \frac{1}{2}\int_{V_0} \boldsymbol{M}\cdot\boldsymbol{B}_0\ \mathrm{d}^3r = \frac{1}{2}\ \boldsymbol{m}\cdot\boldsymbol{B}_0, \tag{13.60}$$

where $\boldsymbol{m}$ is the total magnetic moment associated with the superconductor. Due to the diamagnetic properties, $\boldsymbol{m}$ is opposite to the external field, and therefore ΔW is negative. According to the London theory, the spontaneous transition of a material to the superconducting state is therefore energetically favorable. An appropriate explanation of the thermodynamical aspects of this transition requires a study of the free energy of the system, which constitutes the starting point of the Landau–Ginzburg theory.

13.7 Present Relevance of Superconductivity

In the years 1987 and 1988 intensive research activities developed in the area of superconductivity, motivated by the discovery of materials that are superconducting at high temperatures. Once the initial enthusiasm and excitement calmed down, this gave place to more systematic experiments. In fact, the new superconducting materials exhibit quite unusual mechanical and superconducting properties. Several technical problems appeared that reduced the possibility of immediately using these materials in the wide range of applications initially foreseen.

However, after having overcome the obstacles, it is now possible to use high temperature superconductors in a variety of applications. These include the construction of conducting cables on an industrial scale and applying them conventionally, such as in electromagnets or in providing energy supply to large cities, where the cost of the installation is compensated by the savings in energy consumption.

Problems

13.1 Determine the boundary conditions for $\boldsymbol{M}$ and $\boldsymbol{H}$ at the limiting surface between a superconductor and vacuum.

13.2 Consider a superconducting slab of thickness $2d$ and of infinite extension in the other dimensions. Determine the magnetic field inside the slab assuming that the external field is B_1 at one side of the slab and B_2 at the other side (Fig. 13.7). Calculate the field $\boldsymbol{H}$, the local magnetization $\boldsymbol{M}(\boldsymbol{r})$ and the total magnetic moment per unit volume.

Hint: Notice that the boundary conditions imply a net current flowing within the slab.

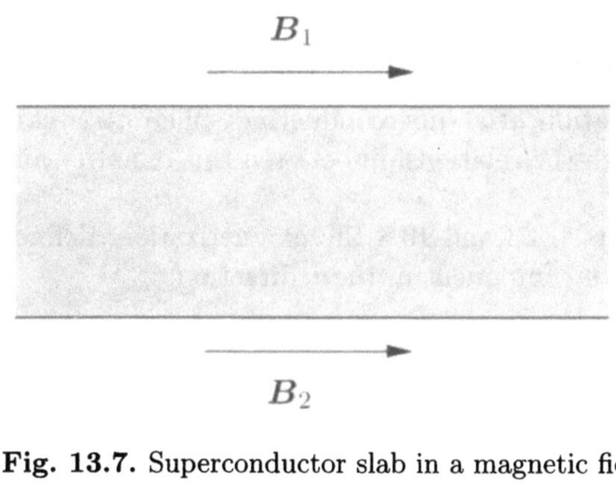

Fig. 13.7. Superconductor slab in a magnetic field

13.3 In the previous problem, determine the total force acting on the slab by direct calculation. Consider the energy per unit volume and calculate the force as the derivative of energy with respect to a coordinate.

13.4 Show that the total magnetic moment $\boldsymbol{m}$ of a superconductor is independent of whether the magnetization density is defined through the relation $c\nabla \times \boldsymbol{M} = \boldsymbol{j}_\mathrm{s}$ or as $\boldsymbol{M} = \boldsymbol{r} \times \boldsymbol{j}_\mathrm{s}/2c$.

Hint: Note that both the magnetization $\boldsymbol{M}$ and the superconducting current $\boldsymbol{j}_\mathrm{s}$ vanish at the boundary.

13.5 Determine the explicit form of the energy for a superconducting sphere in an external field.

13.6 Calculate explicitly the variation in magnetic energy ΔW_1 for the case of a superconducting sphere in a field.

Hint: Express the superconducting current in terms of the velocity.

13.7 In order to discuss the form of the energy in a London superconductor we can consider the total energy of currents and field including magnetic and kinetic energy. Show that if this is expressed in terms of the magnetic field using Maxwell's equations, the London equation follows as a consequence of minimizing the total energy.

13.8 Determine the Green's function for the London equation. Using Green's theorem, prove that the magnetic field at any interior point of a spherical region of superconducting material is always smaller than the average of this field on the surface of the limiting sphere.

Hint: Use Green's theorem (5.15), identifying $\psi(\boldsymbol{r})$ with the function $|\boldsymbol{r}-\boldsymbol{r}'|^{-1}\exp(-|\boldsymbol{r}-\boldsymbol{r}'|/\lambda_\mathrm{L})$. Notice that $|\boldsymbol{r}-\boldsymbol{r}'|^{-1}\exp(+|\boldsymbol{r}-\boldsymbol{r}'|/\lambda_\mathrm{L})$ is also a solution.

13.9 Study the solution of the London equation for the case in which the vorticity is different from zero at a point $\boldsymbol{r}_0$, such that

$$\boldsymbol{Q} = \frac{q}{c}\Phi_0\delta(\boldsymbol{r}-\boldsymbol{r}_0)\hat{\boldsymbol{z}}.$$

This condition defines the existence of a vortex in the London theory.

13.10 Determine the magnetization of a superconducting sphere in a constant magnetic field and establish the relationship between this quantity and the applied field.

Solution: According to Sects. 13.3 and 13.4 the magnetization, defined through $c\nabla \times \boldsymbol{M} = \boldsymbol{j}_\mathrm{s}$ has only a component in the z direction:

$$M_z = \frac{A_1}{4\pi\lambda_\mathrm{L}}\left[\frac{\sinh(r/\lambda_\mathrm{L})}{r/\lambda_\mathrm{L}} - \frac{\sinh(a/\lambda_\mathrm{L})}{a/\lambda_\mathrm{L}}\right].$$

If $\lambda_\mathrm{L} \ll a$, M_z is practically uniform and is given by

$$M_z = -\frac{3}{2}\frac{B_0}{4\pi}.$$

Using the relationship for uniformly magnetized media discussed in Sect. 11.4 we have for a superconductor of arbitrary ellipsoidal shape

$$\boldsymbol{B} - 4\pi \boldsymbol{M} = \boldsymbol{B}_0 - 4\pi \mathcal{N} \cdot \boldsymbol{M},$$

where $\mathcal{N}$ is the demagnetization tensor. In the above mentioned limit $\boldsymbol{B} \to 0$ inside the superconductor so that the previous relationship implies

$$\boldsymbol{M} = -\frac{1}{4\pi} \left[1 - \mathcal{N}\right]^{-1} \cdot \boldsymbol{B}_0.$$

This corresponds to an intrinsic diamagnetic susceptance $\xi_{\mathrm{M}} = -1/4\pi$ due to superconductivity modified by a geometrical factor. For the case of the sphere we obtain

$$M_z = -\frac{1}{4\pi}(1 - 1/3)^{-1} B_0.$$

If we had made the same calculation for a long cylinder we would have obtained a null factor coming from the demagnetization factor of an infinite cylinder instead of 1/3.

References

13.1 H. Kamerlingh Onnes: *Further Experiments with Liquid Helium: The Disappearence of the Resistance of Mercury*, Leyden Commun. **122b** (1911).
13.2 W. Meissner and R. Ochsenfeld: *Ein neuer Effekt bei der Eintritt der Supraleitfähigkeit*, [2] Die Naturwissenschaften **21**, 787 (1933).
13.3 B.S. Deaver and W.M. Fairbank: *Experimental Evidence for Quantized Flux in Superconducting Cylinders*, Phys. Rev. Lett. **7**, 43 (1961).
13.4 R. Doll and M. Naebauer: *Experimental Proof of Magnetic Flux Quantization in a Superconducting Ring*, Phys. Rev. Lett. **7**, 51 (1961).
13.5 J.G. Bednorz and K.A. Müller: *Superconductivity in the Ba-La-Cu-O System?*, Z. Phys. B **64**, 189 (1986).

Further Reading

W.J. Carr Jr.: *Macroscopic Theory of Superconductors*, Phys. Rev. B **23**, 3208 (1981).
P.G. de Gennes: *Superconductivity of Metals and Alloys* (Addison–Wesley, Reading, 1989).
C.J. Gorter: *Superconductivity until 1940 in Leyden and as seen from there*, Rev. Mod. Phys. **36**, 1 (1964).
L.D. Landau and E.M. Lifschitz: *Electrodynamics of Continuous Media* (Pergamon, New York, 1960).
F. London: *Superfluids* (Dover, New York, 1961).

[2] *A New Effect at the Setup of Superconductivity*

A.M. Portis: *Electromagnetic Fields: Sources and Media* (Wiley, New York, 1978).

Shu-Ang Zhou: *Electrodynamic Theory of Superconductors* (Peter Peregrinus, London, 1991).

Appendix
A. The Dirac Delta Distribution

The *Dirac delta* is an element of the set of mathematical objects called *distributions.* These are a generalization of functions. Some of them are particularly suitable for describing physical entities in an approximated way. For instance, the Dirac delta can be used to represent highly concentrated distributions of mass or charge. The theory of distributions was developed in the first half of the twentieth century by the French mathematician L. Schwartz, and its name is due precisely to its connection with mass and charge distributions.

Under appropriate conditions, the Dirac delta distribution $\delta(x)$ can be treated, *cum grano salis*, as a function of the variable $x \in (-\infty, \infty)$, defined by the conditions

$$\begin{cases} \delta(x) = 0 & \text{for } x \neq 0, \\ \int_{-\infty}^{\infty} \delta(x)\, \mathrm{d}x = 1. \end{cases} \tag{A.1}$$

This is an even distribution, $\delta(x) = \delta(-x)$, and it is infinitely concentrated at $x = 0$.

The Dirac delta distribution can be approximated by functions, called *approximant kernels* of $\delta(x)$, in appropriate limits. Some of these approximant kernels are:

$$\begin{aligned} f(x) &= (2\pi\sigma)^{-1/2} \exp(-x^2/2\sigma) && \text{for } \sigma \to 0 \text{ (Gaussian)}, \\ f(x) &= a/\pi(x^2 + a^2) && \text{for } a \to 0 \text{ (Lorentzian)}, \\ f(x) &= \lambda \exp(-2\lambda|x|) && \text{for } \lambda \to \infty, \\ f(x) &= \sin(\alpha x)/\pi x && \text{for } \alpha \to \infty. \end{aligned} \tag{A.2}$$

The main property of the Dirac delta of interest for applications is that for any function $f(x)$

$$f(x_0) = \int_a^b f(x)\, \delta(x - x_0)\, \mathrm{d}x, \tag{A.3}$$

where $a < x_0 < b$, if $f(x)$ is continuous at $x = x_0$. Integrating by parts, the "derivatives" of $\delta(x)$ can be introduced. They are also distributions, defined by the property

$$f^{(n)}(x_0) = (-1)^n \int_a^b f(x)\, \delta^{(n)}(x - x_0)\, \mathrm{d}x, \tag{A.4}$$

where the labels (n) indicate nth-order differentiation.

When the argument of the delta distribution is a function $\phi(x)$, a variable change makes it possible to prove that

$$\int_a^b f(x)\, \delta[\phi(x)]\, \mathrm{d}x = \sum_i f(x_i) \left| \frac{\mathrm{d}\phi}{\mathrm{d}x}(x_i) \right|^{-1}, \tag{A.5}$$

where x_i are the zeros of $\phi(x)$ in the interval (a, b). In particular,

$$\int_a^b f(x)\, \delta[\alpha(x - x_0)]\, \mathrm{d}x = |\alpha|^{-1} f(x_0), \tag{A.6}$$

or, formally, $\delta(\alpha x) = \delta(x)/|\alpha|$.

The product of Dirac deltas allows us to define the multidimensional version of this distribution. For instance, in three dimensions,

$$\delta(\boldsymbol{r}) = \delta(x)\delta(y)\delta(z), \tag{A.7}$$

where (x, y, z) are the Cartesian components of $\boldsymbol{r}$. In spherical coordinates (r, θ, φ) we find

$$\delta(\boldsymbol{r}) = \frac{1}{r^2 \sin\theta}\, \delta(r)\delta(\theta)\delta(\varphi) = \frac{1}{r^2}\, \delta(r)\delta(\cos\theta)\delta(\varphi). \tag{A.8}$$

The multidimensional delta satisfies

$$\int_V f(\boldsymbol{r})\, \delta(\boldsymbol{r} - \boldsymbol{r}_0)\, \mathrm{d}^3 r = f(\boldsymbol{r}_0), \tag{A.9}$$

if $\boldsymbol{r}_0$ is in the integration volume V.

Other important properties of the Dirac delta are the following:

$$\delta(x) = \frac{\mathrm{d}}{\mathrm{d}x}\theta(x), \tag{A.10}$$

where $\theta(x)$ is the *Heaviside function*

$$\theta(x) = \begin{cases} 0 & \text{if } x < 0, \\ 1 & \text{if } x > 0, \end{cases} \tag{A.11}$$

also called the *step function*. Moreover,

$$\delta(\boldsymbol{r}) = -\frac{1}{4\pi} \nabla^2 \frac{1}{r}, \tag{A.12}$$

which, according to Gauss's law, implies that $\delta(\boldsymbol{r})$ is the charge density to be associated with a unit point charge at the origin. Finally,

$$\delta(x) = \frac{1}{2\pi} \int_{-\infty}^{\infty} \exp(\mathrm{i}kx)\, \mathrm{d}k, \tag{A.13}$$

which relates the Dirac delta to the Fourier tranform of a constant function.

B. Legendre Polynomials and Spherical Harmonics

From Laplace's equation,

$$\nabla^2 \phi = 0, \tag{B.1}$$

the separation of variables in spherical coordinates, $\phi(r,\theta,\varphi) = R(r)Y(\theta,\varphi)$, produces for $Y(\theta,\varphi)$ the differential equation

$$\sin\theta\ \partial_\theta(\sin\theta\ \partial_\theta Y) + \partial^2_{\varphi\varphi} Y = -\lambda Y, \tag{B.2}$$

where λ is a separation constant.

Under conditions of azimuthal symmetry, the solution of Laplace's equation does not depend on φ. Requiring regularity conditions at $\theta = 0$ and $\theta = \pi$, the solutions of the angular part of the equation are the *Legendre polynomials* $P_l(x)$, $l = 0, 1, 2, \ldots$, evaluated in terms of $x = \cos\theta$: $Y(\theta,\varphi) = Y(\theta) = P_l(\cos\theta)$.

The lth-degree Legendre polynomial $P_l(x)$ is defined by

$$P_l(x) = \frac{1}{2^l l!}\,\frac{\mathrm{d}^l}{\mathrm{d}x^l}(x^2-1)^l. \tag{B.3}$$

Legendre polynomials satisfy the orthogonality relation

$$\int_{-1}^{1} P_{l'}(x)\ P_l(x)\ \mathrm{d}x = \frac{2}{2l+1}\ \delta_{ll'}, \tag{B.4}$$

and form a set of orthogonal functions in the interval $(-1,1)$. Any well-behaved function can thus be expanded in a series of Legendre polynomials within that interval, and such an expansion is unique.

The generating function of $P_l(x)$,

$$|\boldsymbol{r} - \boldsymbol{r}'|^{-1} = \sum_{l=0}^{\infty} \frac{r_<^l}{r_>^{l+1}} P_l(\cos\gamma), \tag{B.5}$$

where γ is the angle determined by $\boldsymbol{r}$ and $\boldsymbol{r}'$, and $r_< = \min(r,r')$, $r_> = \max(r,r')$, turns out to be particularly important in applications to electrostatics (see Chap. 4).

Some relevant properties of the Legendre polynomials are the following:

$$\begin{aligned} &P_l(1) = 1, \qquad P_l(-1) = (-1)^l, \\ &P_{2n}(0) = (-1)^n (2n-1)!!/(2n)!!, \qquad P_{2n+1}(0) = 0, \\ &(l+1)P_{l+1}(x) = (2l+1)xP_l(x) - lP_{l-1}(x), \\ &(x^2-1)\frac{\mathrm{d}}{\mathrm{d}x}P_l(x) = lxP_l(x) - lP_{l-1}(x). \end{aligned} \tag{B.6}$$

The first few Legendre polynomials are

$$\begin{aligned} &P_0(x) = 1, \\ &P_1(x) = x, \\ &P_2(x) = \tfrac{1}{2}(3x^2-1), \\ &P_3(x) = \tfrac{1}{2}(5x^3-3x). \end{aligned} \tag{B.7}$$

The solutions of the angular part of Laplace's equation when there is no azimuthal symmetry are the *spherical harmonic* functions of order (l, m), $Y_{lm}(\theta, \varphi)$, with $l = 0, 1, 2, ...$ and $m = -l, -l+1, ..., l-1, l$. For $m \geq 0$,

$$Y_{lm}(\theta, \varphi) = \sqrt{\frac{2l+1}{4\pi}\frac{(l-m)!}{(l+m)!}}\; P_l^m(\cos\theta)\; \exp(im\varphi), \tag{B.8}$$

and $Y_{l,-m} = (-1)^m Y_{lm}^*$, where $*$ indicates complex conjugation. The functions $P_l^m(x)$ are the *associated Legendre functions*, defined by

$$P_l^m(x) = \frac{1}{2^l l!}\,(1-x^2)^{|m|/2}\,\frac{\mathrm{d}^{l+|m|}}{\mathrm{d}x^{l+|m|}}(x^2-1)^l. \tag{B.9}$$

Note that $P_l^0(x) = P_l(x)$.

Spherical harmonics satisfy the following orthogonality and completeness relations:

$$\int_0^{2\pi} \mathrm{d}\varphi \int_0^{\pi} \mathrm{d}\theta\, \sin\theta\; Y_{l'm'}^*(\theta, \varphi) Y_{lm}(\theta, \varphi) = \delta_{ll'}\delta_{mm'}, \tag{B.10}$$

$$\sum_{l=0}^{\infty} \sum_{m=-l}^{l} Y_{lm}^*(\theta', \varphi') Y_{lm}(\theta, \varphi) = \delta(\cos\theta - \cos\theta')\delta(\varphi - \varphi'), \tag{B.11}$$

so that they constitute a complete set of functions in $0 \leq \theta \leq \pi$, $0 \leq \varphi \leq 2\pi$.

The following addition theorem, which is useful for several applications, holds:

$$P_l^0(\cos\gamma) = \frac{4\pi}{2l+1} \sum_{m=-l}^{l} Y_{lm}^*(\theta', \varphi') Y_{lm}(\theta, \varphi), \tag{B.12}$$

where γ is the angle between the directions defined by the angular coordinates (θ, φ) and (θ', φ'). This theorem can be used to write (B.5) as

$$|\boldsymbol{r} - \boldsymbol{r}'|^{-1} = 4\pi \sum_{l=0}^{\infty} \sum_{m=-l}^{l} \frac{1}{2l+1} \frac{r_<^l}{r_>^{l+1}} Y_{lm}^*(\theta', \varphi') Y_{lm}(\theta, \varphi), \tag{B.13}$$

where (θ, φ) and (θ', φ') are the angular coordinates of the vectors $\boldsymbol{r}$ and $\boldsymbol{r}'$, respectively, and $r_>$ and $r_<$ are defined as in (B.5).

The first few spherical harmonics are

$$\begin{aligned}
Y_{00} &= (4\pi)^{-1/2}, \\
Y_{10} &= (3/4\pi)^{1/2} \cos\theta, \\
Y_{1,\pm1} &= \mp(3/8\pi)^{1/2} \sin\theta \ \exp(\pm \mathrm{i}\varphi), \\
Y_{20} &= (5/16\pi)^{1/2} (3\cos^2\theta - 1), \\
Y_{2,\pm1} &= \mp(15/8\pi)^{1/2} \sin\theta \ \cos\theta \exp(\pm \mathrm{i}\varphi), \\
Y_{2,\pm2} &= \mp(15/32\pi)^{1/2} \mathrm{sen}^2\theta \ \exp(\pm 2\mathrm{i}\varphi).
\end{aligned} \tag{B.14}$$

C. Covariant Notation and Tensor Calculus

Tensor calculus is the formalism used in special and general relativity theory for operating in the four-dimensional space with spatial coordinates (x, y, z) and time coordinate ct. Generally, four-dimensional tensor calculus involves mathematical objects defined in a space with generalized coordinates (x^0, x^1, x^2, x^3). Such objects, called *tensors*, are characterized by transforming in a well-defined way under a coordinate change to the new variables (x'^0, x'^1, x'^2, x'^3), which are related to the original coordinates through certain functions

$$\begin{cases} x'^0 = x'^0(x^0, x^1, x^2, x^3), \\ x'^1 = x'^1(x^0, x^1, x^2, x^3), \\ x'^2 = x'^2(x^0, x^1, x^2, x^3), \\ x'^3 = x'^3(x^0, x^1, x^2, x^3). \end{cases} \tag{C.1}$$

A zeroth-rank tensor or *scalar*, s, is a quantity whose value does not change under the coordinate transformation,

$$s' = s. \tag{C.2}$$

A first-rank contravariant tensor or, *contravariant four-vector*, is a set of four quantities (v^0, v^1, v^2, v^3), whose elements are generically denoted by v^μ and which transform according to

$$v'^\mu = \sum_{\alpha=0}^{3} \frac{\partial x'^\mu}{\partial x^\alpha}\, v^\alpha = \frac{\partial x'^\mu}{\partial x^\alpha}\, v^\alpha. \tag{C.3}$$

We have introduced here the *Einstein convention* of summation over repeated indices. According to this convention, when an index appears twice in an expression it is implicitly assumed that the expression is summed over the repeated index, from 0 to 3. The summation over repeated indices is called *contraction* with respect to those indices.

A first-rank covariant tensor, or *covariant four-vector*, u_μ, is a set of four quantities (u_0, u_1, u_2, u_3) which transform according to

$$u'_\mu = \frac{\partial x^\alpha}{\partial x'^\mu}\, u_\alpha. \tag{C.4}$$

Four-vector physical quantities have, in general, both a contravariant and a covariant representation.

Higher rth-rank tensors are sets of 4^r quantities labeled by r indices, which transform as products of r four-vector components. Each index can be of contravariant or covariant character, according to the corresponding transformation rule. For instance, any of the sixteen components of a twice-contravariant second-rank tensor $t^{\mu\nu}$ transforms according to

$$t'^{\mu\nu} = \frac{\partial x'^{\mu}}{\partial x^{\alpha}} \frac{\partial x'^{\nu}}{\partial x^{\beta}} \, t^{\alpha\beta} . \tag{C.5}$$

In general, we have

$$T'^{\mu\nu\cdots}_{\rho\sigma\cdots} = \frac{\partial x'^{\mu}}{\partial x^{\alpha}} \frac{\partial x'^{\nu}}{\partial x^{\beta}} \cdots \frac{\partial x^{\delta}}{\partial x'^{\rho}} \frac{\partial x^{\epsilon}}{\partial x'^{\sigma}} \cdots T^{\alpha\beta\cdots}_{\delta\epsilon\cdots} . \tag{C.6}$$

It is worthwhile stressing that the contraction operation preserves the covariant properties of the result. This implies that, in general, the contraction of two tensors over an arbitrary number of indices produces another tensor. For instance, given a four-vector v^{μ} its *modulus* $v^{\mu}v_{\mu}$ is a scalar, i.e. is invariant under coordinate transformations. The transformed coordinates of a four-vector also form a four-vector, as they are obtained from the contraction of the Jacobian tensor associated with the transformation, $J^{\mu}_{\ \nu} = \partial x'^{\mu}/\partial x^{\nu}$, and the original four-vector, $v'^{\mu} = J^{\mu}_{\ \nu} v^{\nu}$.

When the four-dimensional space has metric properties, it is possible to define a (squared) length element corresponding to an infinitesimal displacement $(\mathrm{d}x^0, \mathrm{d}x^1, \mathrm{d}x^2, \mathrm{d}x^3)$ as

$$\mathrm{d}s^2 = g_{\mu\nu} \mathrm{d}x^{\mu} \mathrm{d}x^{\nu} , \tag{C.7}$$

where $g_{\mu\nu}$ is a symmetric tensor, $g_{\mu\nu} = g_{\nu\mu}$, called the *metric tensor*, which characterizes the metric geometry of space. The metric tensor defines the transformation between the contravariant and the covariant components of a tensor. For instance, for a four-vector we have

$$v_{\mu} = g_{\mu\nu} v^{\nu} . \tag{C.8}$$

In the case of special relativity, the four-dimensional space to be considered is Minkowski space, where the coordinates of a point – or *event* – are

$$x^0 = ct, \quad x^1 = x, \quad x^2 = y, \quad x^3 = z . \tag{C.9}$$

In this space, the most interesting coordinate transformation is the Lorentz transformation, which relates the coordinates of an event in two inertial reference frames S and S′ as

$$x'^{\nu} = a^{\nu}_{\ \mu} x^{\mu} , \tag{C.10}$$

where the tensor $a^{\nu}{}_{\mu}$ is defined in (3.18). For the Lorentz transformation, which is linear, the general expressions analyzed above are considerably simpler. In fact, $\partial x'^{\nu}/\partial x^{\mu} = a^{\nu}{}_{\mu}$ does not depend on the coordinates. Consequently, any set of four quantities that transform between reference frames with the same rules as the coordinates of an event,

$$v'^{\nu} = a^{\nu}{}_{\mu} v^{\mu}, \tag{C.11}$$

constitutes a four-vector. As stated above, higher-rank tensors transform as products of four-vector components. For instance, for a twice-contravariant second-rank tensor, we have

$$t'^{\alpha\beta} = a^{\alpha}{}_{\mu} a^{\beta}{}_{\nu} t^{\mu\nu}. \tag{C.12}$$

In the framework of special relativity, the metric tensor defines the *Minkowski metric*, and is

$$g_{\mu\nu} = \begin{pmatrix} 1 & 0 & 0 & 0 \\ 0 & -1 & 0 & 0 \\ 0 & 0 & -1 & 0 \\ 0 & 0 & 0 & -1 \end{pmatrix}. \tag{C.13}$$

The length element associated with the Minkowski metric is then

$$\mathrm{d}s^2 = c^2 \mathrm{d}t^2 - \mathrm{d}x^2 - \mathrm{d}y^2 - \mathrm{d}z^2. \tag{C.14}$$

Finally, we point out that in Minkowski space the operation of differentiation with respect to the contravariant components can be associated with the product with a covariant four-vector, whose components are denoted by

$$\frac{\partial}{\partial x^{\mu}} = \partial_{\mu}. \tag{C.15}$$

These four derivatives are the covariant components of the *four-gradient*. Of course, its contravariant components ∂^{μ} are given by the derivatives with respect to x_{μ}. The definition of the gradient four-vector makes it possible to extend some of the three-dimensional vector operations to four-dimensional space. The four-divergence of a vector V^{μ}, for instance, is the scalar obtained from the contraction $\partial_{\mu} v^{\mu}$. In Chaps. 3 and 9, several applications of these operations are studied.

D. Vector Identities, Theorems and Operators

In the following, ϕ and ψ represent scalars, and $\boldsymbol{a}$, $\boldsymbol{b}$ y $\boldsymbol{c}$ are vectors.

Three-vector products

$$\boldsymbol{a} \cdot (\boldsymbol{b} \times \boldsymbol{c}) = \boldsymbol{b} \cdot (\boldsymbol{c} \times \boldsymbol{a}) = \boldsymbol{c} \cdot (\boldsymbol{a} \times \boldsymbol{b})$$

$$\boldsymbol{a} \times (\boldsymbol{b} \times \boldsymbol{c}) = \boldsymbol{b}(\boldsymbol{a} \cdot \boldsymbol{c}) - \boldsymbol{c}(\boldsymbol{a} \cdot \boldsymbol{b})$$

Product rules

$$\nabla(\phi\psi) = \phi\nabla\psi + \psi\nabla\phi$$

$$\nabla(\boldsymbol{a} \cdot \boldsymbol{b}) = \boldsymbol{a} \times (\nabla \times \boldsymbol{b}) + \boldsymbol{b} \times (\nabla \times \boldsymbol{a}) + (\boldsymbol{a} \cdot \nabla)\boldsymbol{b} + (\boldsymbol{b} \cdot \nabla)\boldsymbol{a}$$

$$\nabla \cdot (\phi\boldsymbol{a}) = \phi\nabla \cdot \boldsymbol{a} + \boldsymbol{a} \cdot \nabla\phi$$

$$\nabla \cdot (\boldsymbol{a} \times \boldsymbol{b}) = \boldsymbol{b} \cdot (\nabla \times \boldsymbol{a}) - \boldsymbol{a} \cdot (\nabla \times \boldsymbol{b})$$

$$\nabla \times (\phi\boldsymbol{a}) = \phi\nabla \times \boldsymbol{a} - \boldsymbol{a} \times \nabla\phi$$

$$\nabla \times (\boldsymbol{a} \times \boldsymbol{b}) = (\boldsymbol{b} \cdot \nabla)\boldsymbol{a} - (\boldsymbol{a} \cdot \nabla)\boldsymbol{b} + \boldsymbol{a}\nabla \cdot \boldsymbol{b} - \boldsymbol{b}\nabla \cdot \boldsymbol{a}$$

Second-order derivatives

$$\nabla \cdot (\nabla \times \boldsymbol{a}) = 0$$

$$\nabla \times (\nabla\phi) = 0$$

$$\nabla \times (\nabla \times \boldsymbol{a}) = \nabla(\nabla \cdot \boldsymbol{a}) - \nabla^2\boldsymbol{a}$$

Fundamental theorems

$$\int_{\boldsymbol{r}_1}^{\boldsymbol{r}_2} (\nabla\phi) \cdot \mathrm{d}\boldsymbol{l} = \phi(\boldsymbol{r}_2) - \phi(\boldsymbol{r}_1)$$

$$\int_V (\nabla \cdot \boldsymbol{a})\mathrm{d}^3r = \int_{S(V)} \boldsymbol{a} \cdot \mathrm{d}\boldsymbol{s}$$

$$\int_{S(C)} (\nabla \times \boldsymbol{a}) \cdot \mathrm{d}\boldsymbol{s} = \oint_C \boldsymbol{a} \cdot \mathrm{d}\boldsymbol{l}$$

Vector operators

Cartesian coordinates

$$\mathrm{d}\boldsymbol{l} = \mathrm{d}x\ \hat{\boldsymbol{x}} + \mathrm{d}y\ \hat{\boldsymbol{y}} + \mathrm{d}z\ \hat{\boldsymbol{z}}$$

$$\mathrm{d}^3r = \mathrm{d}x\ \mathrm{d}y\ \mathrm{d}z$$

$$\nabla\phi = \partial_x\phi\ \hat{\boldsymbol{x}} + \partial_y\phi\ \hat{\boldsymbol{y}} + \partial_z\phi\ \hat{\boldsymbol{z}}$$

$$\nabla\cdot\boldsymbol{a} = \partial_x a_x + \partial_y a_y + \partial_z a_z$$

$$\nabla\times\boldsymbol{a} = (\partial_y a_z - \partial_z a_y)\ \hat{\boldsymbol{x}} + (\partial_z a_x - \partial_x a_z)\ \hat{\boldsymbol{y}} + (\partial_x a_y - \partial_y a_x)\ \hat{\boldsymbol{z}}$$

$$\nabla^2\phi = \partial_x^2\phi + \partial_y^2\phi + \partial_z^2\phi$$

Spherical coordinates

$$\mathrm{d}\boldsymbol{l} = \mathrm{d}r\ \hat{\boldsymbol{r}} + r\ \mathrm{d}\theta\ \hat{\boldsymbol{\theta}} + r\sin\theta\ \mathrm{d}\varphi\ \hat{\boldsymbol{\varphi}}$$

$$\mathrm{d}^3r = r^2\sin\theta\ \mathrm{d}r\ \mathrm{d}\theta\ \mathrm{d}\varphi$$

$$\nabla\phi = \partial_r\phi\ \hat{\boldsymbol{r}} + r^{-1}\partial_\theta\phi\ \hat{\boldsymbol{\theta}} + (r\sin\theta)^{-1}\partial_\varphi\phi\ \hat{\boldsymbol{\varphi}}$$

$$\nabla\cdot\boldsymbol{a} = r^{-2}\partial_r(r^2 a_r) + (r\sin\theta)^{-1}\partial_\theta(\sin\theta\ a_\theta) + (r\sin\theta)^{-1}\partial_\varphi a_\varphi$$

$$\nabla\times\boldsymbol{a} = (r\sin\theta)^{-1}[\partial_\theta(\sin\theta\ a_\varphi) - \partial_\varphi a_\theta]\ \hat{\boldsymbol{r}} + r^{-1}[(\sin\theta)^{-1}\partial_\varphi a_r - \partial_r(r a_\varphi)]\ \hat{\boldsymbol{\theta}}$$
$$+ r^{-1}[\partial_r(r a_\theta) - \partial_\theta a_r]\ \hat{\boldsymbol{\varphi}}$$

$$\nabla^2\phi = r^{-2}\partial_r(r^2\partial_r\phi) + (r^2\sin\theta)^{-1}\partial_\theta(\sin\theta\partial_\theta\phi) + (r^2\sin^2\theta)^{-1}\partial_\varphi^2\phi$$

Cylindrical coordinates

$$\mathrm{d}\boldsymbol{l} = \mathrm{d}r\ \hat{\boldsymbol{r}} + r\ \mathrm{d}\varphi\ \hat{\boldsymbol{\varphi}} + \mathrm{d}z\ \hat{\boldsymbol{z}}$$

$$\mathrm{d}^3r = r\ \mathrm{d}r\ \mathrm{d}\varphi\ \mathrm{d}z$$

$$\nabla\phi = \partial_r\phi\ \hat{\boldsymbol{r}} + r^{-1}\partial_\varphi\phi\ \hat{\boldsymbol{\varphi}} + \partial_z\phi\ \hat{\boldsymbol{z}}$$

$$\nabla\cdot\boldsymbol{a} = r^{-1}\partial_r(r a_r) + r^{-1}\partial_\varphi a_\varphi + \partial_z a_z$$

$$\nabla\times\boldsymbol{a} = [r^{-1}\partial_\varphi a_z - \partial_z a_\varphi]\ \hat{\boldsymbol{r}} + [\partial_z a_r - \partial_r a_z]\ \hat{\boldsymbol{\varphi}} + r^{-1}[\partial_r(r a_\varphi) - \partial_\varphi a_r]\ \hat{\boldsymbol{z}}$$

$$\nabla^2\phi = r^{-1}\partial_r(r\partial_r\phi) + r^{-2}\partial_\varphi^2\phi + \partial_z^2\phi$$

E. Operation of PhysicSolver

PhysicSolver is a computer program designed to solve two-dimensional problems in electrostatics and magnetostatics using the finite element method described in Chap. 5. The problems may involve three-dimensional geometries with either planar or axial symmetry. A problem with planar symmetry is solved in the (x, y) plane. A problem with axial or cylindrical symmetry is solved in one half of the (r, z) plane.

To use the software it is recommended that a simple sketch of the problem is first made on paper. Regions of the plane where materials with different physical properties are present should be identified. Mirror symmetries should be determined and the plane must be divided accordingly. The solution domain must be defined. If the domain extends to infinity, an artificial boundary must be fixed at some distance from the region of interest. Finally, the boundary conditions at all parts of the boundary must be defined.

Units of length (mm, cm, m, inch or feet) must be selected to obtain the solution. Though they can be specified or changed at any time before the solution is calculated, it is recommended that the units of length be specified as early as possible. All other units are fixed and cannot be changed. The program uses a mixed unit system. Results are displayed in ampere, volt, coulomb, gauss and oersted. The unit of force is the newton and the unit of energy is the joule.

To proceed with the solution, points, lines, and arcs have to be created to define the geometry of the problem. These steps lead to the definition of regions. In each one of these regions the material is assumed to be homogeneous. It is also possible to divide a region where the material is homogeneous into two or more subregions where different behaviors of the solution are expected, for instance, in the vicinity of a sharp point or of a region where material properties change.

The symmetry of the problem – planar or cylindrical – must be selected. One must also of course tell the program whether the problem to be solved is one of electrostatics or magnetostatics.

To apply the finite element method, the software creates a mesh. This mesh can be refined by regions before creating it. This is done to adapt the size of the mesh to the expected variations of the solution in each region. In a region where a smooth variation is expected the mesh can be coarser than in a region where the solution is likely to have more abrupt variations.

To completely define the problem, boundary conditions must be fixed once the regions have been created. They must be applied at every line or arc on the outside boundary, but cannot be applied at any internal line or arc. Only one type of boundary condition can be applied at each line or arc. Boundary conditions cannot be changed after a solution has been obtained.

To apply a boundary condition to a line or arc, select one of the Boundary Condition Tools. The cursor becomes an arrow with a legend, indicating which type of boundary condition tool has been selected. Click on the lines or arc where that boundary condition applies. The lines or arcs change color, indicating that the boundary condition has been accepted.

The available boundary conditions are the following.

Field Normal: The resulting field is normal to the boundary. In electrostatics, this is a Dirichlet boundary condition for the potential, and the boundary is an equipotential line at zero potential. In magnetostatics, this is equivalent to a Neumann boundary condition for the vector potential.

Field Confined: The resulting field is parallel to the boundary. In electrostatics, this is a Neumann boundary condition for the scalar potential. In magnetostatics, this is a Dirichlet boundary condition for the vector potential. The vector potential at such a boundary vanishes.

Fixed Potential: In this case a value for the potential is fixed at the boundary. In electrostatics, it applies to the scalar potential and the resulting field is normal to the boundary. In magnetostatics, it applies to the vector potential, and the field is parallel to that portion of the boundary.

Infinite Elements: These approximate an infinite boundary, and the net result is an improvement of the accuracy of the solution in the finite domain.

Material properties must be assigned to the different regions. In electrostatics the regions can correspond to vacuum, or to a conducting or dielectric material. For conductors one must fix the value of the potential whereas for dielectrics the value of the dielectric constant must be given.

In magnetostatics, the regions can correspond to vacuum, to coils carrying electric current, to linear magnetic materials or to permanent magnets. Electric currents can be given as total currents or as current densities. For linear magnetic materials one must specify the relative permeability, a dimensionless number introduced in Chap. 11.

A permanently magnetized material is described by assuming a fixed direction of magnetization in the plane. When an external magnetic field is applied to such a material, the magnetic field has two components: one is parallel to the magnetization and the other is perpendicular. The external field causes additional magnetization to develop in the material, and our concern is how to calculate the total magnetization at each point in the material when the total magnetic field is given at that point. To do this, we separate the problem into its two components. In the direction perpendicular to the magnetization, we simply assume $\boldsymbol{B} = \mu_0 \boldsymbol{H}$. This approximation is used only for simplicity. It is good if the perpendicular field component is small compared with the parallel component, a condition which holds for most common applications of permanent magnets. If this condition is not met, large errors can result.

For the parallel component, we describe the behavior by a linear interpolation in the $(\boldsymbol{H}, \boldsymbol{B})$ plane, passing through the points $(-H_\mathrm{c}, 0)$ and $(0, B_\mathrm{r})$, where B_r is the residual inductance and H_c is the coercive force. Manufacturers of permanent magnets usually supply H_c and B_r for each material. The approximation is good in the range $-H_\mathrm{c} < H < 0$, since the materials are strongly non-linear outside that range.

The solution consists of the set of values of the potential at the nodes of the mesh and a linear interpolation inside each finite element. Once the solution has been obtained, additional results or plots of the electrostatic equipotentials or magnetostatic lines of force can be obtained from the Reports menu.

Table E.1 has been included to facilitate the conversion between cgs and SI units. The coefficients 3 and 9 appearing in this table are respectively related to the value of the speed of light and its square. To evaluate the conversion factors with higher precision the coefficient 3 has to be replaced by 2.9979246.

Table E.1. SI and cgs unit conversion

Quantity	SI unit	Conversion factor	cgs unit
length	meter (m)	10^2	centimeter (cm)
mass	kilogram(kg)	10^3	gram (g)
time	second (s)	1	second (s)
force	newton (N)	10^5	dyne (dyn)
energy, work	joule (J)	10^7	erg
power	watt (W)	10^7	erg/s
electric charge	coulomb (C)	3×10^9	statcoulomb (ues)
charge density	C/m^3	3×10^3	statcoulomb/cm^3
electric current	ampere (A)	3×10^9	statampere (statA)
current density	A/m^2	3×10^{-2}	statA/cm^2
potential	volt (V)	1/300	statvolt (statV)
electric field	V/m	$(1/3) \times 10^{-4}$	statV/cm = gauss (G)
polarization	C/m^2	3×10^5	G
displacement	C/m^2	3×10^5	G
magnetic field	tesla (T)	10^4	G
magnetization	A/m	10^{-3}	G
H-field	A-loop/m	$4\pi \times 10^{-3}$	oersted (Oe)
conductivity	mho/m	9×10^9	1/s
resistance	ohm (Ω)	$(1/9) \times 10^{-11}$	s/cm
capacity	farad (F)	9×10^{11}	cm
magnetic flux	weber (Wb)	10^8	G cm^2
induction	henry (H)	$(1/9) \times 10^{-11}$	stathenry (statH)

Index

MIX
Papier aus verantwortungsvollen Quellen
Paper from responsible sources
FSC® C105338

If you have any concerns about our products,
you can contact us on
ProductSafety@springernature.com

In case Publisher is established outside the EU,
the EU authorized representative is:
Springer Nature Customer Service Center GmbH
Europaplatz 3, 69115 Heidelberg, Germany

Printed by Libri Plureos GmbH
in Hamburg, Germany